S0-BVP-151

SOME LESSONS IN MATHEMATICS

SOME LESSONS IN MATHEMATICS

by members of the

ASSOCIATION OF TEACHERS OF MATHEMATICS

EDITED BY

T. J. FLETCHER

CAMBRIDGE
AT THE UNIVERSITY PRESS
1964

PUBLISHED BY
THE SYNDICS OF THE CAMBRIDGE UNIVERSITY PRESS
Bentley House, 200 Euston Road, London, N.W.1
American Branch: 32 East 57th Street, New York 22, N.Y.
West African Office: P.O. Box 33, Ibadan, Nigeria

Printed in Great Britain at the University Printing House, Cambridge
(*Brooke Crutchley, University Printer*)

To G.

WHO WILL SEE WHAT IS WRONG

FOREWORD

BY PROFESSOR BRYAN THWAITES

Director of the School Mathematics Project

It has sometimes seemed that we, in England, have in some perverse fashion preferred to turn our backs on the pedagogical problems which are posed by changed mathematical courses and which have been faced with such vigour and resolution in other countries. Certainly, at the secondary level (for there has been some remarkable work, here, at the primary level) it has been left to a few individuals and small groups to begin to answer some of the classroom difficulties inherent in any reform of mathematical syllabuses.

However, whatever the truth may have been in the past, suddenly the situation is utterly altered. *Some Lessons in Mathematics* must, surely, mark the beginning of a new era in the history of English school mathematics; in an instant, a new standard is set by which other efforts are bound to be judged. It is a tremendous leap forward.

No need here to explain why. The reader has only to open this book at random to recognize instantly the authority with which it is written, the insight which it gives into the classroom situation and the comprehension which it displays of the rightful impact on school teaching of modern mathematics (that innocent name which, as the introduction says, is now so highly charged with emotional associations).

Thus, though this book is neither a pupils' text nor a teachers' guide for some special syllabus, it is at once compulsory reading for everyone concerned in mathematical education: and one is left to congratulate the group of authors who have produced such a very remarkable contribution to the cause of the reform of school mathematical teaching.

CONTENTS

CONTENTS

ACKNOWLEDGEMENTS

This book was written by members of the Association of Teachers of Mathematics, and very largely compiled at a Writing Week in Leicester, in September 1962. The week's meeting was made possible by a grant from the Nuffield Foundation, whose kindness we most gratefully acknowledge. We also thank the Cambridge University Press for the care they have given to the production of a book, whose varied texture and many points of detail presented special problems.

T. J. F.

The following shared in the writing of this book:

T. J. FLETCHER, *editor*	BERYL FLETCHER
G. P. BEAUMONT	I. HARRIS
A. W. BELL	E. H. LEATON
JOAN BLANDINO	M. D. MEREDITH
BRENDA BRIGGS	T. D. MORRIS
T. H. F. BRISSENDEN	A. G. SILLITTO
W. M. BROOKES	W. O. STORER
A. P. K. CALDWELL	D. G. TAHTA
D. E. CONWAY	NORA USHER
D. S. FIELKER	D. H. WHEELER

NOTE

Figures in square brackets refer to the bibliography on p. 357

1

INTRODUCTION

If the teaching of mathematics is to remain healthy it must be continually refreshed. This is particularly true at a time when mathematical knowledge is advancing rapidly. In general, people do not think that the physics and chemistry which were taught to the last few generations of schoolchildren will be good enough for coming decades; this book is written because some of us feel the same about mathematics.

Broadly speaking, we are in favour of what is, for want of a better name, called 'modern mathematics'. This innocent name is by now highly charged with emotional associations, and we must try to explain what we understand by it, acknowledging that our interpretation does not necessarily agree with that of others who have just as much right to use the label. For us 'modern mathematics' involves several things. The content of the school course should be modified by the introduction of fresh material, some of this being comparatively new in mathematical history, and some being much older but now having an importance which it lacked before. Some traditional material needs rethinking, and teaching in a new way; sometimes because contemporary technology uses the material in new ways, sometimes because of changing perspectives within mathematics itself, and sometimes because the material now has to be taught to large numbers of pupils for whom it was previously considered too difficult.

The teaching of mathematics also needs to be better integrated with contemporary applications in industry and research; indeed these applications should often be the vehicle by which the subject is taught—to regard them merely as ancillary illustration, as a sauce to be added after the meal is cooked, is to misunderstand their role in teaching. But it is not the applicability of the material which is the main point: it is its power to evoke a mathematical response from the pupil.

The last ingredient of 'modern' mathematical teaching, as we see it, is a proper understanding of the relevance of recent psychological investigations. The days when a teacher of mathematics could shut his eyes to psychology, and dismiss it as well-meaning advice of which he had no need, or as opinion which he did not share, are gone. It is the responsibility of teachers to seek to understand this new knowledge and to use it as a basis for a technology of teaching. This technology they will have to create; and it is our aim to assist in the work of creating it. Although teaching should acquire a technology, it will remain an art; but an artist cannot develop to the full unless he is master of the available techniques.

We are concerned then with new mathematics, new ways with old mathematics, contemporary applications of the subject, and the psychology of teaching it. Our theme will be developed largely by means of 'lessons', that is to say by discussions of specific items as much as possible in a way that could be taken into the classroom. But this is most certainly not a textbook for pupils. The 'lessons' are part of a larger context which we seek to explain by commenting to the teacher as we go along. We are not drawing up a syllabus or giving a balanced course of study, and the omission of some piece of traditional work from this book is by no means an indication that we wish to see it disappear. The book was written by a limited number of people in a very limited time, and we could not discuss everything. The reform of the curriculum and the development of suitable methods of teaching the new ideas are research tasks for teachers, and here we aspire to produce a guide for future research. Some of our suggestions are still experimental, and we do not mind if we occasionally go too far. The general consensus of teachers must achieve the synthesis between the old and the new later.

The lessons which follow show a variety of styles and approach, but it seems to us that behind many there is a common strategy. Mathematics does not start with the finished theorem in the textbooks; it starts from situations. Before the first results are achieved there is a period of discovery, creation, error, discarding and accepting. This period is notoriously difficult to discuss, or even to describe in a convincing way, but in this book we have tried to make this period our concern.

The 'situations' from which mathematics starts are rich and varied. The start may be a puzzle, an unsolved problem from the previous lesson, a deliberately contrived piece of teaching apparatus, a portion of the textbook, or a problem from the world outside. Mathematics involves the process of abstraction, starting with the concrete situation, recognizing corresponding structures, and using one structure to solve problems presented by the other. Our concern is with the whole process and not merely with the techniques of the internal operations within one system.

At many places in this book explanatory comment sets the ideas in a more advanced perspective. This is because we want the teacher to recognize the important ideas of advanced mathematics in their primitive forms, and to introduce elementary aspects of these ideas much earlier than is usually the case. We might quote as examples the primitive notion of a vector, or the idea of closure in a group. This seeding process seems to us essential to good teaching.

From all of this it should be quite clear that we repudiate any suggestion that 'modern mathematics' is to be equated with the axiomatic method. A mathematician is not fully educated today unless he understands something of the axiomatic approach, and we believe that the type of material we suggest is at least as good as any other in developing the maturity which can eventually appreciate this approach; but we regard axioms as one of the fruits of a course of study, and not as the starting-point. In teaching, a set of axioms can only be understood after a long initial, quasi-experimental investigation.

Our principal concern is to show the development of modern ideas at stage A, with something of stage B. There is an important difference between the teaching which is possible today and much of the teaching of the past. It is now possible for the teacher to have a far better idea of what stage C really should be. This is especially the case with some of the systems of algebra and geometry which we approach in an intuitive and concrete way. The stage C axiomatic formulations may be found in a variety of university textbooks and in other references which we give. This was scarcely the case with traditional Euclidean geometry (for many years the mathematician's principal example of an axiomatic system), because Euclid's axioms were deficient, and a proper axiomatic approach to this part of mathematics is in any case extremely awkward. The axiom systems involved in some modern algebraic structures are much more simple.

In this book we give lessons covering a wide range of ability. There is no wish to modernize the course of instruction only for those students who will subsequently go on to university; we are writing for all. Further we do not think that university entrance requirements should be a major influence in determining the school course. Only a very small percentage of those studying mathematics at school proceed to study it at university, and it is quite wrong to educate the majority by a course designed for the few. At the same time there is no conflict with university requirements, because we believe that if the schools really teach mathematics, and lead their pupils to think in a mathematical way about the problems with which they are confronted, then the few who go on to university cannot fail to benefit.

A recent inquiry among university teachers showed great concern about the preparation which present school mathematics gives for a university course. There was general agreement that far too much time was spent on learning tricks by rote to solve certain types of examination question, and that there was too little understanding either of what was involved in the tricks or of more general mathematical issues. University teachers were not agreed on the remedies needed to overcome these deficiencies; indeed there were two almost completely opposed schools of thought on the best solution. In this absence of agreement among those most concerned, we cannot think that the type of work we propose will make this situation any worse for the specialists than it is at the moment, and it is our opinion that it offers good hopes of making things better.

In answer to the objection that we cannot have change because university entrance requirements demand that we adhere to the old routine, we must point out that extensive change is already taking place in this country, and that it is taking place in a number of schools which have always been considered strongly traditional, and in which the need to equip boys for entry to the older universities in the face of strong competition is as great as anywhere, and greater than it is in most state schools. The case for reform does not rest on the desire to force potential experts into quicker development; it rests on the desire to produce a classroom environment in which as many pupils as possible may work creatively, with ideas which they find exciting and with ideas which refer to the world of today. This book is mainly concerned with children above the age of eleven, but we would like to deal with corresponding changes in the primary school in another book, as the reform which we desire must extend to the very youngest pupils in our schools.

The present time is one of the rare occasions in the history of mathematics when there has been a revolution which affects not merely the advanced levels of the subject but also its elementary foundations. In the past it was scarcely ever possible to be abreast of important changes in the way in which it is possible at the moment. The elementary foundations of mathematics are changed because whereas in the past elementary mathematics was mainly about number, and had to be, this is no longer the case. In a word it is now about 'structure' instead. Number is still important because number systems certainly have interesting structures, and they may be used to describe structures of many kinds in the world around us. But number is not all, and it is no longer where mathematical instruction should begin.

In neighbouring countries a number of writers [1, 2, 40–2, 98] feel this reorientation to have philosophical and pedagogical implications far beyond the specialized regions of mathematics in which it first occurred, and we wish to see these ideas properly discussed in this country where the tradition is more empirical and pragmatic. We are also extremely aware of the work that has been done recently, and that still needs to be done, on the nature of thinking and the child's development of mathematical concepts. The mathematical models that Piaget uses [96, 97] for the central structures which he describes are such that, if he is right, then some of the non-traditional topics which we discuss, with their emphasis on structure, are more relevant to the understanding of the developing mind than any others.

Had we been able to give these last two items the attention they deserve our book would be even more controversial than many may find it at the moment. But it should be remembered that in advocating the study of these controversial ideas we do not imply that we believe them to be correct in every point; we only believe that in confronting their challenge our own ideas may benefit. These matters must be pursued elsewhere, and throughout this book we give many suggestions for further reading. No one book can tell a teacher just the 'modern mathematics' which he needs to know. This knowledge can only be acquired by constant study, and it is never complete.

We would draw especial attention to the thought-provoking

publications of the OECD [89, 90]. With the principles underlying these documents we are in substantial agreement, and we believe our book to be a practical step towards the implementation of these important proposals; proposals which have not yet received a proper discussion by those in this country to whom they should be of most concern.

We also draw attention to the textbooks which are being written by participants in the programme of reform under the leadership of Professor Thwaites at the University of Southampton. We hope that these books and our own will be complementary; but they are the products of independent minds and neither group is under obligation to uphold the opinions of the other.

At no time in history did the teaching of mathematics present more problems, make more demands, or offer the promise of richer satisfaction.

2

BINARY SYSTEMS

The rapid advance in recent years in the use of computers and similar machines has led inevitably to the adoption of the binary and other scales of notation. But the binary system is not to be considered as merely a tool in a specialized field. It can also be used as a language and symbolism in both elementary and advanced mathematics and other subjects. Nor does its usefulness end here. Many teachers have found that an understanding of it, together with other-based systems, has resolved many, if not all, of the difficulties which children experience in dealing with decimals. If introduced when children start at secondary school (or even earlier) the binary system provides a new experience for them, enjoyable, exciting and shedding new light on number.

The chapter begins with three different treatments; the use of familiar objects (already in binary scale), by a game based on computer operating, and the historical approach. The subject is afterwards developed by means of codes and coding to an advanced lesson on delay networks.

1. INTRODUCING THE BINARY SYSTEM

This arithmetic lesson has been given mainly to less able pupils in the lower half of the secondary school. It is suitable for upper half primary or lower half secondary pupils. Pairs of scales and their weights would be useful with less able pupils, but are not essential.

Lesson

What weights do we get with a pair of scales?

1 oz., 2 oz., 4 oz., 8 oz., 1 lb., 2 lb., 4 lb.

If we rewrite them all as ounces

1, 2, 4, 8, 16, ...

what do we notice?

How else can we write these numbers?

$$1, 2, 2^2, 2^3, 2^4, \ldots$$

What do we call them? Powers of two. *This index notation may be omitted if it is not already familiar.*

How would we weigh:

11 oz.? 2 oz.? 3 oz.? 5 oz.?

If we restrict answers to weights up to 16 oz., is there any weighing we cannot make? What is the most we can weigh? Can we weigh *every* weighing up to 31 oz.?

Every positive integer can be expressed as a sum of distinct powers of 2.

Check this in your notebooks.

What weighings can we make with just the 1 oz.? 1 oz. and 2 oz.? Up to 4 oz.? Up to 8 oz.? Up to 16 oz.? So what is the next weight after 16 oz.? After 32 oz.? After .... Need we check any more?

$$(1+2+2^2+2^3+\ldots+2^{n-1} = 2^n-1).$$

How many *different* weighings can we make with the 1 oz.? 1 oz. and 2 oz.? 1 oz. and 2 oz. and 4 oz.? etc.? Is this reasonable? Can we work out how many weighings we can make with, say, eight weights?

The number of ways of combining any number of articles from n articles is 2^n-1.

How can we weigh 5 oz.? Is there any way other than (4+1) oz.? How many ways can we weigh 3 oz.? 7 oz.? 31 oz.? Any weighing? *This demonstrates uniqueness.*

Make a list of numbers from 1 to 31 in the margin, and tick which weights are necessary, thus:

Number	16	8	4	2	1
1					✓
2				✓	
3				✓	✓
4			✓		
5			✓		✓
etc.					

With the insertion of noughts and removal of column lines and headings we now have the binary notation.

Further work

There are two paths to be followed from this and the next lesson, the first to binary arithmetic as an example of using a base other than ten (and hence to multibase arithmetic), the second to particular applications of the binary system. The first is indicated in the section on a historical approach and in the lesson on binary fractions. Some of the specific applications which follow are to binary codes, delay networks, punched cards and use in computers. In addition the ideas in this lesson lead on to permutations and combinations and to laws of indices, slide-rule work and logarithms. Powers 0 and 1 follow automatically from the table:

Number	32	16	8	4	2	1
Power	5	4	3	2	1	0

and negative indices may be introduced immediately. If we do this, we have what can be described from a more sophisticated point of view as a mapping of the integers into the rationals; which is later extended to a mapping of the real numbers on to the positive real numbers when logarithms are introduced. The present notation applies to various puzzles, games and tricks, to multiplication by duplation (p. 12) and 'Russian' or 'peasant' multiplication.

One might also try another weight problem such as the one in which weights may be put on both scales. This uses addition and subtraction of powers of three [47].

The next lesson is based on an idea by Engelbart [39].

Lesson

Five children are arranged in line in front of the class from right to left, and told that their right hand must, in what follows, be clearly UP or DOWN. The child on the extreme right is instructed to change position (UP to DOWN or DOWN to UP) whenever he receives a signal, a handclap, say, from the teacher. The others change their position when the hand of the child on their LEFT moves DOWN.

Practice of the two instructions may be necessary before the teacher begins to clap. When eventually all the hands are up, the teacher may ask what will happen at the next clap. All the hands go down.

Now the children are given cards, labelled 1, 2, 4, 8 and 16, beginning with 1 on the right. The clapping begins again, and is interrupted to ask how many claps have been given. When the question is repeated at different stages the class will notice that the number of claps given is equal to the sum of the numbers on the cards held up.

All hands are lowered, and one may ask how to show such numbers as 8, 10, 13 and 17. The class can write down how to do this, and the 'machine' can present the correct cards.

Are there any numbers that cannot be made? Yes, fractions, and numbers higher than 31. A machine has been constructed, then, which will count up to ...? 31. How can it count more than 31? What is noticed about the numbers

1, 2, 4, 8, 16, ...?

The lesson can then proceed along the lines of the previous lesson, or, if a sequel to it, be used to revise the material in a new light.

Further work

The 'machine' can be turned into a 'human turnstile' if the signal is someone walking through a 'gate', e.g. the door of the room. A mechanical model could be made from meccano.

A 'decimal counter' can be formed if each child counts on its fingers, nudging the next child and 'returning to zero' after reaching ten. (Use one hand only for a 'quinary counter'.) This shows the decimal system in a new light. A mechanical decimal counter can be made, and an inspection made of milometers, cyclometers, gas and electric meters, etc.

The advantages of the binary system can be discussed. Each part of the 'machine' has only two positions—up or down, backwards or forwards, on or off. The last is most suitable for electric machines. This can be demonstrated during the last lesson on a winter's afternoon with torches on or off replacing or supplementing hands up or down. A semi-mechanical model can be made from lamp switches, or a complete electrical binary counter can be wired up using the appropriate relays. Binary numbers may also be represented on punched cards, a hole corresponding to one and a slot corresponding to nought ([48] and §5.2).

Decimal analogues can be demonstrated by various primitive methods of counting, beginning with finger-counting and leading through pebbles, notches and knots to forms of the abacus.

One may also notice the pattern in the binary notation. This can be seen by writing the numbers 1 to 64 in this notation. Alternatively, the children who are part of the 'machine' soon become aware of it.

The child who is '4' will notice that he is alternately UP and DOWN for four successive signals. This pattern is also apparent from a set of punched cards arranged in order; this helps to explain the method of putting the set into numerical order by stabbing the holes in turn with a needle.

The object of Engelbart's games is to give insight into the workings of computers and a more complicated game is described in §5.6.

2. A HISTORICAL APPROACH

The history of numbers and of the recording of numbers provides another way of introducing the binary system. A short discussion on the need for primitive man to record number should precede an investigation into methods of recording number, methods such as by drawing pictures, by knotting and by simple markings, as for example on a tally stick. Some of these methods have remained in use until comparatively recently. Knotting on cords when taking a census in a remote part of India and the use in England of the tally stick for government accounts until 1826 are examples.

The Babylonian cuneiform symbols, and the base of 60 (still evident in time and angle measures), the Egyptian hieroglyphics, the Greek use of the alphabet and the more familiar Roman notation could all be discussed. Wall charts illustrating the various notations are helpful.

The main points to be noted in these notations are the base used, the lack of a symbol for zero and the lack (or limited use) of the idea of 'place' value.

These points can be brought out by attempting a few examples with the various notations, without changing to the familiar 1, 2, 3, etc. For example, to use an alphabetic notation similar to that of the Greeks we could take:

A = 1	D = 4	G = 7	J = 10	M = 40	P = 70	S = 100
B = 2	E = 5	H = 8	K = 20	N = 50	Q = 80	T = 200
C = 3	F = 6	I = 9	L = 30	O = 60	R = 90	etc.

The number 12 is composed of 10 and 2, or 2 and 10, and can be written as either JB or BJ. Even the simplest questions in addition seem difficult with this notation (what is G+LC?) and the awkwardness of the notation is quickly appreciated.

The more familiar Roman notation is fairly easy to use for

addition, but difficulties soon arise when multiplication by the usual method is attempted. Here it is useful to introduce the method of multiplication by duplation, which not only overcomes most of the difficulties but also gives experience in decomposing a number into a sum of powers of two, which is of value when the binary system is introduced.

The following example in both Roman and Hindu–Arabic notations is included to illustrate this.

Multiply XXXIX	by XXI		Multiply 39×21	
XXXIX	(I times)		39 (1 times)	
XXXIX			39	
LXXVIII	(II times)		78 (2 times)	
LXXVIII			78	
CLVI	(IV times)		156 (4 times)	
CLVI			156	
CCCXII	(VIII times)		312 (8 times)	
CCCXII			312	
DCXXIV	(XVI times)		624 (16 times)	
CLVI	(IV times)	add	156 (4 times)	add
XXXIX	(I times)		39 (1 times)	
DCCCXIX	(XXI times)		819 (21 times)	
			for $21 = 16+4+1$	

A brief history of our usual notation (the Hindu–Arabic) could follow, paying particular attention to the use of the ideas of position and zero, whereby ten symbols suffice to express any number, however large or small, in the system with ten as base (the denary system).

Index notation can conveniently be introduced when discussing position and very easily extended to include both zero and negative exponents. The sufficiency of ten symbols for the base ten system, and the question of how many symbols have to be used for systems with other bases, 60, 20, 12, 2, 5, 3, can then be considered.

The concepts of position and zero will be seen to be applicable to systems of any base, the number of symbols used being the same as the number of the base. If the system base two (the binary system) is chosen, the symbols 0 and 1 are used; for the base three system 0, 1 and 2, and so on.

It is not suggested that the work in the foregoing summary should be covered in one lesson; five at least would be needed. Many interesting discoveries can be made about number, its composition and properties, by investigating the various notations systematically.

Other methods for the four basic operations, particularly multiplication, can be a fruitful field to explore. For example, 'Russian' multiplication, which is usually carried out in this way:

27×93		
	27	93
	13	186
	~~6~~	~~372~~
	3	744
	1	1488
		2511

The method is to multiply one number by 2 and divide the other by 2. Remainders (if any) are neglected. When the division is complete, cross out any numbers in the 'multiples' column which are on the same line as even numbers in the division column. The sum of the remaining multiples is the required answer.

Let the children investigate whether or not it matters if one number rather than the other is chosen for the 'division' or the 'multiples' columns. Does it make any difference to the result? Is there any other reason for making a particular choice?

An alternative way of writing the same question can also be profitable. Here the number which is divided is kept in a bracket until the division is completed.

$$
\begin{aligned}
(27) \times 93 &= (13+\tfrac{1}{2}) \times 186 \\
&= (6+\tfrac{1}{2}+\tfrac{1}{4}) \times 372 \\
&= (3+\tfrac{1}{4}+\tfrac{1}{8}) \times 744 \\
&= (1+\tfrac{1}{2}+\tfrac{1}{8}+\tfrac{1}{16}) \times 1488 \\
&= 1488+744+186+93 \quad \text{(Expanding).}
\end{aligned}
$$

This method leads to the idea of binary fractions (p. 15).

Numbers can be converted from one scale to another. This is usually done in one of two ways: by expressing the number as a sum of the powers of the base used, or by 'reduction'.

In the case of binary numbers the first method has already been suggested.

Example

$$\begin{aligned} 23 &= 16+7 \\ &= 16+4+3 \\ &= 16+4+2+1 \end{aligned}$$

$\therefore$ 23 (base ten) = 10111 (base two).

This method is easy for the binary scale but increases in difficulty with larger numbers and larger bases. The 'reduction' method has the advantage that it is just as easy to use for any multibase or mixed base system (by 'mixed' base is meant one where the columns each have a different base, as in £ *s. d.*).

Example

To bring 234 (base ten) to a number base seven: divide by 7 (the new base) until the quotient is zero and the required number (base seven) is given by the remainders reading upwards. *Discuss why.*

Thus

$$\begin{array}{r|l} 7 & 234 \\ \hline 7 & 33 \quad \text{rem. } 3 \\ \hline & 4 \quad \text{rem. } 5 \\ \hline & 0 \quad \text{rem. } 4 \end{array}$$

$\therefore$ 234 (base ten) = 453 (base seven).

To convert back to denary form the reverse process of either method can be used.

Some experience in working in different bases should be gained before converting numbers from one base to another where neither is ten.

Multiplication or division by powers of two in the binary system is analogous to the same operations with powers of ten in the denary system. This can be generalized to all systems, whatever the base.

Reference to the octal system (base eight) might also be made as, combined with a binary notation, it is used in many computers.

So binary numbers	101	111	011	011101	001111 etc.
are written in octals	5	7	3	35	17

References

Bowman [18], Cajori [22], Dantzig [34], Newman [88].

3. A LESSON ON BINARY FRACTIONS

An arithmetic lesson which has been given to bright first-year secondary children. Suitable for bright top primary children or any pupils of secondary age familiar with the manipulation of vulgar fractions and with integral binary notation.

Preliminary work

Apart from a special symbol for $\frac{2}{3}$ the ancient Egyptians only had a notation for fractions with unit numerator, and other fractions had to be expressed as sums of these. Thus:

$$\tfrac{3}{4} = \tfrac{1}{2}+\tfrac{1}{4}.$$

The actual Egyptian notation is irrelevant, but would of course add interest to this particular lesson if Egyptian numerals were already known.

It becomes a simple exercise in addition and subtraction of vulgar fractions to express the simpler of them in this way. It also becomes a matter of honour to do this as briefly as possible, and to avoid repetitions such as

$$\tfrac{3}{4} = \tfrac{1}{4}+\tfrac{1}{4}+\tfrac{1}{4}.$$

Some systems may be evolved, such as trying successive Egyptian fractions in order ($\frac{1}{2}, \frac{1}{3}, \frac{1}{4}, \frac{1}{5}, \ldots$) thus:

$$\tfrac{7}{9} = \tfrac{1}{2}+\tfrac{5}{18} = \tfrac{1}{2}+\ldots\ (\tfrac{1}{3} = \tfrac{6}{18} > \tfrac{5}{18})$$
$$= \tfrac{1}{2}+\tfrac{1}{4}+\tfrac{1}{36}.$$

The solutions are by no means unique, for example,

$$\tfrac{7}{8} = \tfrac{1}{2}+\tfrac{1}{4}+\tfrac{1}{8} = \tfrac{1}{2}+\tfrac{1}{3}+\tfrac{1}{24} = \tfrac{1}{3}+\tfrac{1}{4}+\tfrac{1}{6}+\tfrac{1}{8}.$$

This may be the pupil's first introduction to lack of uniqueness of an answer and he can be asked for the complete set of non-repetitive solutions.

He may soon hit on something like

$$\begin{aligned}\tfrac{1}{2} &= \tfrac{1}{3}+\tfrac{1}{6}\\ &= \tfrac{1}{3}+(\tfrac{1}{3}\cdot\tfrac{1}{2})\\ &= \tfrac{1}{3}+\tfrac{1}{3}(\tfrac{1}{3}+\tfrac{1}{6})\\ &= \tfrac{1}{3}+\tfrac{1}{9}+\tfrac{1}{18}\\ &= \tfrac{1}{3}+\tfrac{1}{9}+\tfrac{1}{9}(\tfrac{1}{3}+\tfrac{1}{6})\\ &= \tfrac{1}{3}+\tfrac{1}{9}+\tfrac{1}{27}+\tfrac{1}{54}\ldots\end{aligned}$$

which would indicate the endless possibilities.

Lesson

A recurrent puzzle is that of the frog in the centre of a circular lily-pond who, gradually tiring, jumps half-way out, then a quarter of the way, then an eighth, etc., halving his jump each time. Will he ever reach the edge? *The class will have to decide that the size of the pond is irrelevant.* Let us write down how far he goes in terms of a fraction of the radius:

$$\tfrac{1}{2}$$
$$+\tfrac{1}{4} = \tfrac{3}{4}$$
$$+\tfrac{1}{8} = \tfrac{7}{8}$$
$$+\tfrac{1}{16} = \tfrac{15}{16}$$
$$+\tfrac{1}{32} = \tfrac{31}{32} \text{ etc.}$$

Without actually adding, can we continue the sequence?

$$\tfrac{1}{2}, \tfrac{3}{4}, \tfrac{7}{8}, \tfrac{15}{16}, \tfrac{31}{32}, \tfrac{63}{64}, \ldots.$$

Both the sequences of denominators and of numerators should be familiar from p. 8.

The binary notation may be introduced here. The distances gone are now

·1

·11

·111

·1111

·11111 etc.

Will this sequence ever reach one? If the numerator is always one less than the denominator then it cannot. How close to one can we get? Let us inspect the fraction of the radius left at each jump:

$$\tfrac{1}{2}, \tfrac{1}{4}, \tfrac{1}{8}, \tfrac{1}{16}, \ldots$$

(·1, ·01, ·001, ·0001, ...).

The distance to the edge of the pond halves each time, and we cannot get to zero by halving anything (except zero). Can we get as close as we like? Nearer than $\frac{1}{10}$? $\frac{1}{16}$. Nearer than $\frac{1}{100}$? $\frac{1}{128}$. Nearer than $\frac{1}{1,000,000}$?

It is obvious, then, that the frog can get as close to the bank as he likes without actually getting there.

Now suppose that he wishes to reach a leaf $\frac{5}{8}$ of the distance from the centre in a similar way. Can he do it? What if we let him omit some of his jumps?

$$\tfrac{1}{2}+\tfrac{1}{8} = \tfrac{5}{8}$$

so he can get there! What about $\frac{7}{8}$? $\frac{13}{16}$? $\frac{5}{32}$? $\frac{17}{32}$?

A slightly more sophisticated approach is to introduce the hypothetical tribe living, shall we say, further up the Nile than the Egyptians, who only have notation for $\frac{1}{2}$, $\frac{1}{4}$, $\frac{1}{8}$, $\frac{1}{16}$, etc., and are allowed to use each only once for any particular fraction; or the problem can be posed merely as an extension of the method suggested above for Egyptian fractions. It can be pointed out that any whole number can be expressed as a sum of distinct powers of two, so why may we not express a fraction as distinct negative powers of two? Here negative index notation may be introduced, and the binary notation extended.

The 'binary point' is a convenient way of denoting where whole numbers end and fractions begin. We can now write, for example,

$$\tfrac{7}{8} = \cdot 111$$
$$\tfrac{13}{16} = \cdot 1101$$
$$\tfrac{5}{32} = \cdot 00101.$$

How can we write $\frac{1}{3}$? We may relate the problem to the frog. As before

$$\begin{aligned}\tfrac{1}{3} &= \ldots(\tfrac{1}{2} \text{ is too big})\\ &= \tfrac{1}{4}+\tfrac{1}{12}\\ &= \tfrac{1}{4}+\ldots(\tfrac{1}{8} \text{ is too big})\\ &= \tfrac{1}{4}+\tfrac{1}{16}+\tfrac{1}{48}\\ &= \tfrac{1}{4}+\tfrac{1}{16}+\ldots\\ &\quad\ldots\\ &= \tfrac{1}{4}+\tfrac{1}{16}+\tfrac{1}{64}+\tfrac{1}{256}+\ldots\\ &= \cdot 01010101\ldots\end{aligned}$$

Discussion is necessary here. The frog, if that is the problem, can again get as close to his leaf as he wishes. If we are not worried about, say, $\frac{1}{1,000,000}$, we can stop at the appropriate place. Otherwise a row of *dots will indicate that we can go on for as long as we like, and get as near as we like. The original problem can now be stated as*

$$1 = \cdot 11111\ldots$$

meaning that by continuing to add smaller fractions we can get as close as we like to 1.

Can we now write in binary notation $\frac{1}{7}$, $\frac{1}{15}$, $\frac{1}{31}$, ...? Notice what happens.

$$\tfrac{1}{7} = \cdot 001001\ldots$$
$$\tfrac{1}{15} = \cdot 00010001\ldots$$
$$\tfrac{1}{31} = \cdot 0000100001\ldots$$

Why does this happen? *No answer is really expected yet, but with faith the sequence may be continued! These examples are special cases of*

$$\frac{1}{n-1} = \frac{1}{n}+\frac{1}{n^2}+\frac{1}{n^3}+\ldots.$$

What about $\frac{2}{3}$?

$$\tfrac{2}{3} = \cdot 101010\ldots$$

Has this any connection with $\frac{1}{3}$? *Work on doubling and halving with whole numbers in binary notation should have been covered beforehand.* Now we can tell immediately that

$$\tfrac{1}{6} = \cdot 0010101\ldots$$

Or, if manipulation of common fractions is still necessary,

$$\tfrac{1}{3} = \tfrac{1}{4}+\tfrac{1}{16}+\tfrac{1}{64}+\ldots$$

so

$$\tfrac{2}{3} = \tfrac{2}{4}+\tfrac{2}{16}+\tfrac{2}{64}+\ldots$$
$$= \tfrac{1}{2}+\tfrac{1}{8}+\tfrac{1}{32}+\ldots$$

and

$$\tfrac{1}{6} = \tfrac{1}{2}\cdot\tfrac{1}{4}+\tfrac{1}{2}\cdot\tfrac{1}{16}+\tfrac{1}{2}\cdot\tfrac{1}{64}+\ldots$$
$$= \tfrac{1}{8}+\tfrac{1}{32}+\tfrac{1}{128}+\ldots$$

Can *any* fraction be written in binary notation? Return to a discussion of the frog. In the diagram, O is the centre of the pond and A the edge, P the point he wishes to reach.

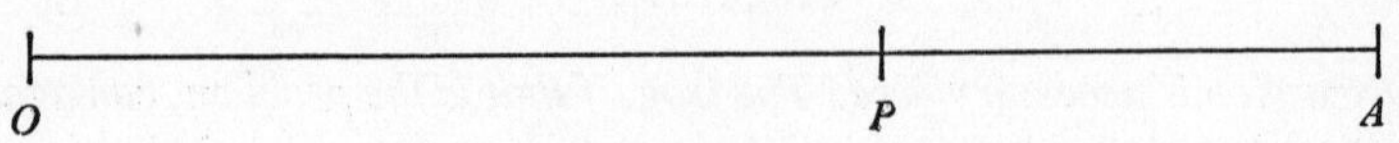

Fig. 2.1

Halve OA at B. P is to the left or right of B, say right. Halve BA at C. P is left or right of C, say left.

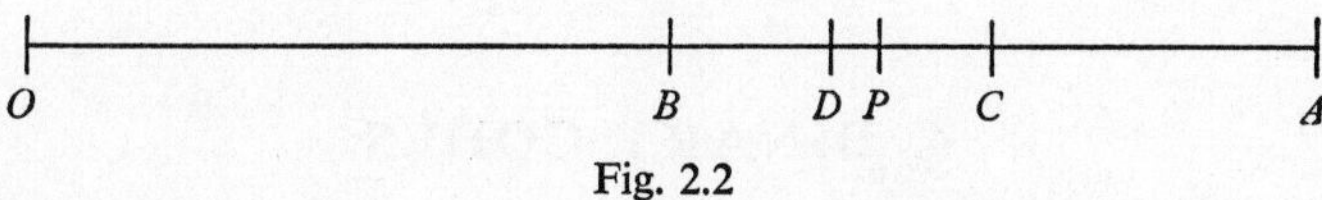

Fig. 2.2

Halve BC at D, etc. Clearly we are including or excluding 2^{-n} according as P is to the right or left of the nth mid-point, and we can always get as close to P as we like. In terms of the frog, if his leaf is over half the distance left, he jumps half that distance; otherwise he tries one quarter of the distance; and so on.

Further work

The human, mechanical and electrical computers may be adapted to calculations involving fractions.

General manipulations of numbers in the binary system involving the four rules can be considered.

A discussion of approximations would be beneficial considering, for example, whether we need for practical purposes anything less than $\frac{1}{32}$.

Work in bases other than 2 and changing from one base to another may also be done. Here the conversion of vulgar fractions to binaries by division would give more insight into recurring binaries.

Comment

An important line of development from here is the generalization to bases other than two, and then specialization to decimal fractions, so that they are seen against this multibase background. This is Dienes' idea of abstracting the concept of place value from a variety of concrete situations—the different bases used [36].

The simplicity of the binary notation in introducing place-value fractions will be obvious; just as we use the system for computers because of the 'on or off' feature of the two symbols, the fraction ideas are easily comprehensible because either we use a place (a negative power of two) or we don't. Recurring decimals often get obscured in the symbols, and any idea of convergence is lost. We are content to show that 'it goes on and on for ever' without explaining 'for ever' or using concrete illustrations. It is so difficult to divide an interval into ten, let alone a hundred, parts, and the intervals get too small too quickly! Halving has always been a simple operation.

4. BINARY CODES

A simple lesson for beginners, which could be developed in a sophisticated way. Copies of codes and some punched tape would be useful.

Preliminary work

Knowledge of the binary system and modular arithmetic.

Lesson

In teleprinters and other electrical machinery we need signalling systems which depend on the use of two symbols only, which we can denote by 0 and 1. Using only these symbols, can we construct a code to send the ordinary alphabet? One way is to construct five-digit groups, based on binary numbers 0 to 31, to represent letters:

00000, 00001, 00010, etc.

We will henceforth concern ourselves with codes in which each character always contains the same number of digits. *Examples may be shown*, *see pp.* 27–8. How many letters can we construct with an n-digit code?

Errors may arise through a symbol being incorrectly transmitted. What is the best way to overcome this? If we sent everything twice some random errors would be detected, although we might not know how to correct them, but to be more certain of how to correct the error we would have to send everything three times, and take a majority vote. Is this a very efficient way of correcting occasional errors? Can we find a better one?

The main principle of error detection and correction is the use of parity checks. We may take our five-digit code and increase each letter to a six-digit symbol, introducing the last digit in such a way as to make the number of 1's appearing always an even number.

Thus the character 01000 is now to be coded as 010001 and 00011 is now to be coded as 000110. A character containing a single mistake can now be recognized.

As an exercise complete the code and set up a simple message in it. Change messages with your neighbour, and decode.

Now set up a message, and introduce errors by shutting your eyes, marking a few places at random with a pencil and changing the digits. Hand the faulty message to your neighbour and see if he can decode it correctly, pointing out the errors.

We now wish to construct a code which will not only indicate errors, but which will also enable them to be corrected automatically. We can often spot errors by common sense, but is this enough? No! the message may sometimes consist of information, such as reference numbers or names with an unfamiliar spelling, which we cannot check by common sense. Assuming that errors are not too frequent, these difficulties can be overcome by using a very ingenious coding principle due to Hamming[60].

We will first take a simple example with an alphabet of only 16 characters. Set up the 16 characters 0000, 0001, 0010, 0011, etc. Now increase each character to 7 digits X_1, X_2, X_3, X_4, X_5, X_6, X_7 in the following way. Positions 3, 5, 6 and 7 are to be *information positions* and 1, 2 and 4 are to be *check positions.*

$$\left.\begin{aligned} X_4 \text{ is chosen to make } X_4+X_5+X_6+X_7 &= 0\\ X_2 \text{ is chosen to make } X_2+X_3+X_6+X_7 &= 0\\ X_1 \text{ is chosen to make } X_1+X_3+X_5+X_7 &= 0 \end{aligned}\right\} \pmod 2. \quad (1)$$

Thus with 1001 we have $X_3 = 1$, $X_5 = 0$, $X_6 = 0$, $X_7 = 1$, so X_4 has to be 1, X_2 has to be 0, and X_1 has to be 0; and so 1001 is now written as 0011001.

Construct the entire alphabet of 16 characters according to this rule.

The code now detects errors 'automatically' in a most remarkable way. When a letter is received calculate

$$\left.\begin{aligned} X_4+X_5+X_6+X_7 &= \alpha \text{ (say)}\\ X_2+X_3+X_6+X_7 &= \beta \text{ (say)}\\ X_1+X_3+X_5+X_7 &= \gamma \text{ (say)} \end{aligned}\right\} \pmod 2.$$

If the letter has been received correctly, what should the values of α, β and γ be? $\alpha = \beta = \gamma = 0$, by equations (1).

If any one of the seven digits in the letter is incorrect one or more of the α, β, γ will differ from zero, and this will indicate an error. Do α, β and γ tell us anything else?

Suppose we wish to send 0011001. An error in the first position results in 1011001, and then

$$\left.\begin{aligned} \alpha &= X_4+X_5+X_6+X_7 = 0 \\ \beta &= X_2+X_3+X_6+X_7 = 0 \\ \gamma &= X_1+X_3+X_5+X_7 = 1 \end{aligned}\right\} \quad (\text{mod } 2).$$

Work out α, β and γ for errors in the other six positions. This gives us:

Position	α	β	γ
1	0	0	1
2	0	1	0
3	0	1	1
4	1	0	0
5	1	0	1
6	1	1	0
7	1	1	1

What do we notice? Of course, the binary number $\alpha\beta\gamma$ gives us the position in which the error occurs.

Those familiar with matrix notation will see that the checking operations can be put into a very striking form as a matrix product:

$$[X_1, X_2, X_3, X_4, X_5, X_6, X_7]\begin{bmatrix} 0 & 0 & 1 \\ 0 & 1 & 0 \\ 0 & 1 & 1 \\ 1 & 0 & 0 \\ 1 & 0 & 1 \\ 1 & 1 & 0 \\ 1 & 1 & 1 \end{bmatrix} = [\alpha, \beta, \gamma].$$

Further work and notes

The system may be explained, and compared with packs of magic cards for guessing numbers. A nine-digit code could be constructed with five information positions and four check positions for an alphabet of 32 letters.

There are now many exercises. Code and decode messages. Code messages and insert errors by some random process, such as reading numbers from a telephone directory and inserting an error every time a 7 is read, and then let someone else try to decode the faulty message correctly.

Note that errors escape detection if more than one occurs in a letter. We may attach the idea of a *distance* between two letters in this code as follows. The *distance* between two letters is defined as the number of differences in the digits. Thus 0011001 and 0101101 differ in three positions, so the distance between them is 3. This definition may seem very artificial; but note that this distance function has the following properties. If $D(x, y)$ denotes the distance between letter x and letter y then

$$D(x, y) = 0 \quad \text{if and only if } x = y.$$

$$D(x, y) = D(y, x),$$

$$D(x, y) + D(y, z) \geqslant D(x, z).$$

Cases of this may be examined.

Distance in ordinary geometry has these three properties. The study of metric spaces is concerned with mathematical systems in which there is a distance function with these properties. We have here an elementary example of a metric space. In this system all combinations of digits with a distance 1 from the character intended can be corrected, but errors which take the combination further away cannot.

The symbols with which we are dealing form a vector space over the field GF(2) (see §11.6, 3.4 and [14], [109]). This system provides simple illustrations of many key ideas of linear algebra. If we are dealing with a seven-digit code the set of all seven-digit symbols forms a space of dimension 7. The particular subset of 16 symbols (with error-correction) defined above forms a subspace of dimension 4. Why is it a subspace?

We are only concerned with the additive properties of the space. A vector space is always an additive (Abelian) group (by its definition). The symbols here provide an example of an additive group. All the seven-digit symbols form a group of order 2^7; the 16 particular symbols form a subgroup. Given a group and a subgroup there is always a corresponding decomposition into cosets (p. 61). This decomposition has a relevance in the theory of these codes [60].

The reduction of the seven-digit symbol to the three-digit symbol $\alpha\beta\gamma$ is a *linear mapping*. In any such mapping the set of elements in the object space which are mapped on the zero element of the image space is called the *kernel*. Here the kernel consists of those seven-digit combinations which are free of errors.

Exercise

If a symbol has n binary digits, of which m are information positions and $n-m$ are checks, show that

$$2^m \leqslant \frac{2^n}{n+1}.$$

As another example of automatic error detection we may quote the problem of identifying the dud penny out of the twelve [56].

Chain codes

Indian drummers remember the word yamátárájabhánasalagám. It is a mnemonic for drumming rhythms. The essential pattern is 0111010001 where 1 and 0 denote strong and weak accents. All possible three-digit combinations occur in sequence. If we remove the last two digits the eight left may be arranged in a closed cycle.

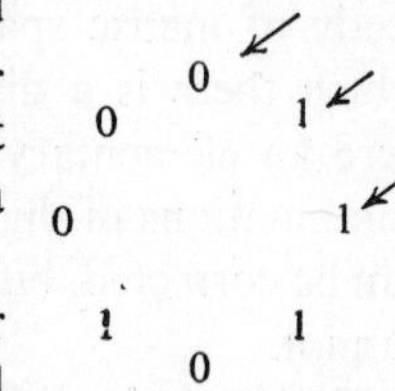

Fig. 2.3

This has uses in electrical machinery. For example, suppose we wish to arrange for all eight possible combinations of 0 and 1 to appear on three wires. These three wires can be attached to three successive positions on a rotating switch and turned to the eight positions above. We show the combination 011; if we move one position clockwise we pick up 111. Such a sequence of symbols is called a *chain code*.

In 1895 Baudot used a five-digit chain code in telegraphy

$$1001010110100011001111101110000.$$

Note that the combination 00000 is deliberately omitted. Can such codes always be constructed?

A heuristic discussion of the lines on which such a proof might be discovered is given by Stein [113], and if Euler's results on the universal description of figures (in particular the application to the domino problem, p. 233) are fresh in the pupils' minds they might well discover for themselves the method devised by Good [55].

This method can be appreciated by considering the construction of a four-digit binary code. We regard a letter such as 0011 as arising from a transition from 001 to 011. The four-digit chain code can then be constructed if we can write down all three-digit combinations and connect them up by a unicursal path, such that the links are of the type considered. This leads to the figure:

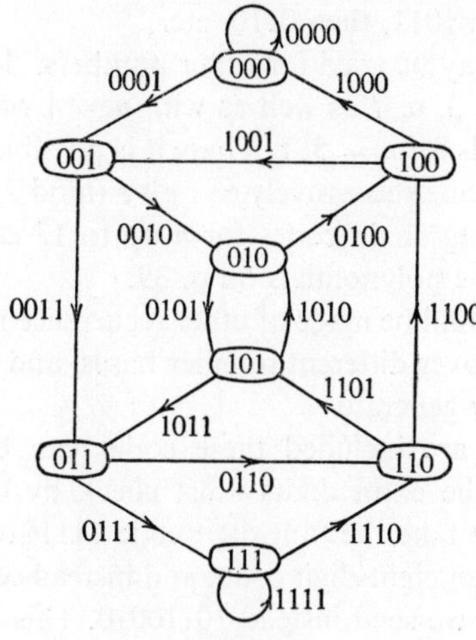

Fig. 2.4

It is clear that there are just two paths into and out of each node, and so the figure can be described unicursally. It is equally clear that the same method will apply to a code containing letters composed of any other number of binary digits. Also it extends to digits taken over any other base.

Rees [104] has given an alternative method for constructing these codes which depends on using polynomials to construct recurring 'decimals' over Galois fields; and so this basically simple classroom problem leads into some intricate considerations in the theory of polynomials over finite fields. The above diagram can be compared with others in chapters 6 and 8.

An account of these codes is given by Heath and Gribble [61] who give applications to frequency dividing in electrical circuitry, where the basic problem is to produce switching cycles which repeat after an assigned number of operations, and the method of solution is to produce chains like this, either of full length or shortened by a 'skip'.

There are sometimes simple rules which will generate chain codes. For example, if we start to build a four-digit binary code with the combination *abcd* and get the next by putting $e = a+d \pmod 2$ leading to the combination *bcde*, this is a recurrence relation which generates the whole cycle. Thus suppose we start with 1010 we get in turn 0101, then 1011, then 0110, etc.

This method may be tried for other numbers. It will be found to work with $n = 2, 3, 6, 7$ as well as with $n = 4$ which has just been considered. It fails for $n = 5$; but here it is possible to start with any *abcde* and then define successively $f = c+e \pmod 2$, etc. Recurrence formulae producing chain codes for n up to 13 can be derived immediately from the polynomials on p. 39.

Investigation could be made of other recurrence relations for codes employing digits over different number bases, and the lengths of the chains which they generate.

If extra digits are included these codes can be used for error detection, since the extra digits must check by the recurrence relation. Thus if we take the four-digit code 1111010110010000 it can be lengthened to an eight-digit code; and instead of sending say 1011 to denote a letter we send instead 10110010. This means that

one error in a letter can always be corrected;

two errors in a letter can sometimes be corrected, and always detected;

three errors can sometimes pass undetected.

The ideas of vector spaces, Abelian groups and the metric discussed on p. 23 also apply here.

Examples of binary codes

The essence of a binary code is that it uses only two symbols. The best known is perhaps the Morse code, each 'signal' being a dot or a dash; but this is really a ternary code as the space signal plays an important part. Without knowing where the spaces come a Morse message cannot be understood.

Braille is a binary code, each of six places in each section being raised or not. This gives $2^6-1 = 63$ different symbols, which are used for frequently occurring words and pairs of letters and for punctuation, as well as the alphabet which is given in Fig. 2.5.

Baudot's five-digit code for telegraphy was based on an idea by Gauss and Weber, and is similar to codes used on punched tape in computers. J. B. Moore invented a seven-digit code, modified by Van Duuren, in which out of the $2^7 = 128$ possible arrangements were used only the 35 which had three of one symbol and four of the other; reception of any of the others indicated errors.

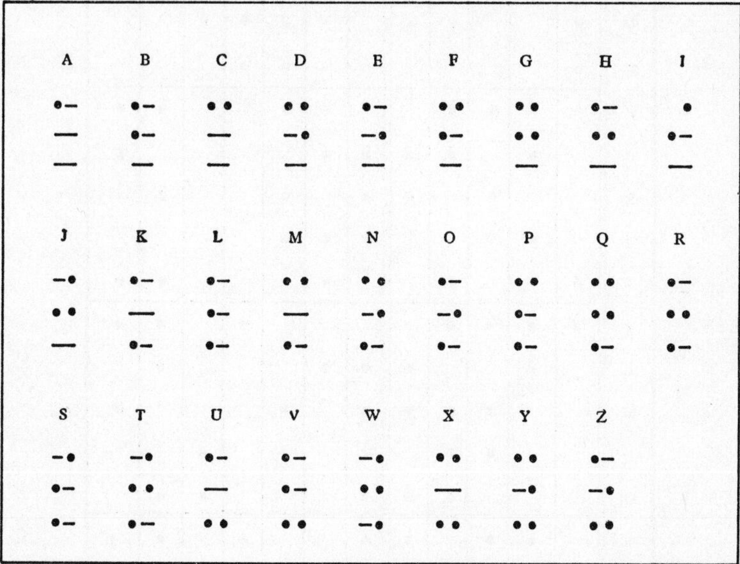

Fig. 2.5. The Braille alphabet.

Both alphabets are shown in Fig. 2.6. Other arrangements are used for punctuation, etc.

5. THE DESIGN OF DELAY NETWORKS

An advanced lesson leading from a practical problem to many of the ideas of modern abstract algebra.

Preliminary work

Boolean algebra of circuitry useful but not essential; earlier binary lessons; polynomial theory.

Lesson

We are going to study some of the processes which can be used in computers to transform one signal into another. We will be

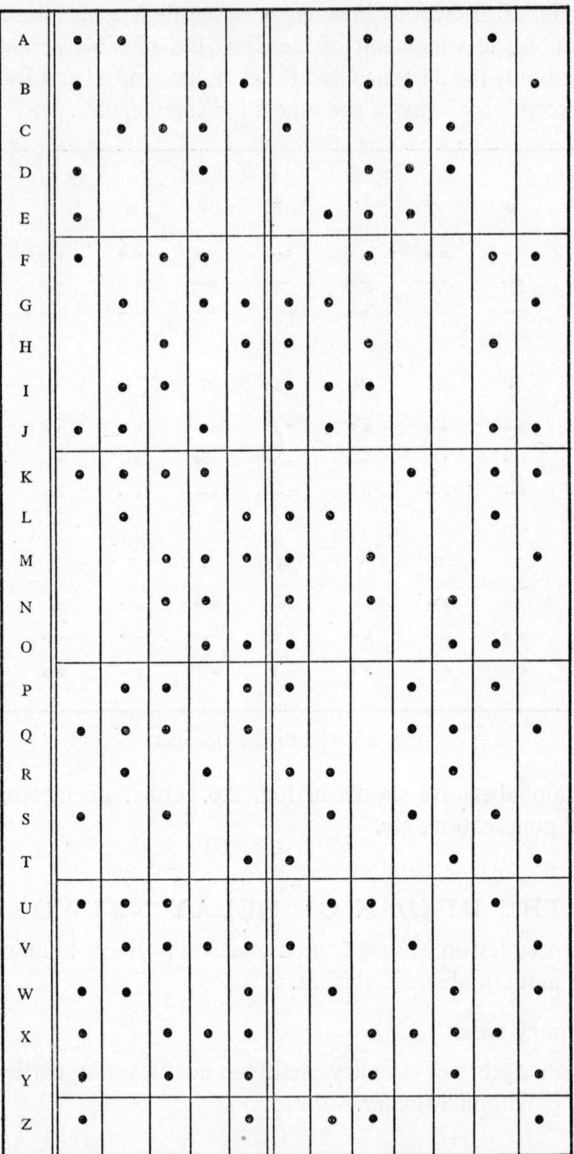

A	•	•						•	•		•	
B	•			•	•			•	•			•
C		•	•	•		•			•	•		
D	•			•				•	•	•		
E	•						•	•	•			
F	•		•	•				•			•	•
G		•		•	•	•	•					•
H			•		•	•		•			•	
I		•	•			•	•	•				
J	•	•		•			•				•	•
K	•	•	•	•					•		•	•
L		•			•	•	•				•	
M			•	•	•	•		•				•
N			•	•		•		•		•		
O				•	•	•				•	•	
P		•	•		•	•			•		•	
Q	•	•	•		•				•	•		•
R		•		•		•	•			•		
S	•		•				•		•		•	
T					•	•				•		•
U	•	•	•				•	•			•	
V		•	•	•	•	•			•			•
W	•	•			•		•			•		•
X	•		•	•	•			•	•	•	•	
Y	•		•		•			•		•		•
Z	•				•		•	•				•

Fig. 2.6 International Telegraph Alphabet No. 2 (left); Van Duuren Alphabet (right). Note the tendency in the five-digit code, as in the Morse code, to use fewer symbols for the letters more frequently used.

concerned with binary systems which operate in sequence; for example, some of the previous coding systems where the message is sent as a sequence of symbols. Consider a sequence

$$X = 0\ \ 0\ \ 1\ \ 1\ \ 1\ \ 0\ \ 1\ \ 0\ \ 0\ \ 1\ \ 1\ \ 1\ \ \ldots$$

Let us put

$$DX = .\ \ 0\ \ 0\ \ 1\ \ 1\ \ 1\ \ 0\ \ 1\ \ 0\ \ 0\ \ 1\ \ 1\ \ 1\ \ \ldots$$

and

$$D^2X = .\ \ .\ \ 0\ \ 0\ \ 1\ \ 1\ \ 1\ \ 0\ \ 1\ \ 0\ \ 0\ \ 1\ \ 1\ \ 1\ \ \ldots \quad \text{etc.}$$

D is called a *delay operator.* Adding sequences modulo 2 we may form such functions as

$$(D+1)X = DX+X = .\ \ 0\ \ 1\ \ 0\ \ 0\ \ 1\ \ 1\ \ 1\ \ 0\ \ 1\ \ 0\ \ 0\ \ \ldots$$
$$(D^2+1)X = D^2X+X = .\ \ .\ \ 1\ \ 1\ \ 0\ \ 1\ \ 0\ \ 0\ \ 1\ \ 1\ \ 1\ \ 0\ \ \ldots$$

Since a shift operator leaves blanks at the beginning of a sequence it is often convenient to fill these in with noughts; this convention, which corresponds later to switching on the apparatus in a quiescent state, will be adopted henceforth unless it is explicitly stated otherwise. *Note. For recurring sequences corresponding to steady state operation we may wish to have the convention of shifting in the recurring figures. There is a theory of transients and steady states as in the theory of differential equations.* With these conventions, sequences form a vector space (see p. 321). Readers should consider why. Note that the space is of infinite dimension.

The operators are *linear*, that is $D(X+Y) = DX+DY$, and similarly for any power of D, or any polynomial in D. The operators form a *ring* (p. 61), and we later extend them to a *rational domain.*

Express the recurrence relations on p. 26 and the equations on p. 21 in D notation. For example, the chain code previously defined by $e = a+d$ can be written

$$x_n = x_{n-4}+x_{n-1}$$

in terms of individual elements, or

$$X = D^4X+DX = (D^4+D)X$$

in terms of the sequence as a whole.

We are working with two operations, D and addition modulo 2. These can be realized by electrical components called a *delay box* and a *binary adder*, which will be shown diagrammatically as in Figs. 2.7 and 2.8.

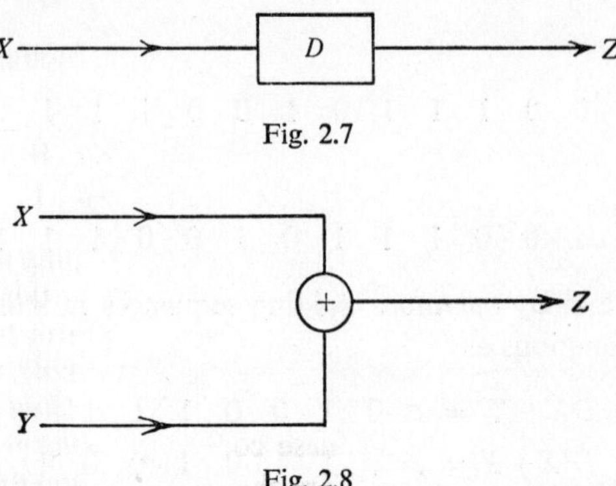

Fig. 2.7

Fig. 2.8

We assume that the delay box operates over discrete intervals of time (dits). The output, Z, is always what the input X was 1 dit earlier. With these two elements we can construct circuit counterparts of polynomials.

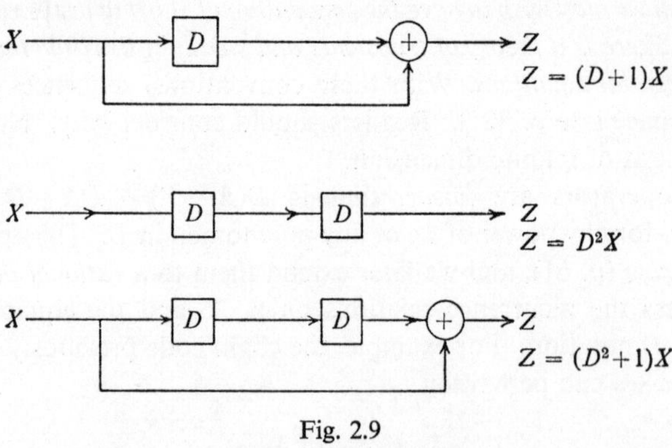

Fig. 2.9

Circuits such as these should be studied in some detail; design various circuits, put in different inputs and find the output. Where the polynomial may be factorized the circuit may be constructed in different ways corresponding to the factorizations.

In the classroom the circuits may be constructed by using the students 'Engelbart fashion' (p. 9 and p. 150).

There is a symmetry in a binary adder which we have not noticed so far.

$$x+y = z \Leftrightarrow x+y+z = 0 \pmod{2},$$

so any two elements may be regarded as the input and the third as the output. With a piece of physical equipment this may mean rewiring; our concern is with logical structure, and we will have to reverse the arrows on some of our diagrams as we change our point of view. We can alter the connections to an adder at will, provided that two go in and one comes out. On the other hand a D box is not symmetric. Going the opposite way through a D box shifts a sequence in the opposite direction.

Consider the circuit $Z = (D+1)X$

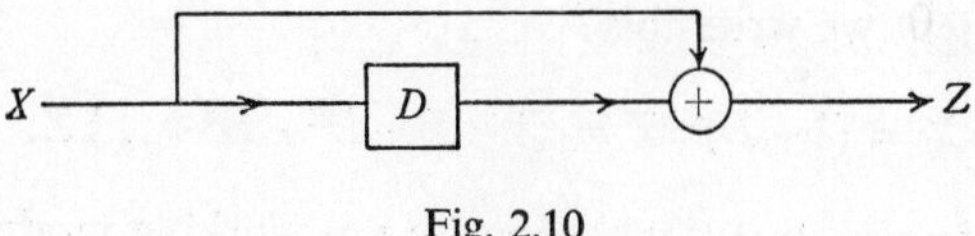

Fig. 2.10

Reverse the directions of the input and output

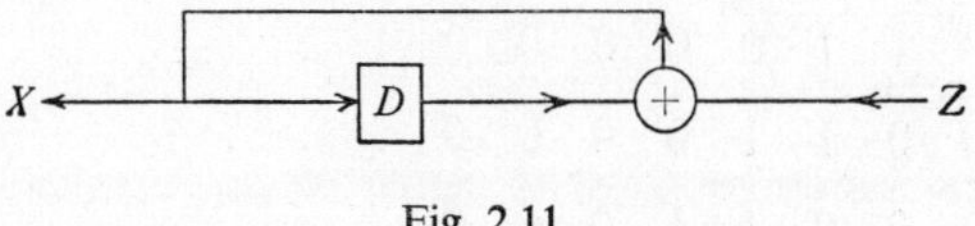

Fig. 2.11

How does this circuit behave? Notice that we are now introducing 'feed-back'. This is a most important idea in electrical machinery.

Let us consider a particular case, putting $X = 0, 1, 0, 0, 0, \ldots$ into the circuit in Fig. 2.10.

X	0	1	0	0	0	.	.
Z	0	1	1	0	0	.	.

The method of construction is indicated.

Now consider the circuit of Fig. 2.11, and feed in this same Z as the input. The output X must now be found as follows:

Z	0	1	1	0	0	.	.
X	0	1	0	0	0	.	.

Note very carefully the direction of the arrows and the way the new table is built up. In this case we find that

> a given sequence fed into the first circuit produces a certain output, this output fed into the second circuit produces the original input.

Whenever the directions of input and output of a circuit are reversed it will be found that this happens. This being so, how may we denote the second circuit?

By
$$X = \frac{1}{D+1} Z.$$

How else might we write this?

$$X = (1+D)^{-1} Z = (1-D+D^2-D^3+\ldots) Z.$$

Need we have negative signs when we are working modulo 2? No: so
$$X = (1+D+D^2+D^3+\ldots) Z.$$

So if

$$Z = 0\ 1\ 1\ 0\ 0\ 0\ 0\ 0\ \ldots$$
$$DZ = 0\ 0\ 1\ 1\ 0\ 0\ 0\ 0\ \ldots$$
$$D^2Z = 0\ 0\ 0\ 1\ 1\ 0\ 0\ 0\ \ldots$$

etc.

then
$$(1+D+D^2+D^3+\ldots) Z = 0\ 1\ 0\ 0\ 0\ 0\ 0\ 0\ \ldots$$

Try this with other inputs and outputs and also with other circuits. The circuit obtained by reversing the input and output is called the *inverse* circuit.

We can now extend the whole theory to rational functions, for example,

$$Z = \frac{D}{D^2+1} X = \left[D.\frac{1}{D^2+1}\right] X = \frac{1}{D^2+1}\cdot[DX].$$

$\frac{D}{D^2+1}$ is got by wiring up D^2+1

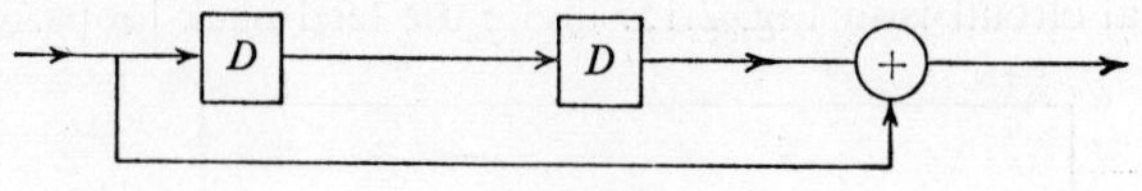

Fig. 2.12

and reversing,

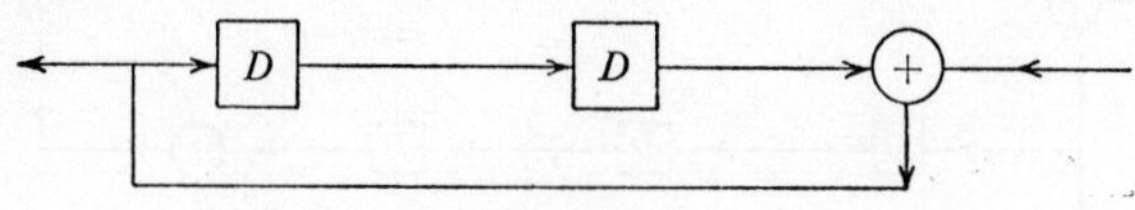

Fig. 2.13

then multiplying by D (front or back does not matter).

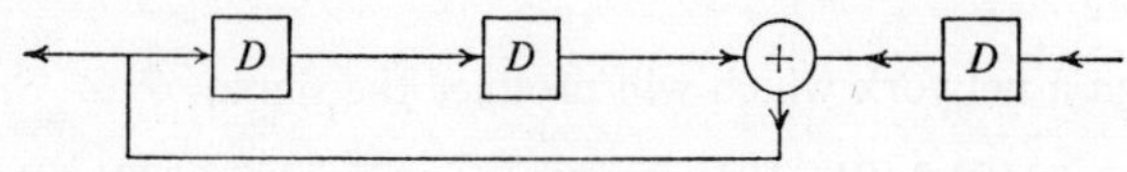

Fig. 2.14

Find Z for various X's

(*a*) by 'following round' the circuit;

(*b*) by algebra;

(*c*) by Engelbart's method.

Note that in this work expressions of the form $1/D$ in isolation should be avoided as they are not physically realizable; but something like $\frac{1}{D}+1$ is.

The following two examples show why we need to introduce rational functions:

Example

Design a network such that when

$$X = 1 \quad 0 \quad 0 \quad 0 \quad 0 \quad 0 \quad 0 \quad 0 \quad 0 \quad . \quad .$$

$$Z = 0 \quad 1 \quad 1 \quad 0 \quad 1 \quad 1 \quad 0 \quad 1 \quad 1 \quad . \quad .$$

$$\begin{aligned} Z &= (D+D^2+D^4+D^5+D^7+D^8+\ldots)\,X \\ &= [(1+D+D^2+D^3+\ldots)+(1+D^3+D^6+\ldots)]\,X \\ &= \left[\frac{1}{1+D}+\frac{1}{1+D^3}\right]X. \end{aligned}$$

The final circuit is in Fig. 2.15. Note the feed-back loops.

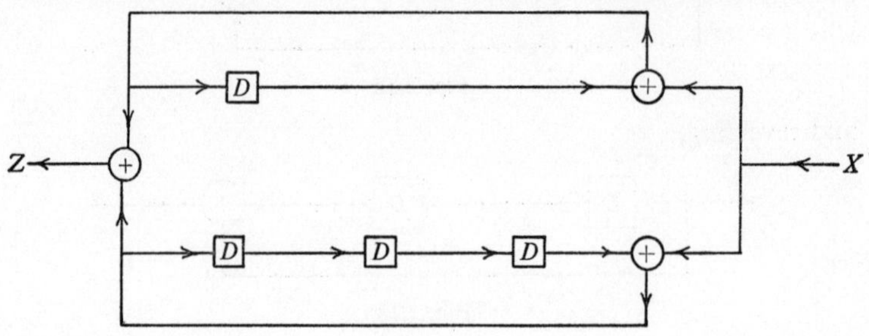

Fig. 2.15

Example

Design a network which will produce the signal

1101 1101 1101 ... from 1000 0000 0000...

$$Z = \frac{D^3+D+1}{D^4+1}\,X.$$

If we wire this up as it stands we get a circuit which needs eight D-boxes and three adders. Rearranging the algebra may do better.

Putting the expression into partial fractions in the usual way, and using such relations as $(D+1)^4 = D^4+1$, which are valid when the coefficients are taken modulo 2, we have

$$Z = \frac{D^3+D+1}{D^4+1} = \frac{1}{D^4+1}+\frac{1}{D^2+1}+\frac{1}{D+1}.$$

This requires seven D-boxes and three adders.

We may, however, take the opportunity for an exercise on Euclid's algorithm and put the expression into continued fraction form:

$$Z=\cfrac{1}{D+\cfrac{1}{(D+1)+\cfrac{1}{(D+1)+\cfrac{1}{D}}}}.$$

Wiring this calls for only four D-boxes, but five adders. There is a better method still, suggested by the arrangement of polynomials which is most convenient for numerical computation,

$$(D^4+1)\,Z = (D^3+D+1)\,X$$
$$Z = X+D[X+D^2(X+DZ)].$$

This is wired in Fig. 2.16.

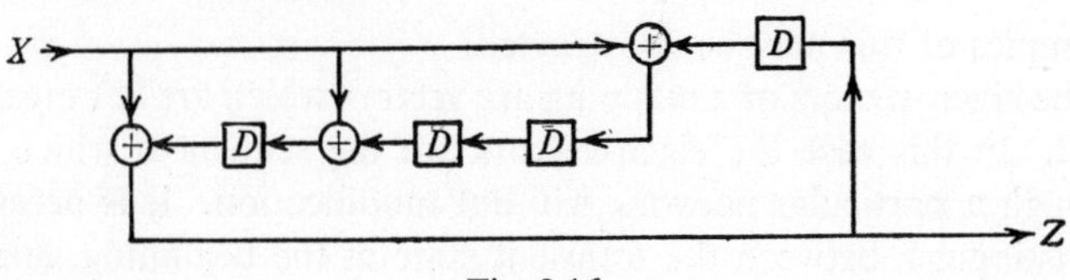

Fig. 2.16

Exercise

Design networks to produce signals corresponding to the recurring binary fractions on p. 18.

Further work and notes

From the point of view of modern algebra we may regard our input sequences as forming a space $\mathfrak{I}$. Likewise the output sequences also form a space $\mathfrak{O}$. In any vector space there are two operations, addition (which we have defined) and scalar multiplication. As our ground field is GF(2) the only scalar multipliers at our disposal are 0 and 1. Multiplying by the operators involving D maps the input space into the output space. As we have observed before, this mapping is a linear operation (that is addition in $\mathfrak{I}$ corresponds to addition in $\mathfrak{O}$). These spaces are of infinite dimension.

Example

Recurring sequences of period n form subspaces of dimension n.

When spaces are mapped on one another certain ideas are nearly always important. We may say that a sequence, or vector, X is annihilated by an operator $F(D)$ if $F(D)\,X = 0$. The set of vectors which are annihilated by an operator forms a subspace. Why? This subspace is called the *kernel* space. The set of vectors Z arising from the vectors X in $\mathfrak{I}$ also forms a space. Why? This space is called the *image* space. Note carefully that we are here concerned with the Z's which actually do arise from X's, bearing in mind the particular operator $F(D)$ concerned. There may be other output sequences which cannot be realized by any X because of the particular operator $F(D)$ involved; but more of this later. We have here an example of the distinction between a mapping *into* and a mapping *onto*.

If we start with a space of dimension n, then there is a general theorem which states that

$$\dim(\text{kernel space}) + \dim(\text{image space}) = n.$$

Examples of this may be constructed.

The eigen-vectors of a mapping are vectors which are left invariant by it. In this case the eigen-vectors are the sequences which pass through a particular network without modification. It is necessary to distinguish between the transient state at the beginning, and the steady state when the delay boxes have 'filled up'. *Compare recurring decimals; especially in this instance recurring decimals in the scale of two.*

The teacher who is interested in these topics may construct numerical examples for himself on the following lines.

What inputs in the steady state give a zero output?

There is always the trivial solution $X = 0, 0, 0, 0, \ldots$, and so this is of no further interest.

Consider the circuit

$$Z = (D^2 + D + 1)X.$$

If $Z = 0$, where by 0 is meant the sequence $0, 0, 0, 0, \ldots$,

$$(D^2 + D + 1)X = 0,$$

so the elements of the sequence X satisfy

$$x_{n-2} + x_{n-1} + x_n = 0$$

$$x_n = x_{n-1} + x_{n-2}.$$

Here we are concerned with sequences which extend to infinity in *both* directions. We may decide one pair of adjacent digits arbitrarily, and the rest are then determined. This gives three non-trivial solutions, starting with (1, 0), (0, 1) and (1, 1).

. . . . 1 0 1 1 0 1 1 0 1 1 . . . .
. . . . 0 1 1 0 1 1 0 1 1 0 . . . .
. . . . 1 1 0 1 1 0 1 1 0 1 . . . .

The third sequence is the sum of the first two, modulo 2, so there are two independent solutions.

Consider other circuits in the same way:

$$D+D^2, \quad D^4+D+1, \text{ etc.}$$

Investigate some cases where the polynomial factorizes:

$$D^3+D^2+D+1 = (D+1)(D^2+1), \text{ etc.}$$

Eigen-vectors of the mapping form the kernel space of $F(D)-1$. Likewise, in this context, it is of interest to ask what sequences pass through a particular network with a simple delay of one, or more, dit; that is, to find sequences satisfying $F(D)X = DX$, etc.

Example

$$F(D) = D^3+D^2+1.$$

For a kernel vector $x_n+x_{n-2}+x_{n-3} = 0$, so $x_n = x_{n-2}+x_{n-3}$. There are three independent solutions:

$$\ldots\ \dot{1}\ 0\ 0\ 1\ 0\ 1\ \dot{1}\ \ldots$$
$$\ldots\ \dot{0}\ 1\ 0\ 1\ 1\ 1\ \dot{0}\ \ldots$$
$$\ldots\ \dot{0}\ 0\ 1\ 0\ 1\ 1\ \dot{1}\ \ldots$$

This assumes suitable initial conditions on the D-boxes, i.e. recurrencc is fcd in as wc *shift*. Also we use dots over the top to denote recurrence, as in recurring decimals. Any linear combination of these is a kernel vector. Including the trivial case of the zero vector, there are eight kernel vectors in all. They form a space of dimension 3. It is a subspace of what might conveniently be called $\mathfrak{S}$, the space of all sequences of period 7. These vectors are always, because of the recurrence relation, chains of the type discussed on p. 24.

Some networks produce chains of the maximum possible length; others do not. The chain aspect is important, but cannot be discussed fully here. (See the note at the end of the chapter.) All of these eight inputs result in zero output from the circuit (check). If the circuit is driven backwards these outputs can rise with zero input, i.e. they are self-maintaining oscillations, which may arise by accident or because of initial states of the D-boxes.

It is also appropriate in this context to employ the language of group theory. It is part of the definition of a vector space that, with respect to its additive properties, it forms an Abelian group. The kernel space is a subgroup. Whenever we have a group and a subgroup there is a corresponding decomposition into cosets. What significance do the cosets have in this application?

The cosets are produced by taking some other element and adding (additive terminology is conventional in an Abelian group) to the elements of the kernel. In other words members of the same coset differ from one another by an element of the kernel. Since any element from the kernel gives a zero output the physical significance of cosets is that all members of the same coset produce the same output.

Applying this to the present circuit consider the sequence

$$X = \ldots\ldots\ldots\ 1\ 1\ 1\ 1\ 1\ 1\ 1\ \ldots\ldots$$

It is easy to verify that the output $Z = (D^3+D^2+1)\,X$ is the same. The general theory about decomposition into cosets tells us that there are seven other sequences which give the same output; all of these are of the form $\ldots\ldots\ 0\ 1\ 1\ 0\ 1\ 0\ 0\ \ldots\ldots$, but they occur at different phases, corresponding to different initial states of the delay boxes.

Continuing with the same example we may say that the circuit maps the input space on to the output space *homomorphically*—meaning that many vectors of the one correspond to one vector of the other. What happens then in the case of the inverse circuit? If the circuit is driven the other way clearly only one output can arise from a given input, so what has become of the many-one relation? The answer to this is that, remembering that we are dealing now with recurring sequences of signals, one input can maintain a number of different outputs according to the values of certain arbitrary terms (the initial states of the delay boxes). If these are specified then we know which of the possible coset of outputs will actually arise. This problem is of interest if we switch from one input to another at some arbitrary point in a cycle. Consider, for example, the circuit

$$Z = \frac{D^2+D+1}{D^3+D^2+1}\,X.$$

An expression such as this may be written in the form

$$(D^3+D^2+1)\,Z = (D^2+D+1)\,X = Y\ \text{(say)},$$

and the circuit may be wired accordingly.

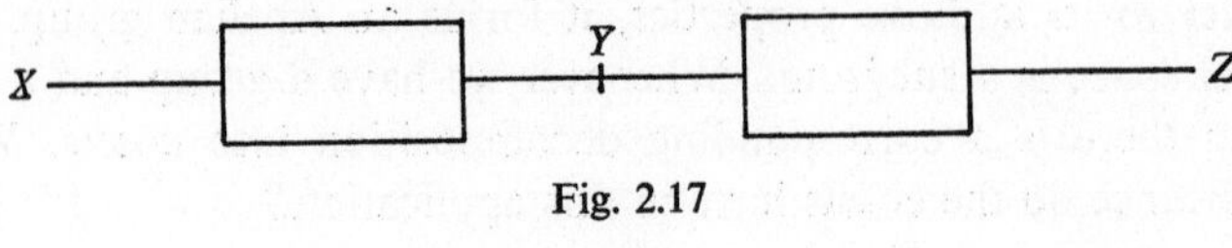

Fig. 2.17

This will not usually be the most convenient arrangement electrically, but it clarifies the discussion to follow. What inputs correspond to what outputs? We find the respective kernels when the X space and the Z space are mapped on the Y space. Cosets in these spaces are then mapped on individual elements of the Y space. If the circuit is driven from one side then we may say that any of a given coset of inputs will produce any of a given coset of outputs, and that this is reversible. Which output comes from which input depends on the initial states of the delay boxes.

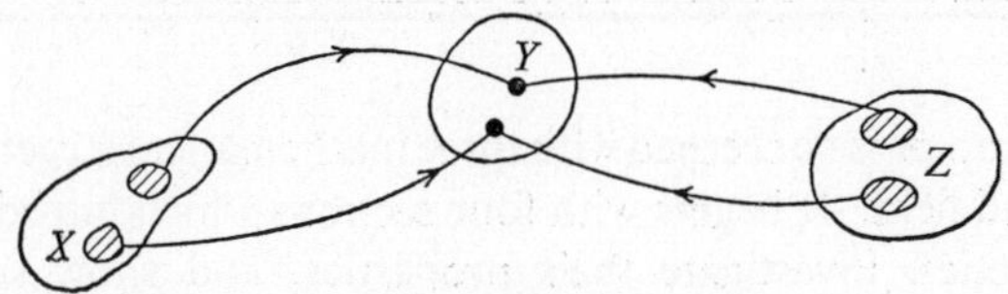

Fig. 2.18

The Z kernel is of dimension 3, with 8 vectors.
The X kernel is of dimension 2, with 4 vectors.

Polynomials producing chains of maximum length (Albert [5])

Digits	Length of chain	Polynomial
4	15	D^4+D+1
5	31	D^5+D^2+1
6	63	D^6+D+1
7	127	D^7+D^3+1
8	255	$D^8+D^4+D^3+D^2+1$
9	511	D^9+D^4+1
10	1023	$D^{10}+D^3+1$
11	2047	$D^{11}+D^2+1$
12	4095	$D^{12}+D^6+D^4+D+1$
13	8191	$D^{13}+D^4+D^3+D+1$

Reference

Huffmann [65].

3

FINITE ARITHMETICS AND GROUPS

This chapter is concerned with three mathematical structures—group, ring and field. It begins with four sections which introduce the finite arithmetics, investigate their properties, and show some of their simple applications. These arithmetics are either rings or fields. The section on isomorphism, on p. 51, introduces an idea which is a main theme of the whole chapter—the idea that two mathematical systems with different elements may be structurally the same; the same, that is, in the way the elements combine to give other elements under operations such as addition and multiplication. In section five, starting a new line of development, we study symmetrical figures since it is in this context that the idea of the group, perhaps the most fundamental structure of all, most naturally arises. Subsequent sections illustrate different aspects of group structure, notably the subgroup and the coset, and also show the same structure appearing within the finite arithmetics. The last two sections explore the field structure of the arithmetics, indicating some unusual but important applications.

The material of the earlier sections is intended to be usable more or less directly for lessons; the later sections aim to give a brief survey of a wider field. Notes for teachers are interpolated at what seemed the appropriate points. The question is often asked: how does this fit in to the course I am teaching? In the first place, most teachers will probably want to try some of these topics out as individual lesson-sequences, and for this purpose much of the material is capable of adaptation to suit a wide range of secondary ages and abilities. In the longer view, we would regard the material of this chapter as being an outline which could be filled in to make one part of a developing course of modern algebra, continuing from the first secondary year to the sixth form. It would fit into the whole mathematics course in much the same way as the traditional work in algebra, some of which it would replace.

Teachers unfamiliar with this material may find some of it difficult at first reading; where this is so the sections should be omitted. This applies, for example, to all sections which discuss GF(p^n).

1. TIMETABLES, CALENDARS AND CLOCKS

At a certain school, which works five days a week, morning assembly is held every second day. So if there is an assembly on a certain Wednesday, succeeding ones are on Friday, then Tuesday, Thursday, Monday and then Wednesday again. Thus in any sequence of five assemblies there is just one on every possible day. Another school which works six days a week also has assemblies on every second day. In this case there is a choice: the assemblies may be arranged to take place on Mondays, Wednesdays and Fridays, or else on the other three days. The two possible sets of days do not overlap at all. We might represent these situations thus:

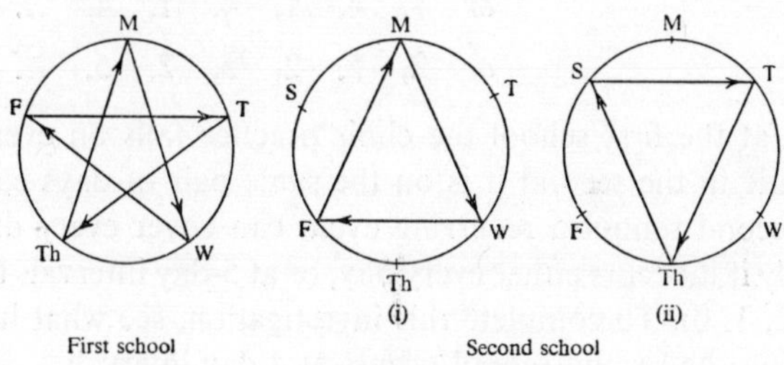

Fig. 3.1

Alternatively, we might give numbers 0, 1, 2, ..., 4 or 5 to the days, in which case the sequences of days would be as follows (the reason for preferring to start at 0 and not at 1 will appear immediately).

First school		2,	4,	1,	3,	0,	2,	...
Second school	(i)	0,	2,	4,	0,	2,	4,	...
or	(ii)	1,	3,	5,	1,	3,	5,	...

These sequences are arithmetic progressions of common difference 2, but the numbers are kept within the set 0–4 or 0–5 by subtracting multiples of 5 or 6. These numbers, operated on in this way, are said to form the arithmetic 'modulo 5' or 'modulo 6'.

Now suppose that at each of these two schools choir practices are held during the lunch hour every third day. This situation could be represented thus:

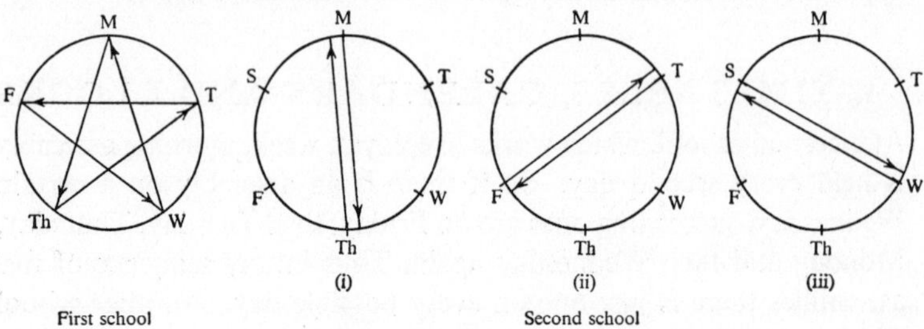

Fig. 3.2

or thus:

First school	1,	4,	2,	0,	3,	1,	...
Second school	0,	3,	0,	3,	0,	3,	...
or	1,	4,	1,	4,	1,	4,	...
or	2,	5,	2,	5,	2,	5,	...

Again at the first school the choir practice falls on every day in turn, while in the second it is on the same pair of days each week. In the second school a recurring event can cover every day of the week only if it occurs either every day, or at 5-day intervals (sequence 5, 4, 3, 2, 1, 0). To complete this investigation, see what happens if an event at the second school occurs at 4-day intervals.

What we have been considering here are arithmetic progressions in the arithmetics modulo 5 and modulo 6. Here are some more arithmetical progressions in different arithmetics:

(1) (mod 8, difference 3) 1, 4, 7, 2, 5, 0, 3, 6, 1, 4, ...

(2) (mod 8, difference 6) 3, 1, 7, 5, 3, 1, ...

(3) (mod 8, difference 2) 2, 4, 6, 0, 2, 4, ...

(4) (mod 7, difference 2) 2, 4, 6, 1, 3, 5, 0, 2, 4, ...

Make up some more. Try to formulate statements of the conditions under which the sequence contains the full set of possible numbers; and, in the other cases, how many different subsets the numbers are divided into.

Problems

(1) Which day is half-way between Monday and Tuesday? Show that if there is such a day, it is necessarily Friday.

(2) A faulty clock strikes correctly up to eleven strokes but then reverts to one. If it is striking the correct hours on Monday morning, when will it next be striking the right hours?

Extend your solution to deal with the case in which the clock strikes only up to ten; then nine; can you formulate a general result?

(3) If an event occurs every n months, for what values of n can the event fall, over a suitable period, in every different month of the year?

2. CONGRUENCES

Terminology and background for the teacher

We have spoken in the previous section of 'finite arithmetics' to a given modulus. The teacher should be familiar with other modes of expression. If two integers differ by a multiple of an integer n they are said to be *congruent modulo n*. Thus

$$15 \equiv 39 \pmod 6$$

and

$$35 \equiv 0 \pmod 7.$$

Relations of this kind are called congruences.

If a modulus is chosen then all of the integers may be put into *equivalence classes*, two integers being in the same class if they are congruent, with the modulus in question. Thus working modulo 5 we get the classes

$$\begin{array}{l} \ldots\ldots, -15, \quad -10, -5, 0, 5, 10, 15, \ldots\ldots \\ \ldots\ldots, -14, \quad -9, -4, 1, 6, 11, 16, \ldots\ldots \\ \ldots\ldots, -13, \quad -8, -3, 2, 7, 12, 17, \ldots\ldots \\ \ldots\ldots, -12, \quad -7, -2, 3, 8, 13, 18, \ldots\ldots \\ \ldots\ldots, -11, \quad -6, -1, 4, 9, 14, 19, \ldots\ldots \end{array}$$

These classes are called *residue classes*. The reason for the name is obvious if we remember that residue means remainder. The arithmetics described in the previous section can be regarded as arithmetics of these residue classes. On the blackboard it is easy to make a distinction between ordinary numbers and residue classes by writing them in different colours. In print the distinction tends to be obscured, and the five classes listed above are often denoted simply by 0, 1, 2, 3, 4. It is now fashionable to describe this sort of thing as 'an abuse of language'.

In ordinary arithmetic there is a logical distinction between the natural number 2, the integer $+2$, and the rational number $2/1$; but in everyday arithmetic our notation often fails to make the distinction. At a much later stage in the pupil's development than is our concern at the moment the distinction should be made, and then he does not always find it easy to grasp the points at issue. Then it is standard practice to explain that the numbers at successive levels (natural numbers, integers, rational numbers, real numbers, complex numbers) should properly be defined as classes (or sets) of the numbers at the level below. The definition of residue classes in terms of the integers is perhaps the simplest example of this type of definition, and so it has a great value when the pupil comes to reconsider the number system from a strictly logical point of view in the sixth form or in the first year at university.

Here we are only concerned that the teacher should appreciate the logical background; the lessons themselves proceed firmly at an intuitive level. They are taken from Puig Adam [2], and the first develops the idea that a set of residue classes forms a *ring*, and that if the modulus is prime the ring is also a *field*. (These terms are defined in §6 of the present chapter.) This first lesson is particularly remarkable because it shows, among other things, how pupils who have mastered the Cuisenaire rods approach the problem of residue classes using a notation which is *more* abstract than the one which would be used by expert mathematicians. The lesson is only summarized here, but the original account merits careful attention as many of the points are deep, and the whole lesson obviously proceeded at a level of understanding rather different from that which one might expect *a priori*.

Lesson

The lesson was given to 24 pupils of ages 12–13. They were grouped round a large table and they were numbered off and handed out

Cuisenaire rods of five different colours: white, red, green, pink and yellow. The rods were handed out in this order, one to each boy, from 1 to 24, with such questions as:

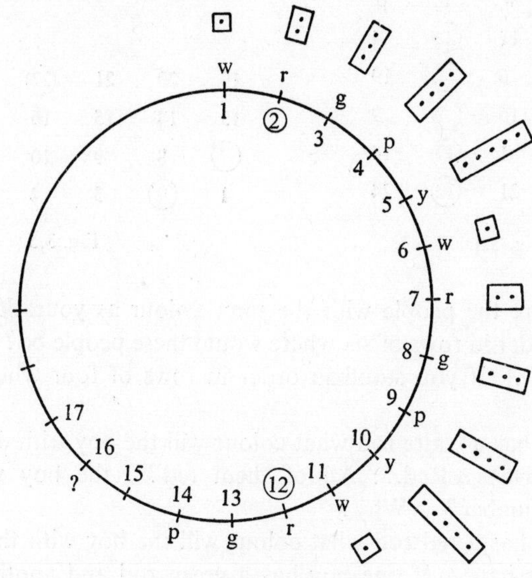

Fig. 3.3

What colour must I give you? How have I given out the rods? List all the numbers with a red rod. What colour has 12? State another number with the same colour as yours. Write down all the numbers with the same colour as yours. Who has the same list as you?

Thus the boys are associated in classes according to the colours of the rods which they have. Boys with the same colour rod have numbers which are 'congruent modulo 5'. This idea is established by question and answer and suitable terminology introduced.

Write in your books some numbers above 50 which are congruent to your own number. Write down some numbers above 85 which are congruent to 12, 28, 64, etc. What do you observe? All of the differences are multiples of 5.

They write in their books, 'The difference between two numbers congruent modulo 5 is a multiple of 5'.

The pupils are then asked to stand in five columns according to the class to which they belong. No. 12 is asked where he should stand, and so on.

g	y	w	r	p
13		11	㉒	4
23	20	6	⑦	19
3	5	16	②	9
8	15	1	⑰	14
18	10	21	⑫	24

Fig. 3.4

19	20	21	㉒	23	24
13	14	15	16	⑰	18
⑦	8	9	10	11	⑫
1	②	3	4	5	6

Fig. 3.5

Where are the people with the same colour as yourself? If you stood in order in rows of six where would these people be? (*Sloping, in diagonals.*) If you stood in order in rows of four where would they be?

If a boy has a white rod what colour will the boy with double his number have?...Red....All of them red?...the boy with four times his number?...Why?

If a boy has a red rod what colour will the boy with three times his number have? If one boy has a green rod and another boy a white rod, what colour will the boy with the sum of their numbers have? If one boy has a yellow rod and another boy has some other colour, what colour will the boy with their sum have? (*The same as the second boy.*)

A boy has a pink rod and another boy has a green one. What colour will the boy with the product of their numbers have? A boy has a white rod and another boy has a rod of some other colour. What colour will the boy with the product of their numbers have? (*The same as the second boy.*)

A boy has a rod of any colour, and another boy has a yellow rod. What colour will the boy with the product of their numbers have?

The pupils had developed the properties of a finite arithmetic modulo 5. Note that their notation was a colour notation and not the 0, 1, 2, 3, 4 that our own mathematical habits might have dictated.

More intelligent pupils were also asked, 'A red number multiplied by another number gives a white number. What colour is the other number?'

Lesson

The second lesson begins by interesting the pupils in a trick. They write a number with a few different digits in their books. They alter the order of the digits and take the smaller number from the larger. They erase one of the digits in the answer. They add the remaining digits together, and the master, on being told the sum, tells them the digit they have crossed out.

The number has to be made up to a multiple of nine since if we consider $513-135$ *we have* $1(10-100)+3(1-10)+5(100-1)$. *One must seek to arouse curiosity as to how the trick is done.*

The explanation is roundabout. First a ring is formed of nine boys (after all, we are dealing with a *ring* of numbers). Numbers are assigned in order to the boys in turn. After a while jumps are made. Who will have 29? Who will have 39, etc.? They gradually understand the pattern.

Who will have 100, 1000, 10000, etc.? Who will have 20, 200, 2000, etc.? Who will have 13, 31, 103, 301, 22, etc.?

Various numbers are assigned and, if it does not occur to someone to do so, the boys are invited to sum the digits. It is then found that 'summing the digits' enables a position to be decided much more quickly.

Who will have 2435? 7264? 4267, etc.? A number x belongs to you. Who will have $x+20$? Who will have $x-20$? If x belongs to you and y belongs to him who will have $x+y$? Who will have $x-y$? Who does 7264 belong to? ...4267...? Who does the difference belong to? Then what do you notice about numbers formed from the difference of two numbers with the same digits? If you sum the digits of such a number what will happen? If you now cross out a digit how do I decide what it is? What would I decide if the sum remaining at the end was a multiple of 9?

While seeking to understand the trick the pupils have discovered the main properties of the ring of residues modulo 9. Of course, other similar tricks can be used in the same way.

The rule of 'Casting out Nines' seems to be taught comparatively little today. It is an extremely useful rule to know, and the reason why it is so little taught must surely be an idea that one should not teach mere rules whose underlying reasons cannot be understood. This shows that the reasons can be understood quite well, and that the process of understanding also gives an introduction to one of the themes of modern algebra—a number ring.

3. CODES

What does the class know about codes? Various systems may be suggested, but the discussion is directed towards substitution codes. Most people will know systems where one letter of the alphabet is substituted for another by using a key. How can we avoid having to memorize, or carry around, a key?

There are simple methods such as replacing each letter by the one which follows. What is to be done with Z? Can we describe the substitution by a formula?

$$x' \equiv x+1 \pmod{26}.$$

After discussing further examples of this type any disadvantages can be considered. For instance such a code is easily discovered.

Can we code using a formula such as

$$x' \equiv 2x \pmod{26}?$$

What goes wrong in this case? To understand better why some formulae will not do consider a case with smaller numbers—

$$x' \equiv 2x \pmod{6}.$$

In order to decode it is convenient to have the multiplication table available.

	1	2	3	4	5
1	1	2	3	4	5
2	2	4	0	2	4
3	3	0	3	0	3
4	4	2	0	4	2
5	5	4	3	2	1

Now how would we decode $x' = 4$? It could have come from either $x = 2$ or $x = 5$, and we are in no position to say which it came from. This sort of breakdown always happens when the mapping of the symbols of one set into another set is not one–one. We notice that even though 6 is not prime had we chosen

$$x' \equiv 5x \pmod{6}$$

then all would have been well.

Could we use $x' \equiv 2x \pmod{27}$?

$x' \equiv 3x \pmod{26}$?

$x' \equiv 4x \pmod{26}$?

When is $x' \equiv nx \pmod{m}$ suitable?

(*m and n must not have any factors in common.*)

Thus
$$x' \equiv kx \pmod{29}, \quad (k < 29),$$
would always be suitable with any choice of k. Also there is room for three extra symbols, which could be 'space', 'stop', 'comma', which is convenient.

How are substitution codes deciphered when the key is not known? There is an interesting account in Edgar Allan Poe's story *The Gold Bug*[99].

Can we devise more effective ways of scrambling the letters? This question is taken up again in §12.1.

4. FINITE ARITHMETICS AS NUMBER-SYSTEMS

We now compare these finite arithmetics with the ordinary number-system, to see which of the usual techniques can be used with them. The addition and multiplication tables for the arithmetics mod 5 and mod 6 are as follows:

mod 5 +	0	1	2	3	4
0	0	1	2	3	4
1	1	2	3	4	0
2	2	3	4	0	1
3	3	4	0	1	2
4	4	0	1	2	3

mod 5 ×	0	1	2	3	4
0	0	0	0	0	0
1	0	1	2	3	4
2	0	2	4	1	3
3	0	3	1	4	2
4	0	4	3	2	1

mod 6 +	0	1	2	3	4	5
0	0	1	2	3	4	5
1	1	2	3	4	5	0
2	2	3	4	5	0	1
3	3	4	5	0	1	2
4	4	5	0	1	2	3
5	5	0	1	2	3	4

mod 6 ×	0	1	2	3	4	5
0	0	0	0	0	0	0
1	0	1	2	3	4	5
2	0	2	4	0	2	4
3	0	3	0	3	0	3
4	0	4	2	0	4	2
5	0	5	4	3	2	1

Exercises

To be worked first in mod 5 arithmetic then in mod 6.

(1) Find $(4+3)+2$ and $4+(3+2)$.

(2) Find $2-4$ (i.e. what has to be added to 4 to make 2) and $5-2$ (mod 6 only); also $4-1$ and $1-4$.

(3) Find $(4.3)2$ and $4(3.2)$.

(4) Find 2^2, 2^3, 2^4, 2^5, 2^6, and 3^2, 3^3, 3^4, 3^5, 3^6; note what you observe.

(5) Solve the following equations and note any differences from ordinary arithmetic. Again work first in mod 5 arithmetic, then in mod 6.

$2x = 4$, $4x = 2$, $2x = 3$, $2x = 1$, $5x = 1$, $2x = 0$, $3x = 0$, $5x = 0$.

(6) If $1/2$ means the number which, when multiplied by 2, gives 1, what is $1/2$ in these two arithmetics? Can we avoid fractions? Consider the equations of Question 5 from this point of view.

(7) Solve $x^2 = 2$, $x^2 = 1$, $x^2 = 4$; $x^4 = 1$; $x^3 = 2$, $x^3 = 3$, $x^3 = 4$; $x^2+3x+2 = 0$; $x^2+1 = 0$. Note what you observe.

Notes

The exercises bring out the following properties of these finite arithmetics.

(1) Subtraction is always possible within the set, without the need of any negative numbers.

(2) In the mod 5 arithmetic division is always possible, without the need of any fractions; except that division by zero is meaningless, as in ordinary arithmetic.

The properties described in (1) and (2) are the essential conditions for a *field.* (See p. 61.)

(3) In the mod 6 arithmetic, division is not always possible, as there are relations such as $2.3 = 0$ and $2.2 = 2.5$. This system is not a field, though it is a *ring*. (See p. 61.)

(4) In the mod 5 arithmetic, the fourth power of every number is 1; no such property exists in the mod 6 system. This is an instance of Fermat's theorem [14].

(5) In the mod 5 arithmetic, quadratic equations have two roots or none; in mod 6, they may have as many as four. (Try substituting values in the equation.) This illustrates how the usual properties of quadratics are true in any field, but not in a ring which is not a field.

Exercise

Investigate the arithmetics mod 7 and mod 9, and see whether these properties exist in them.

The finite arithmetic with any modulus is a ring; those with a prime modulus are also fields; the arithmetic mod p, where p is prime, is called the *Galois field* GF(p). Further remarks and references on rings and fields occur on p. 61.

Isomorphism

If we remove that part of the modulo 6 multiplication table which behaves irregularly, we are left with the left-hand table following, which we compare with the multiplication table for the non-zero numbers modulo 3.

mod 6 ×	1	5
1	1	5
5	5	1

mod 3 ×	1	2
1	1	2
2	2	1

A similar correspondence can be observed between the multiplication tables for 1, 2, 3, 4 (mod 5) and 1, 3, 7, 9 (mod 10). Again we omit those rows and columns of the complete tables which contain zeros.

mod 5 ×	1	2	4	3
1	1	2	4	3
2	2	4	3	1
4	4	3	1	2
3	3	1	2	4

mod 10 ×	1	3	9	7
1	1	3	9	7
3	3	9	7	1
9	9	7	1	3
7	7	1	3	9

1 2 4 3
↕ ↕ ↕ ↕
1 3 9 7

This one–one correspondence between the numbers of the two sets has the property that if two numbers a, b in the first set have product c, then the corresponding two numbers a', b' of the second set have a product c' which is the number corresponding to c. This is the reason underlying the fact that the multiplication tables show the same pattern. We say that this correspondence 'preserves products', and such a one–one correspondence is an example of an 'isomorphism'. The correspondence

$$\begin{array}{cccc} 1 & 2 & 4 & 3 \\ \updownarrow & \updownarrow & \updownarrow & \updownarrow \\ 1 & 7 & 9 & 3 \end{array}$$

is another isomorphism, as may be verified, but

$$\begin{array}{cccc} 1 & 2 & 4 & 3 \\ \updownarrow & \updownarrow & \updownarrow & \updownarrow \\ 1 & 9 & 3 & 7 \end{array}$$

is not, since in the first set $4.4 = 1$ while in the second set $3.3 = 9$, and 9 does not correspond to the 1 of the first set. Further investigation of this situation leads to the following observations: in the modulo 5 system, $1^2 = 4^2 = 1$, while $2^2 = 3^2 = 4$, so that $2^4 = 3^4 = 1$. 4 is said to be of *order* 2, while 2 and 3 are of order 4: the order is the least power of the number which equals 1. In the modulo 10 case, 9 is of order 2, and 3 and 7 are of order 4. To obtain an isomorphism between the two sets, numbers which correspond to each other must be of the same order.

Exercises

Investigate the multiplications of the following sets for isomorphism:

(1) 1, 5, 7, 11 (mod 12). (2) 1, 3, 5, 7 (mod 8).

(3) 1, 2, 4, 5, 7, 8 (mod 9). (4) 1, 3, 5, 9, 11, 13 (mod 14).

Make up and try some further sets.

Isomorphisms also exist between some of the addition tables and certain multiplication tables. For example, the set (0, 1, 2, 3) under addition modulo 4 is isomorphic to (1, 2, 4, 3) under multiplication modulo 5; compare the following addition table with the earlier multiplication table (mod 5).

mod 4 +	0	1	2	3
0	0	1	2	3
1	1	2	3	0
2	2	3	0	1
3	3	0	1	2

$$\begin{array}{ccccl} 0 & 1 & 2 & 3 & +\text{mod } 4 \\ \updownarrow & \updownarrow & \updownarrow & \updownarrow & \\ 1 & 2 & 4 & 3 & \times\text{mod } 5 \end{array}$$

The use of logarithms is based on a similar isomorphism between the set of all real numbers under addition and the set of positive real numbers under multiplication.

Exercise

Find, if possible, additive sets isomorphic to each of the sets considered above.

5. SYMMETRY GROUPS

The letters of Fig. 3.6 possess symmetry of two types, reflection in a line, 'T-symmetry', and reflection in a point, 'S-symmetry'.

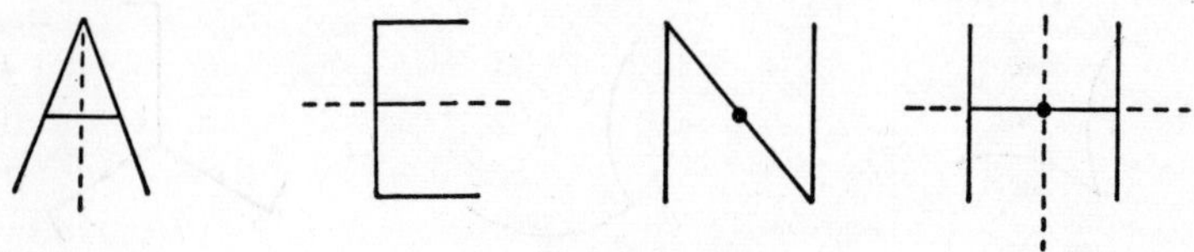

Fig. 3.6

If necessary, a preliminary exercise for this lesson would consist of identifying these two types of symmetry in the capital letters, and in familiar objects. But now consider Fig. 3.7*i*. If this is to be said to possess symmetry—and it seems reasonable to wish to do so—we must adopt a wider definition. What could be suggested? We shall agree that a *symmetry* of the figure is any transformation (strictly, any isometry) which leaves the figure as a whole in the same position, though the position of its constituent parts may be changed. The symmetries of Fig. 3.7*i* then consist of rotations through 120°, 240° and 360°. *Rotating the figure through 360° is equivalent to leaving it where it is. This 'identity operation' is always to be regarded as a symmetry, the main reason being that mathematicians like to have their sets of operations forming closed 'groups', with the resultant of two operations, in this case symmetries, being a third operation of the set.*

Now study the other diagrams of Fig. 3.7 and mark all the symmetries you can find. Are all the figures symmetrical? (*No, h is not.*) Would you say that any of these figures had the same type

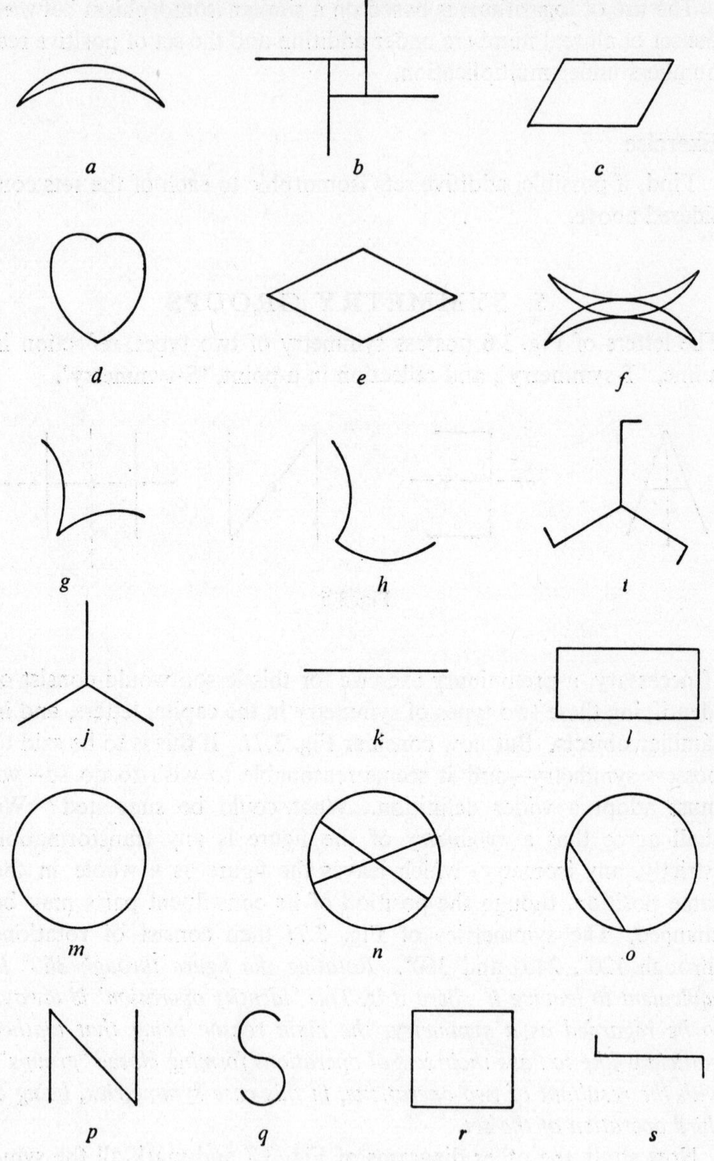

Fig. 3.7

of symmetry? Let us try to see how many different types we can find.

(*a*) Reflection in 1 line	(*b*) Reflection in 2 lines at right angles	(*c*) Rotation through 180° (half-turn)	(*d*) Rotation through 120° and 240°
a	*e*	*c*, *p*, *q*, *s*, and also all those in previous column	*i*
d	*f*		—
g	*k*		—
o	*l*	—	—
—	*n*	—	—

(*e*) Rotations through 120°, 240° and reflection in 3 lines at 120° angles	(*f*) Rotations through 90°, 180°, 270° and no reflections	(*g*) Rotations through 90°, 180°, 270° and reflections in 4 lines
j	*b*	*r*

Collect as many examples as you can of objects or drawings with symmetry, and classify them according to this list.

Can you find any further types?

Now try starting with an irregular shape and making it into patterns. Do we get the same types? Can we get all these types, if we repeat in the right way? *Use tracing and folding.*

Fig. 3.8 is obtained by one fold; Fig. 3.9 by tracing round a cut-out triangle and rotating it four times through 90°.

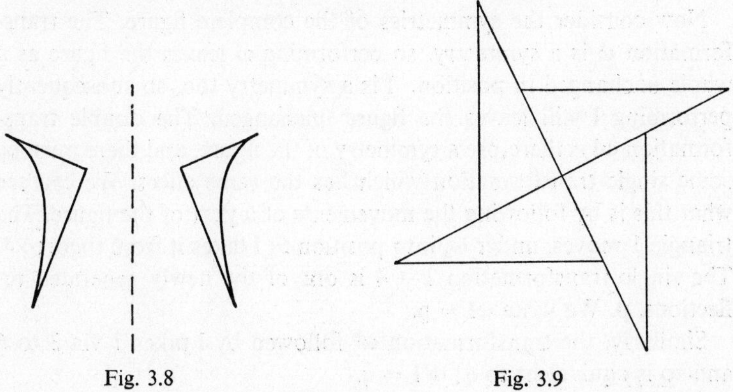

Fig. 3.8　　Fig. 3.9

Make a pattern with the same symmetries as Fig. 3.7*i*. The full-line part of Fig. 3.10 is an example.

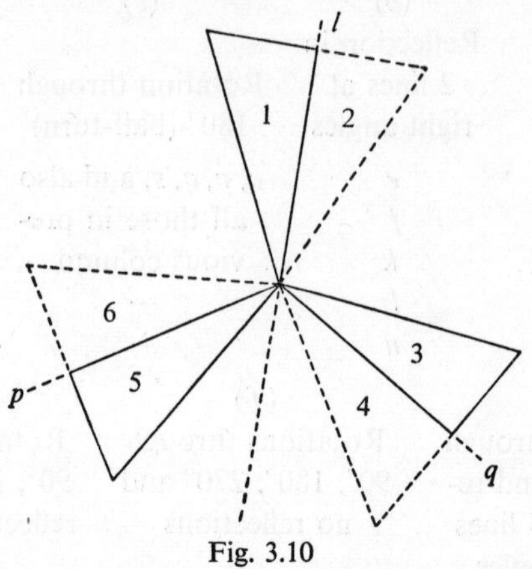

Fig. 3.10

Now, by folding and tracing, make it symmetrical about any line *l* (which need not correspond to a line of the original figure). What do you find? The figure now has two further axes of symmetry, each making 120° with *l*. Let us inquire into this. Denote the transformation of rotating through 120°, anticlockwise, by **ω** and that of reflecting in the line *l*, by **l**. Denote reflections in the two new axes by **p** and **q**; the axes themselves will be referred to as *p* and *q*. Performing transformations **ω** and **l** in succession will be denoted by **ωl**, and this will be called the *product* of **ω** and **l**. A rotation through 240° is then **ω**.**ω** or $\boldsymbol{\omega}^2$.

Now consider the symmetries of the complete figure. The transformation **ω** is a symmetry, so performing **ω** leaves the figure as a whole unchanged in position. **l** is a symmetry too, so subsequently performing **l** still leaves the figure unchanged. The double transformation **ωl** is therefore a symmetry of the figure, and there must be some single transformation which has the same effect. We can see what this is by following the movements of a part of the figure. The triangle 1 moves, under **ω**, into position 5; **l** takes it from there to 4. The single transformation $1 \rightarrow 4$ is one of the newly generated reflections, **p**. We write $\boldsymbol{\omega}\mathbf{l} = \mathbf{p}$.

Similarly, the transformation $\boldsymbol{\omega}^2$ followed by **l** takes 1 via 3 to 6 and so is equivalent to **q**; $\boldsymbol{\omega}^2\mathbf{l} = \mathbf{q}$.

Thus, including the identity transformation **I**, the full set of symmetries of Fig. 3.10 contains six transformations, **I**, **ω**, **ω**^{2}, **l**, **p**, **q**. This set has been generated by starting with **I**, **ω**, **ω**^{2} (the symmetries of the original figure), and combining each of them with **l**.

There is another point worthy of note in this situation. Suppose we perform on this figure the transformation **l** followed by **ω**. This takes triangle 1 to 2 and thence to 6; so **lω** = **q**, which is not the same as **ωl**. It is of interest to display the whole set of combinations of pairs of symmetries in a table.

	I	**ω**	**ω**^{2}	**l**	**p**	**q**
I	**I**	**ω**	**ω**^{2}	**l**	**p**	**q**
ω	**ω**	**ω**^{2}	**I**	**p**	**q**	**l**
ω^2	**ω**^{2}	**I**	**ω**	**q**	**l**	**p**
l	**l**	**q**	**p**	**I**	**ω**^{2}	**ω**
p	**p**	**l**	**q**	**ω**	**I**	**ω**^{2}
q	**q**	**p**	**l**	**ω**^{2}	**ω**	**I**

This table immediately raises further interesting questions. Why, for example, does it divide so neatly into four quarters? In what ways are **l**, **p**, **q** different from **I**, **ω**, **ω**^{2}? Is it just a coincidence that each row contains each letter once and once only? If not, what lies behind this?

Some of these questions will be answered later in this chapter. Others may be pursued in, for instance, Andree [8].

Classification of these symmetry groups

We end this somewhat long 'lesson' by giving in Fig. 3.11 a systematic list of the symmetry types discussed.

Each diagram is built up by starting with the unit hook shown in the first one, and performing on it all the transformations of the symmetry set.

v, **h**, **p**, **q**, **d**$_1$, **d**$_2$ are reflections in the axes indicated.

r, **ω**, **s** are rotations through 180°, 120°, 90° respectively.

C_n, consisting of the n rotations through angles which are the multiples of $2\pi/n$, is called the *cyclic group* of order n; D_n, consisting of these rotations and n reflections in axes making angles of π/n with each other, is called the *dihedral group* of order $2n$.

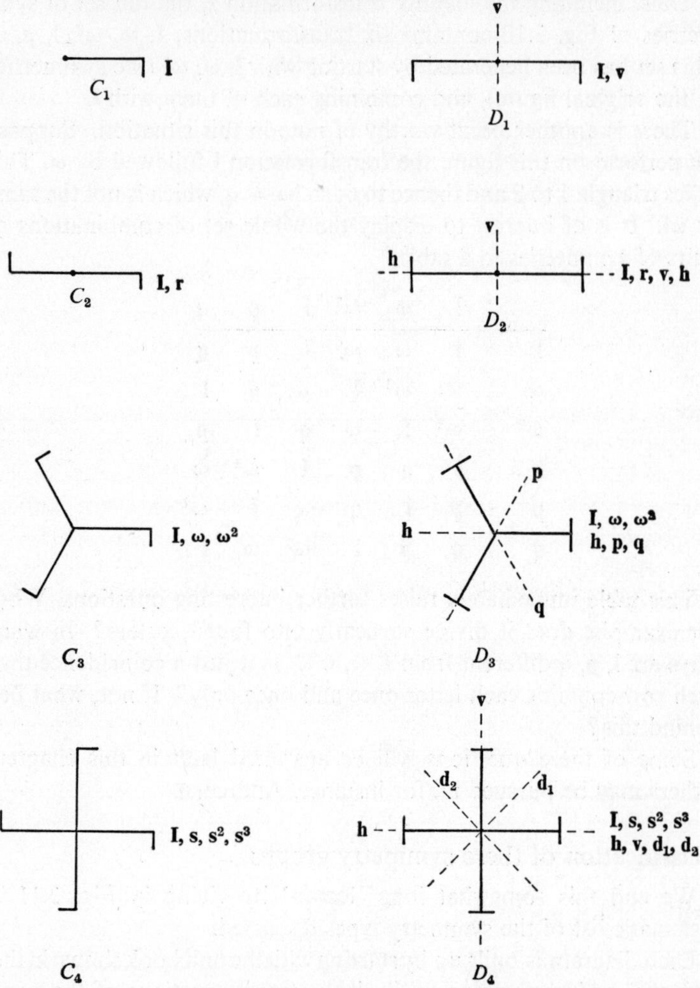

Fig. 3.11. The first eight of the plane symmetry groups with a fixed point. The series continues with C_5, D_5, etc.

Exercises

C_4 and D_2 each contain four symmetries. Write out their multiplication tables, and find systems in the finite arithmetics to which they are isomorphic. (See §3.4.)

Notes

(1) We have defined the product of two transformations **v** and **h** so that **vh** means **v** followed by **h**. Some authors define **vh** to mean **h** followed by **v**. For instance, in chapter 12 successive transformations using matrices are written in the second way.

(2) A slightly different symbolism for reflections is used in chapter 10.

6. GROUPS, RINGS AND FIELDS

Symmetries have been introduced in this chapter not only because of their intrinsic interest but also because the mathematical structures of the symmetry sets show remarkable similarities with finite arithmetics. This similarity is expressed mathematically by defining a structure called a group, which possesses the properties which these systems have in common. The step of abstracting and defining this structure is taken, historically and pedagogically, only after gaining a full working knowledge of the more primitive systems, the finite arithmetics and the symmetry patterns themselves. Most of the lessons of this chapter are therefore designed to give this experience of the systems. Various situations are explored, each of which is intended to bring out some aspect of the group structure, so that at a later stage the group concept will arise and be seen to unify these earlier experiences. This latter would be sixth-form work, whereas many of the lessons here could be taken in the middle school.

The following note on groups, rings and fields is inserted at this point so that the teacher may see the direction in which the lessons are leading.

Definition of a group

A *group* is a set of elements $a, b, \ldots$, which can be combined by an operation o (to be called a 'product') such that:

(i) the result aob, of combining any two elements by the operation o, is itself an element of the group;

(ii) the operation is associative, i.e. $(aob)oc = ao(boc)$ for any elements a, b, c of the group;

(iii) there is an identity element I having the property

$$Ioa = aoI = a$$

for every element a in the group;

(iv) for every element a there is a unique inverse element $\bar{a}$ such that

$$a o \bar{a} = \bar{a} o a = I.$$

Notes

(1) When introducing new operations there is always a problem about symbolism. The number of common symbols is limited and consequently different operations will have to be represented by the same symbol. Students should always realize that a symbol means just what one wants it to mean and nothing else.

Juxtaposition ab is an alternative way of implying the combination of two elements, but it must be remembered that there is not necessarily any connection with multiplication in arithmetic. There is however the advantage with this notation that many of the algebraic expressions occurring in a group structure will seem more familiar.

(2) In most groups the operation o is called 'multiplication' though it is defined in different ways. For example, in the groups of symmetries, the product **vh** means **v** *followed by* **h**. But the finite arithmetic modulo 6 is a group 'under addition', i.e. the operation in this case is $+$. When the operation is written as a product, the inverse of a is written a^{-1}; when it is denoted by $+$, the inverse of a is written $-a$.

(3) aob is not in general equal to boa; for example, in Fig. 3.10, $\mathbf{l}\boldsymbol{\omega} \neq \boldsymbol{\omega}\mathbf{l}$. A group in which the law of combination is commutative (i.e. $aob = boa$ for every pair of elements) is called a *commutative* or *Abelian* group. It is a common convention in such a case to denote the operation by $+$, and to call it an additive group.

(4) The sets of symmetries discussed in this chapter are all groups. (i) and (iii) of the definition have already been dealt with; (ii) is satisfied since there is only one possible meaning for 'a followed by b followed by c'; (iv) is satisfied since, whatever transformation is made, it is possible by a second transformation to return the figure to its original position. (In Fig. 3.10, the inverse of $\boldsymbol{\omega}$ is $\boldsymbol{\omega}^2$, the inverse of **p** is **p**.)

Subgroups and cosets

A *subgroup* H of a group G is a set of elements which is a group, and whose elements are all elements of G. **I**, $\boldsymbol{\omega}$, $\boldsymbol{\omega}^2$ is a subgroup of the full group of Fig. 3.10 and C_n is a subgroup of D_n. The set **l**, **p**, **q** is not a subgroup; it does not contain **I** and so cannot be a group. It is, however, a coset, as will be shown below.

A *coset* is a set formed from a group G and a subgroup H of G as follows. Any element a of the group G is taken, and all the elements of the subgroup H are multiplied by a. The result is called a coset. If the multiplication is with a on the left, it is a left-coset, and vice versa. Thus in the group of Fig. 3.10, using the subgroup $\mathbf{I}$, $\boldsymbol{\omega}$, $\boldsymbol{\omega}^2$ and the element **l** on the left, $\mathbf{l}$, $\mathbf{l}\boldsymbol{\omega}$, $\mathbf{l}\boldsymbol{\omega}^2$ (i.e. **l**, **q**, **p**), is a left-coset. The corresponding right-coset ($\mathbf{l}$, $\boldsymbol{\omega}\mathbf{l}$, $\boldsymbol{\omega}^2\mathbf{l}$) consists in this case (though not always) of the same three elements.

Exercise

Find the left-cosets of all six elements with respect to the subgroup **I**, **p**.

Order of a group and of an element

The *order of a group* is the number of elements it contains.

The *order of an element a* of a group is the least power of a which equals the identity element. In the group of Fig. 3.10, $\boldsymbol{\omega}$ and $\boldsymbol{\omega}^2$ are of order 3, while **l**, **p** and **q** are of order 2.

It is an important theorem that in a group of order n the nth power of every element equals **I** [14, 75].

Rings and fields

A *ring* is a set of elements in which two operations, $+$ and $\times$, are defined. The set is a commutative group with respect to addition; multiplication has the properties (i) and (ii) of a group, and it is distributive over addition, that is

$$a(b+c) = ab+ac, \quad (b+c)a = ba+ca.$$

A *field* is a ring in which multiplication is commutative and has all the group properties (except that the zero element of a field does not have a multiplicative inverse).

These ideas are discussed at length in any text on modern algebra [8, 14, 109].

7. REGULAR POLYGONS

An important class of figures possessing symmetry is the class of regular polygons. One of the easiest to draw is the regular hexagon. Divide the circumference of a circle into six equal parts. Number the points 0–5, and join points 0 to 1, 1 to 2, etc.

Mark all the symmetries possessed by this figure. How many are there? Reflections? *6 reflections, 3 in joins of opposite vertices, 3 in joins of mid-points of opposite sides.* Rotations? *6, including the identity: through*

60°, 120°, ..., 360°.

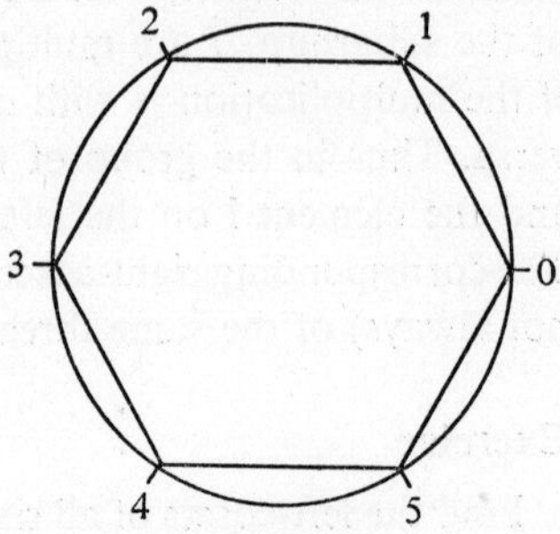

Fig. 3.12

Total 12. Can we classify this group with the ones found earlier? *Generation from the rotations and any one reflection may be discussed if appropriate. The group is* D_6.

Now suppose we join up points whose numbers differ by 2, e.g. 0 and 2, 2 and 4, 4 and 0. We get an equilateral triangle.

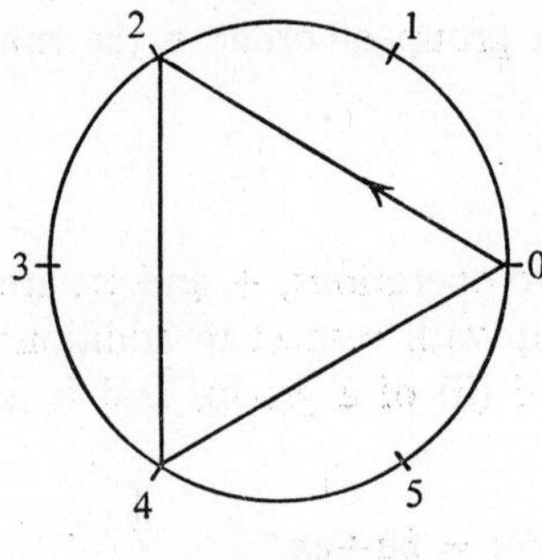

Fig. 3.13

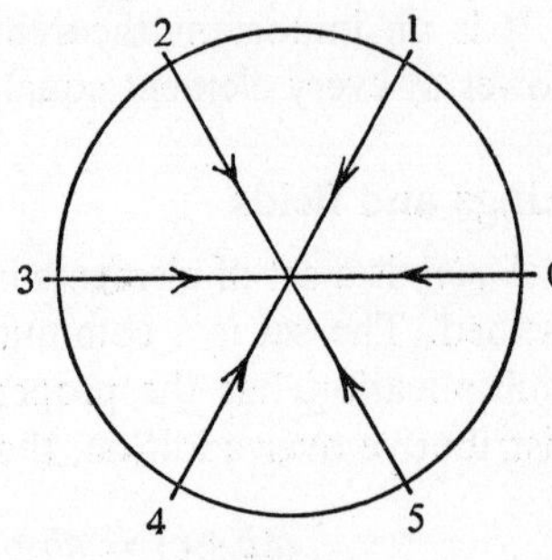

Fig. 3.14

What symmetries have we now? Joining up points 1, 3, 5 gives another equilateral triangle; the hexagon has split into these two triangles. If we join points differing by 3, we get just three separate double line segments.

Try joining points differing by 4 and by 5, and note the results.

These diagrams include, of course, the ones obtained in the lesson on Timetables, Calendars and Clocks, and we note from these that the numbers of the points we join here form arithmetic progressions in the arithmetic modulo 6, with differences 0, 1, 2, ..., respectively.

We now look at the regular heptagon and octagon from the same point of view.

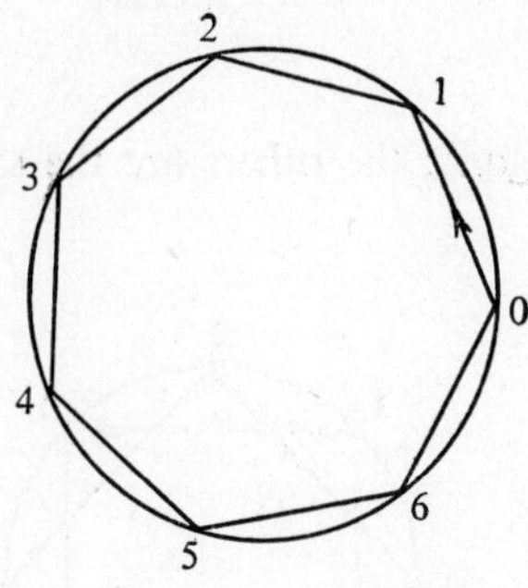

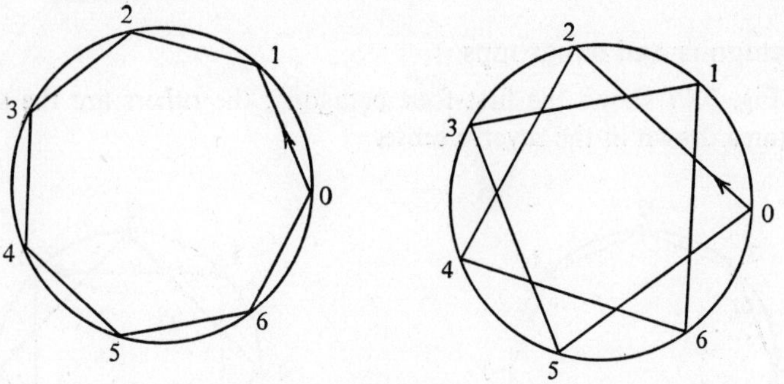

Fig. 3.15

$\{\frac{7}{1}\}$: 0, 1, 2, 3, 4, 5, 6, 0 ... $\{\frac{7}{2}\}$: 0, 2, 4, 6, 1, 3, 5, 0 ...

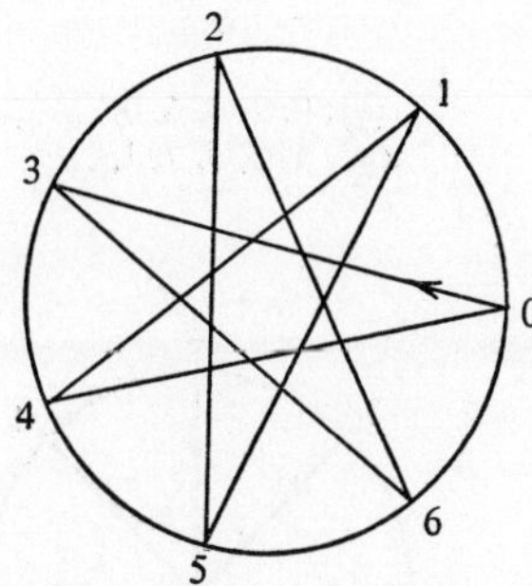

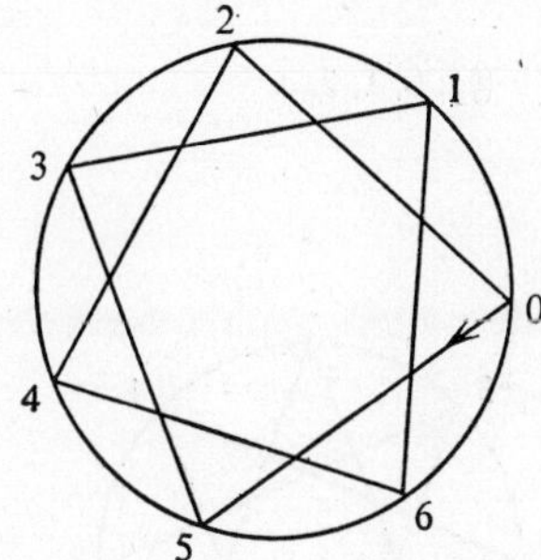

Fig. 3.16

$\{\frac{7}{3}\}$: 0, 3, 6, 2, 5, 1, 4, 0 ... $\{\frac{7}{5}\}$: 0, 5, 3, 1, 6, 4, 2, 0 ...

The second, third and fourth of these are called star polygons [30] and $\{\frac{7}{3}\}$ is the symbol for a polygon with seven vertices, the sides consisting of the joins of every third vertex. The polygon $\{\frac{7}{5}\}$ is $\{\frac{7}{2}\}$ traced the reverse direction, and $\{\frac{7}{4}\}$ and $\{\frac{7}{6}\}$ are similarly related to $\{\frac{7}{3}\}$ and $\{\frac{7}{1}\}$ respectively.

What symmetry groups do these star polygons have?

For teachers

Show that the numbers 0–5 form a group under addition modulo 6, and that this group is isomorphic to the group of *rotations* of the regular hexagon, C_6.

Octagons and subgroups

Fig. 3.17 shows the first four octagons; the others are the same figures drawn in the reverse sense.

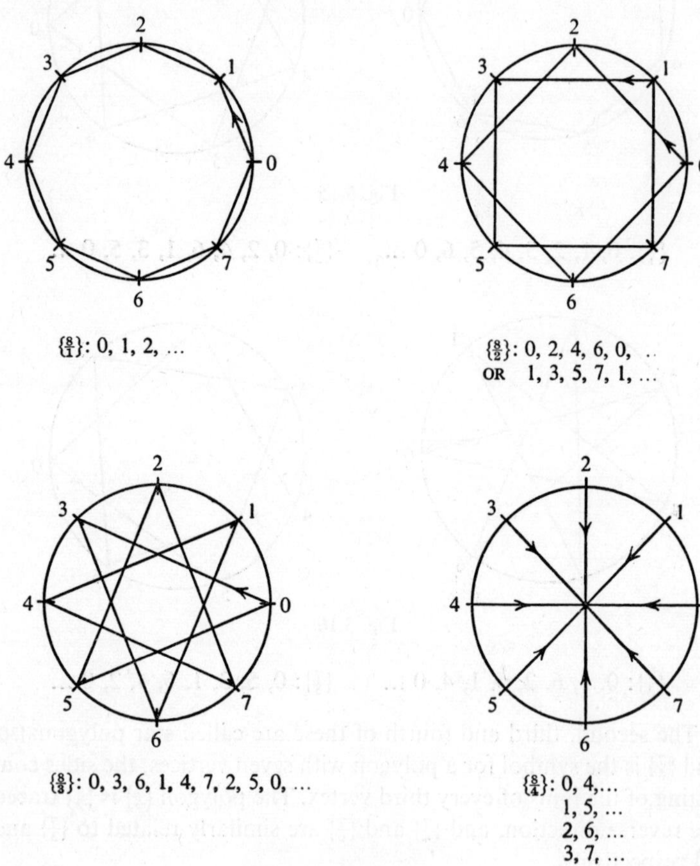

Fig. 3.17. The octagons.

To begin this lesson we are going to look more closely at the symmetries of the second of these, the polygon $\{\frac{8}{2}\}$. If we ignore the two arrows, the figure has the full symmetry group of the octagon D_8.

If we now regard the two squares as being different, what symmetry group does the figure have? If we now take account of the arrows, but ignore the other difference, what group? And if we consider both differences to be significant, what group now? Can we reduce the symmetries of the figure in such ways to get any smaller subgroups of D_8? If not, try the same investigation on the other polygons. Can you, for example, make the polygon $\{\frac{8}{4}\}$ have only the group C_2?

Other exercises

(1) Show that the heptagons all have the symmetry groups either C_7 or D_7.

(2) Find 'hexagons' with groups D_2 and C_3.

(3) Study the dodecagon in a similar way.

For teachers

(1) Show that the numbers 0–7 form a group under addition modulo 8, and that 0, 2, 4, 6 is a subgroup of this group, and 1, 3, 5, 7 the only other coset.

(2) Find similarly the subgroups and cosets of the group 0–5 under addition modulo 6, and note the relation of these to the hexagons (see previous lesson).

Polygons which decompose into separate pieces give a pictorial illustration of the cosets described in the previous section. The nonagon $\{\frac{9}{3}\}$ consists of the triangle 0, 3, 6, which shows D_3 as a subgroup of D_9, and two other triangles which are associated with the cosets.

8. SLIDE-RULES FOR FINITE ARITHMETICS

Lesson

In finite arithmetics the appropriate slide-rules are circular. Two suitable scales can be fixed to the sides of a pillbox, or marked on two concentric circles of card and fixed with a paper fastener at the centre.

Fig. 3.18 shows a suitable device for adding modulo 13. Subtraction can be performed either by having the moving scale marked on both sides and turning it over, or by moving backwards an appropriate number of steps. Subtraction is always possible without the necessity of a set of negative numbers.

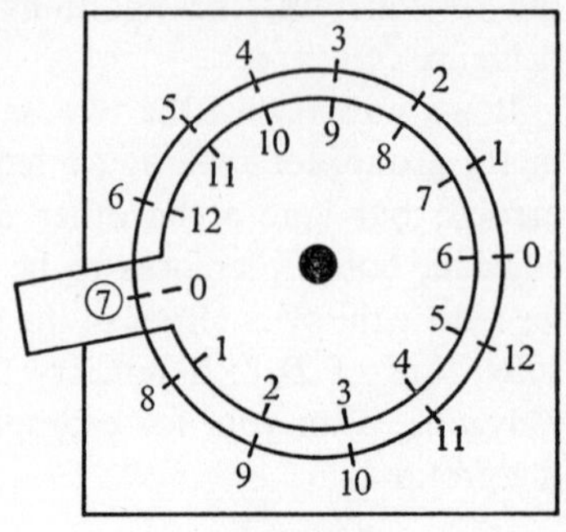

Fig. 3.18

A multiplying scale needs to have equal intervals corresponding to equal multiples. Zero being omitted, there are twelve numbers to be placed on the scale. What scale factor shall we choose—that is, what multiplying factor is to represent one step in the scale? Try 2; then successive numbers must be 1, 2, 2^2, 2^3, 2^4, ..., which, modulo 13, are 1, 2, 4, 8, 3, 6, 12, 11, 9, 5, 10, 7, 1. Every number of the system can be expressed as a power of 2. Other scale factors may be suitable; Fig. 3.19 shows two possibilities.

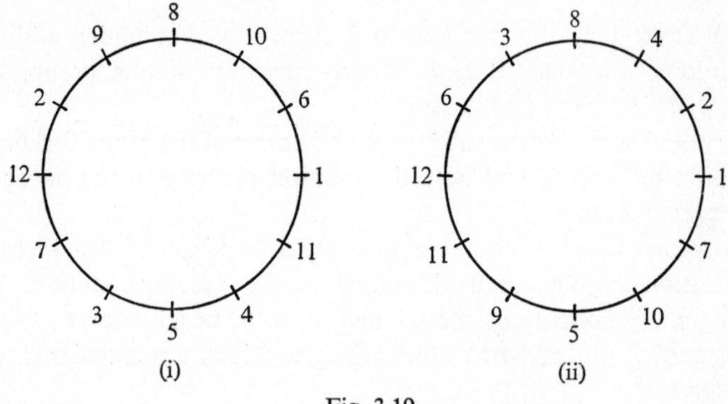

Fig. 3.19

Exercises

(1) Do some multiplications on the scales of Fig. 3.19.

(2) Do some divisions and note that no fractions are required.

Let us take a number of the set, say 11, and multiply it repeatedly by 6, using the scale of Fig. 3.19i. Mark on the scale the numbers obtained (Fig. 3.20i). Now multiply 11 repeatedly by 3. What happens? (Fig. 3.20 ii). Try multiplying another number by threes, say 12. What do you conclude? There are four closed sets of numbers; geometrically they are the four triangles of the polygon $\{\frac{12}{4}\}$.

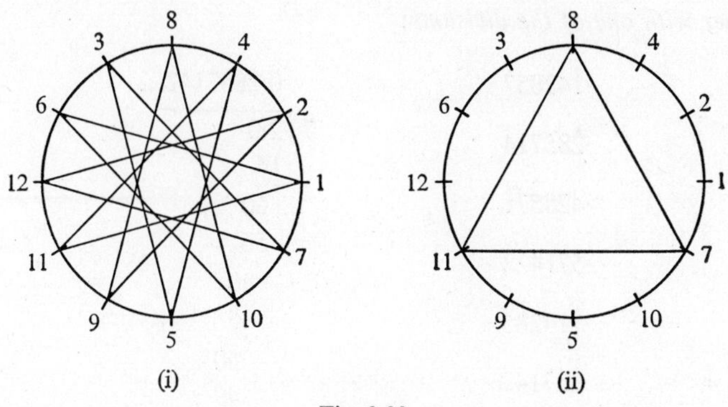

Fig. 3.20

All the dodecagons can be produced from this scale as different sets of multiples. Thus there is a close similarity between the dodecagons and the multiplication relationships of the arithmetic modulo 13; this similarity can be made explicit as follows. The set of numbers 1–12, with the operation of multiplication modulo 13, form a group. *Verify this from the definition, p.* 59. This group is isomorphic (p. 51) to the group of rotations of the regular dodecagons, the group C_{12}. 1, 3, 9 is a subgroup of order 3 (C_3), and 2, 6, 5; 4, 12, 10, and 8, 11, 7 are cosets.

9. RECURRING DECIMALS

This is another application of finite arithmetics, in which the structures of the previous lessons give insight into the properties of recurring decimals.

Having memorized the sequence 142857 the teacher begins: I will guarantee to turn any fraction with denominator 7 into a decimal. Who would like to try me out? Give $\frac{5}{7}$.

With appearance of thought, he chooses the correct starting figure and quotes the decimal. He does several, and asks the class if they can see how to do it. Then he offers to show the class how to do it for themselves.

We observe that there are always six figures in the cycle and that the six figures are always the same.

The pupils now work out, by long division, $\frac{1}{7}$, $\frac{2}{7}$, ..., $\frac{6}{7}$ *and keep their work for reference. The decimals are written on the blackboard, together with one of the divisions*:

$\cdot\dot{1}4285\dot{7}$

$\cdot\dot{2}8571\dot{4}$

$\cdot\dot{4}2857\dot{1}$

$\cdot\dot{5}7142\dot{8}$

$\cdot\dot{7}1428\dot{5}$

$\cdot\dot{8}5714\dot{2}$

$$\begin{array}{r@{}l}
 & 0\cdot2857142\ldots \\
7\,) & 2\cdot0 \\
 & 14 \\
 & 60 \\
 & 56 \\
 & 40 \\
 & 35 \\
 & 50 \\
 & 49 \\
 & 10 \\
 & 7 \\
 & 30 \\
 & 28 \\
 & 20 \\
 & \ldots
\end{array}$$

We now observe further that the first and fourth figures add up to 9, so do the second and fifth, and third and sixth.

Why should there be six figures? And why do we never get 0, 3, 6 or 9? Look at the division. We see that to get 3 as a quotient we need 30 as dividend, but this contains four sevens. Consequently 3 is not a possible quotient. There are only six possible remainders which can appear in the working, and so *at most* six figures in the cycle.

Is the number of figures always one less than the denominator? The class try some examples.

$$\frac{1}{3} = 0\cdot\dot{3}\text{—NO!} \qquad \frac{1}{11} = 0\cdot\dot{0}\dot{9}\text{—NO!}$$

This is strange; why, out of the 10 possible non-zero remainders in the division by 11, do we get only 2?

Instead of pursuing these particular decimals, we leave them for the class to investigate later, and study the thirteenths. We start by working them out—division of labour among the class is advisable—and the results are collected on the blackboard.

We find:

$$\frac{1}{13} = \cdot\dot{0}7692\dot{3} \qquad \frac{2}{13} = \cdot\dot{1}5384\dot{6} \qquad \frac{3}{13} = \cdot\dot{2}3076\dot{9}$$

$$\frac{4}{13} = \cdot\dot{3}0769\dot{2} \qquad \frac{5}{13} = \cdot\dot{3}8461\dot{5} \qquad \text{etc.}$$

What new light does this shed on the problem of finding how many figures there are in the cycle? We have two possible sets of figures; 3 and 6 appear in both sets but they are otherwise distinct. We have, in fact, twelve figures in all in the two cycles. Could this correspond to the twelve remainders possible in division by 13? We see from the division that it does. Why are 3 and 6 repeated? *3 × 13 is the nearest multiple below both 40 and 50.*

Let us make a list of the two sets of remainders. They are:

$$1,\ 10,\ 9,\ 12,\ 3,\ 4$$

and

$$2,\ 7,\ 5,\ 11,\ 6,\ 8.$$

Could these have anything to do with the arithmetic modulo 13? Let us look at the multiplying scale we were using earlier (Fig. 3.19 ii). Where are 1, 10, 9, 12, 3, 4? Do they form a definite set? And the others? What is the relationship between these numbers on the multiplying scale? *1, 10, 9, ..., are the powers of 10 reduced modulo 13.*

What are the others? *They are 2, 2.10, 2.10², etc.* How do these relationships arise in our divisions? Look at the calculation. Successive remainders r, r' are related by

$$10r = k.13 + r',$$

that is

$$10r \equiv r' \pmod{13}.$$

So the remainders, if we start with, say, 3, are 3, 3.10, 3.10², 3.10³, ..., in the arithmetic modulo 13. From the figure we see that these are 3, 4, 1, 10, 9, 12. Who worked out $\frac{3}{13}$? Is this right? Who worked out $\frac{11}{13}$? Read out your remainders. Watch where they appear on the diagram.

How much does this explain about the number of figures in the cycle? Why do the sets of remainders form a hexagon, and not the full 12-sided polygon, or say, a square? *Because* six *multiplications by 10 bring us back to the starting-point. In the language of* §*3.4, 10 is of order 6 in the group of multiplications modulo 13.*

Can you tell how long the cycle would be for $\frac{1}{17}$, or $\frac{1}{19}$? What can you say without doing any working out? *The length of the cycle must be a divisor of 16 or 18 respectively.*

Could you, by writing thirteen to a different base, e.g. 21 in base 6, arrange for the thirteenths to have 'decimals' with a cycle of four digits? Of 2 digits? Of 12 digits?

Exercises

Investigate the elevenths and the thirds in the same way.

Notes

There was another curious fact which we observed but did not investigate; the fact that corresponding digits in the two halves of the thirteenths cycle sum to 9.

This note is added in case the teacher feels it appropriate to pursue with a suitable class.

Let us look at $\frac{2}{13}$, $\cdot\dot{1}5384\dot{6}$. How do the 1 and the 8 arise? As before, we look at the corresponding remainders in the division. 1 leaves the remainder 7, 8 leaves 6; 5 leaves remainder 5, 4 leaves 8; 3 leaves remainder 11 (in this case), 6 leaves 2. Is there any pattern in the remainders? *The pairs sum to 13.* Is this necessarily true? Where do these remainders appear on the diagram (Fig. 3.20)? *At opposite ends of diameters.* Why do they appear there? *They are obtained from each other by multiplying by* 10^3.

Yes, but how does this make the *quotients* (in the division) add up to 9? What is the relation between the quotient 1 and its remainder 7? Look at the division. We have

$$1.13+7 = 20$$

and

$$8.13+6 = 110.$$

Both of these right-hand sides are $\equiv 0 \pmod{10}$; the same is true for all other pairs of quotients and remainders. Let us say that in general, if (q_1, r_1) and (q_2, r_2) are corresponding pairs (so that $r_1 + r_2 = 13$ as shown above) we have

$$q_1 . 13 + r_1 \equiv 0 \pmod{10}$$

$$q_2 . 13 + r_2 \equiv 0 \pmod{10}.$$

Adding $$13(q_1 + q_2) + r_1 + r_2 \equiv 0 \pmod{10}$$

$$13(q_1 + q_2) + 13 \equiv 0 \pmod{10}.$$

Dividing $$q_1 + q_2 + 1 \equiv 0 \pmod{10}$$

i.e. $$q_1 + q_2 \equiv 9 \pmod{10}$$

that is $$q_1 + q_2 = 9.$$

Another possible type of behaviour is displayed by the 'twenty-oneths'.

$\frac{1}{21} = \cdot\dot{0}4761\dot{9}$ and there are five others related.

$\frac{20}{21} = \cdot\dot{9}5238\dot{0}$ and there are five others related.

This gives two families of six in which the property of 'summing to 9' takes a new form; there is another family of six, the already familiar $\frac{3}{21}, \frac{6}{21}, \ldots, \frac{18}{21}$, and a family of two, $\frac{7}{21}, \frac{14}{21}$. The complete theory of recurring decimals is complicated, and may be found in books on the theory of numbers. We do not suggest that it should be studied formally, as part of the syllabus, for its own sake. The point is that the quite simple idea of a group enables the previous geometrical exercises to be given an arithmetical interpretation from which fresh information can be gained. The notions of modern abstract algebra are introduced into the school course so that different parts of the subject reinforce one another. They are unifying principles.

10. FINITE GEOMETRIES, LATIN SQUARES AND MAGIC SQUARES

In this lesson we see how standard work on parallel lines in Cartesian geometry may be used in a different context to provide information which is useful in the design of methods of carrying out statistical experiments in agriculture and quality control in factories. Both of these are activities which at first sight are not in the least geometrical; and this lesson illustrates how the appreciation of mathematics at a sufficiently abstract level increases the range over which the ideas may be applied, and also how the introduction of fresh material into the school course is very often not a matter of competing with traditional material but of enriching it by showing it in action in a greater variety of circumstances.

We assume familiarity with the finite fields GF(p) (see p. 51), define a two-dimensional geometry over such a field, and show how the resulting system has interesting properties of its own which can be applied to practical problems. We give here merely a résumé of the ground which might be covered and do not attempt to cast the work in a discussion form suited to the classroom.

If we are familiar with the field GF(p) we may play a game with ordered pairs of elements from the field which we write (x, y). We refer to these ordered pairs as 'points', and for reasons which appear as we go along we consider the points to form a 'geometry'. The two-dimensional geometry over GF(p) contains p^2 different points. It may be convenient to extend the idea later to spaces of higher dimension, and we will speak of the present system as VS (2, p). (VS stands for vector space. See §11.6[14, 109].) VS (2, 5) is very suitable for classroom demonstration, and elsewhere[45] it has been shown in action illustrating many aspects of sixth-form geometry, including the relation between projective and Euclidean geometry, and the construction of an 'Argand plane' by the introduction of 'complex numbers' over a finite field. Here we will follow a different course.

Consider the array of letters:

$$\begin{matrix} A & B & C \\ C & A & B \\ B & C & A \end{matrix}$$

The main property is immediately obvious. Each letter occurs once and only once in each row, and in each column. An analogous system may be constructed with four letters by experiment (though not immediately by the methods we are going on to discuss). A system of this kind is called a 'Latin square'. A more sophisticated problem with 4×4 squares is often posed in terms of a pack of cards. Can you arrange the sixteen court cards (including aces) so that each row and column contains just one card of each suit and just one of each denomination? The resulting arrangement is an example of what is often called a Graeco-Latin square because the following notation is suitable:

$$\begin{matrix} D & C & B & A \\ C & D & A & B \\ B & A & D & C \\ A & B & C & D \end{matrix} \qquad \begin{matrix} \delta & \beta & \alpha & \gamma \\ \gamma & \alpha & \beta & \delta \\ \beta & \delta & \gamma & \alpha \\ \alpha & \gamma & \delta & \beta \end{matrix}$$

If these two Latin squares are superimposed then each combination of a Latin with a Greek letter appears just once, and this provides a solution to the card-pack problem. Two such squares are said to be *orthogonal*. This overworked word is merely used here in the sense which has just been defined, and one should not seek too long for mysterious connections with its usage elsewhere.

Orthogonal Latin squares have statistical applications. Say, first, that we wish to experiment with four different types of wheat. If these are planted in strips in a field there is a danger that one end of the field may have better soil than the other, and this will militate against a proper comparison. A more satisfactory system is to split the field into 16 plots and use a Latin square arrangement. In this way there is the best chance of avoiding preferential treatment for any of the four varieties. If we wish to repeat the experiment another year, and to redistribute the plots as thoroughly as possible then another orthogonal square may be used.

What methods are there for constructing orthogonal Latin squares of higher order? In the first place it is not always possible to construct such squares. The celebrated problem of Euler concerning the 36 officers [107] is impossible because two orthogonal 6×6 Latin squares cannot be constructed. It can be shown that it is never possible to construct more than $(n-1)$ mutually orthogonal $n \times n$ Latin squares, and we will now describe a method which succeeds in constructing $(n-1)$ such squares whenever n is prime. The method also extends to cases when n is an integral power of a prime, and so it covers a great many cases of practical importance.

To discuss the 5×5 case write down the 25 'points' of VS (2, 5).

04	14	24	34	44
03	13	23	33	43
02	12	22	32	42
01	11	21	31	41
00	10	20	30	40

Fig. 3.21

Here we use, for example, 23 to denote the number pair (x, y), with $x = 2$ and $y = 3$.

We now consider various 'lines'. A line is a set of points whose coordinates satisfy an equation of the first degree.

What points are on the line $y \equiv x \pmod 5$? Label these A.

What points are on the line $y \equiv x+1 \pmod 5$? Label these B.

What points are on the line $y \equiv x+2 \pmod 5$? Label these C.

What points are on the line $y \equiv x+3 \pmod 5$? Label these D.

What points are on the line $y \equiv x+4 \pmod 5$? Label these E.

The result is shown in Fig. 3.22.

E	*D*	*C*	*B*	*A*
D	*C*	*B*	*A*	*E*
C	*B*	*A*	*E*	*D*
B	*A*	*E*	*D*	*C*
A	*E*	*D*	*C*	*B*

Fig. 3.22

Do the same for the lines

$$y \equiv kx, \quad y \equiv kx+1, \quad y \equiv kx+2, \quad y \equiv kx+3, \quad y \equiv kx+4,$$

for $k = 2, 3, 4$. (See Fig. 3.23.)

E	*C*	*A*	*D*	*B*		*E*	*B*	*D*	*A*	*C*		*E*	*A*	*B*	*C*	*D*
D	*B*	*E*	*C*	*A*		*D*	*A*	*C*	*E*	*B*		*D*	*E*	*A*	*B*	*C*
C	*A*	*D*	*B*	*E*		*C*	*E*	*B*	*D*	*A*		*C*	*D*	*E*	*A*	*B*
B	*E*	*C*	*A*	*D*		*B*	*D*	*A*	*C*	*E*		*B*	*C*	*D*	*E*	*A*
A	*D*	*B*	*E*	*C*		*A*	*C*	*E*	*B*	*D*		*A*	*B*	*C*	*D*	*E*

Fig. 3.23

In each case we construct a Latin square. Why?

Furthermore, any two of these are mutually orthogonal. Because if, for example, we superimpose the square of Fig. 3.22 with any one of Fig. 3.23 each combination, such as *CA*, occurs once and once only; that is, the line *C* in the former and the line *A* in the latter meet in exactly one point. In the general case of the superposition of any two of the four possible squares, the condition for 'orthogonality' is that any pair of lines $y \equiv k_1x+c_1$, $y \equiv k_2x+c_2$, with $k_1 \not\equiv k_2$, must meet in one point. This is indeed the case in VS (2, 5), since the x coordinate of the point of intersection is given by

$$(k_2-k_1)\,x \equiv c_1-c_2,$$

whence $$x \equiv (c_1-c_2)/(k_2-k_1).$$

This step is possible in GF (5), by virtue of the field property that each non-zero element has a unique multiplicative inverse. It is at this point that the method breaks down for a 6×6 square; the arithmetic modulo 6 does not have multiplicative inverses for 2, 3 or 4. It is a ring, but not a field. The method does, however, work for $p\times p$ squares, where p is any prime.

It is worthy of note that this division of the 25 points of VS (2, 5) into sets of parallel lines is another example of the use of a subgroup to decompose a group into cosets (p. 61). For the 25 points form a group under the law of vector addition, and the five points of the line $y \equiv kx$ are a subgroup of this group. The five new points formed by adding each of these five points to any other point of the space constitute a coset, which is another line of the set, $y \equiv kx+c$.

These squares illustrate many properties characteristic of the systems studied in modern algebra, and there are many further points which could be followed up. If the fields GF (p^n) are familiar, the above method applies also to these.

(Henceforth, in this chapter, we drop congruence notation and employ the more familiar equality sign. The context will indicate any modulus which is understood.)

Complex finite fields

Texts on modern algebra show that the only finite fields which may be constructed always contain a number of elements equal to p^n, where p is prime and n is a positive integer. We have given many examples using fields GF (p). The fields GF (p^n) are extensions of these, very much as the field of complex numbers is an extension of the real field.

If the finite fields GF(p) become familiar in schools the fields GF(p^n) could be introduced at about the same time as complex numbers. Indeed the process of building up a field like GF(5^2) illuminates the process by which complex numbers are constructed from the reals. For a discussion of the present finite geometry as an 'Argand plane' over GF(5^2) see [45].

The method of constructing extension fields is described in very simple terms in Sawyer [109]. In particular he describes the field GF(2^2), the addition and multiplication tables of which are

+	0	1	m	$m+1$
0	0	1	m	$m+1$
1	1	0	$m+1$	m
m	m	$m+1$	0	1
$m+1$	$m+1$	m	1	0

×	0	1	m	$m+1$
0	0	0	0	0
1	0	1	m	$m+1$
m	0	m	$m+1$	1
$m+1$	0	$m+1$	1	m

Using this field it is a simple exercise to construct the two 4×4 orthogonal squares given above, and also a third square orthogonal to them both [81].

The configurations which arise in finite geometry find many applications in statistics, in particular in the design of experiments and quality control [35, 44]. Not all finite geometries can be constructed by means of Galois fields, and neither can all of the configurations be used in the design of experiments, but these fields enable a great many designs to be constructed and cover most of the practically important cases.

These vector spaces are very fundamental in statistical investigation, but the geometrical nature of the systems is not always made clear by the notation. The famous Twelve-Penny Problem can be solved in VS (3, 3), and this same finite geometry is used in other ways in the design of experiments.

Modern algebra provides a variety of structures which the mathematician may use as frames of reference whenever he recognizes them in the world around.

A note on magic squares

In the square of Fig. 3.22, let the letters A to E be given the number values 0–4 respectively. Do similarly with any square of Fig. 3.23. The square formed by superimposing these two now has at each lattice point a number-pair. Let these now be regarded as two-digit numbers in the scale of 5 (i.e. $21 = 2.5+1$). The resulting square (Fig. 3.24) contains each number from 00 to 44 once only (by the orthogonality property), and the rows and columns all sum to 220 in the scale of 5 (by the 'Latin' property). This is almost a 'magic' square; it fails to be one only because one of the diagonals does not sum to 220. But this square was made from only two of the four Latin squares we have found, those based on $y = x+c$ and $y = 2x+c$. Can we make a fully magic square if we use one or more of the others?

44	32	20	13	01
33	21	14	02	40
22	10	03	41	34
11	04	42	30	23
00	43	31	24	12

Fig. 3.24

44	21	03	30	12
33	10	42	24	01
22	04	31	13	40
11	43	20	02	34
00	32	14	41	23

Fig. 3.25

25	12	4	16	8
19	6	23	15	2
13	5	17	9	21
7	24	11	3	20
1	18	10	22	14

Fig. 3.26

When the letters of the squares are converted into numbers, as described above, the number u occupying the lattice point (x, y) is a function of x and y. Let us call these functions $u_1(x, y)$, $u_2(x, y)$, $u_3(x, y)$ and $u_4(x, y)$, referring to the squares based on $y = x+c$, $y = 2x+c$, $y = 3x+c$, $y = 4x+c$, respectively. Then it can be seen that

$$u_1(x, y) = y-x, \quad u_2(x, y) = y-2x,$$
$$u_3(x, y) = y-3x, \quad u_4(x, y) = y-4x.$$

Now consider what numbers appear on the diagonals of the squares. The equations of the diagonals are $y = x$ and $y = 4x+4$; in the first square the numbers they contain are given by $u_1 = 0$ always for the first diagonal, and $u_1 = 4x+4-x = 3x+4$ for the other. As x takes the values 0, 1, 2, 3, 4, this function takes each possible value once, giving a 'magic' line. In the other squares, all diagonals are of this type, except for the diagonal $y = 4x+4$ in the fourth square, in which $u_4 = y-4x$, making $u_4 = 4$ everywhere on this diagonal.

A fully magic square can thus be formed by combining the second and third of these Latin squares; it is shown in Fig. 3.25. Note that it is 'magic' with respect to all lines parallel to the diagonals as well as to the diagonals themselves. Fig. 3.26 shows the same square, with the numbers written with base ten and 1 added to each.

11. POLYNOMIALS OVER FINITE FIELDS

In the previous section we considered arithmetical operations over finite fields and linear algebraic operations. We now consider algebraic operations of higher degree, and in particular we show how algebraic operations which have been learned in one context may be transferred to another. This is one result of the abstract approach, and one reason why we believe that it should be possible to introduce much material, which at first sight is 'fresh', into the course and still cover the ground in the same time. There is a further pedagogical gain in this approach because when an algebraic operation is seen at work in different contexts, the concept of the 'operation' itself is more easily disentangled from the operands on which it works.

Polynomials over finite fields provide examples of a mapping of one domain into another, and we find tangible applications of the ideas in electrical switching circuits.

Consider the polynomials over GF (5). The table of powers is as follows:

x	0	1	2	3	4
x^2	0	1	4	4	1
x^3	0	1	3	2	4
x^4	0	1	1	1	1
x^5	0	1	2	3	4

It can be seen that $x^5 = x$, $x^6 = x^2$ and so on; in fact, since $x^4 = 1$ for all non-zero x, any index can be reduced modulo 4. There are only five non-identical power-functions, $x^0 \ldots x^4$. A corresponding theorem is true for the general GF (p), in which case $x^{p-1} = 1$ for all x.

This is Fermat's theorem, and is an immediate consequence of the fact that in GF (p) the non-zero elements form a group, under multiplication, of order p−1. Thus in GF (5) all polynomials are reducible to the form $ax^4+bx^3+cx^2+dx+e$, where a, b, c, d, e can take the values 0, 1, 2, 3, 4. There are thus not more than 5^5 different polynomial functions over this field.

Turning now to consider the total number of different functions over GF (5), we note that since any function is specified by its value for each of the five possible values of x, and these values of the function must also be chosen from the set 0...4, the total number of different functions is exactly 5^5. The one step further which must be demonstrated before we can identify these with the 5^5 polynomial functions (which must therefore be all distinct) is provided by the investigation which follows.

We shall in this section use the field GF (3) for illustrations; here we have 3^3 or 27 different polynomials of degree 2, ax^2+bx+c, which, if no two are equal for all values of x, may represent 27 different functions; and 27 is the total number of functions over the field. We might prove the polynomial functions distinct by the use of the factor theorem, leading to the theorem that polynomial functions of degree n which are equal for $n+1$ values of x have equal coefficients; this is true for polynomials over any field. Instead we shall use a method which gives explicitly the polynomial representing any given function, specified by its values for $x = 0$, 1 and 2. This problem is essentially that of finding a polynomial of degree n whose graph passes through $n+1$ given points. The case for $n = 2$ will indicate the general method. The problem is to find $y = f(x)$ taking the values A, B, C when x takes the values a, b, c. If we can find polynomials $f_a(x)$, $f_b(x)$ and $f_c(x)$ which take the following special values the problem is easily solved by adding appropriate multiples of them.

x	a	b	c
$f_a(x)$	1	0	0
$f_b(x)$	0	1	0
$f_c(x)$	0	0	1

By inspection
$$f_a(x) = \frac{(x-b)(x-c)}{(a-b)(a-c)},$$

and $f_b(x)$ and $f_c(x)$ are similar. If we put $F(x) = (x-a)(x-b)(x-c)$ these may be written
$$f_a(x) = \frac{1}{F'(a)}\cdot\frac{F(x)}{x-a}, \text{ etc.}$$
Then the required function
$$y = f(x) = \Sigma \frac{A}{F'(a)}\frac{F(x)}{x-a}. \tag{1}$$

It is clear that this method will extend just as easily to the general case. (1) is called *Lagrange's interpolation formula.*

The calculations which produce this are performed in the field of real numbers. Are they still valid in the finite field GF (3)? Why not?

The operations $+\ -\ \times\ \div$ are still valid, but there is a difficulty with differentiation, which needs a continuum in which to carry out the limiting process. However, over a finite field it is possible to *define* the derivative of x^n as nx^{n-1}. All the familiar laws for the differentiation of polynomials then follow just as before. This is discussed fully in van der Waerden [124]. We shall therefore continue to use the notation as a shorthand here.

Our problem is to find a polynomial over GF (3) with given values $f(0), f(1), f(2)$. Following the method above, and working modulo 3,

$$\begin{aligned} F(x) &= x(x-1)(x-2) \\ &= x(x^2-3x+2) \\ &= x^3+2x. \end{aligned}$$

So

$$\begin{aligned} F'(x) &= 3x^2+2 \\ &= 2. \end{aligned}$$

Then the required function is given by

$$f(x) = \tfrac{1}{2}f(0).(x-1)(x-2)+\tfrac{1}{2}f(1).x(x-2)+\tfrac{1}{2}f(2).x(x-1),$$

which, because we are in GF (3), is

$$\begin{aligned} f(x) &= 2f(0).(x^2+2)+2f(1).(x^2+x)+2f(2).(x^2+2x) \\ &= f(0).(2x^2+1)+f(1).(2x^2+2x)+f(2).(2x^2+x), \end{aligned}$$

which is the required explicit expression for $f(x)$ in terms of $f(0)$, $f(1)$ and $f(2)$.

These formulae may be applied to three-way switching circuits. Consider a three-way switch which is connected to three output channels. We may denote the positions of the switch by $x = 0, 1, 2$ and the output channel which is 'live' by $y = 0, 1, 2$. Throughout the discussion we restrict attention to functions which are one-valued, that is to say different positions of the switch may lead to the same output channel, but each position of the switch leads to only one output channel.

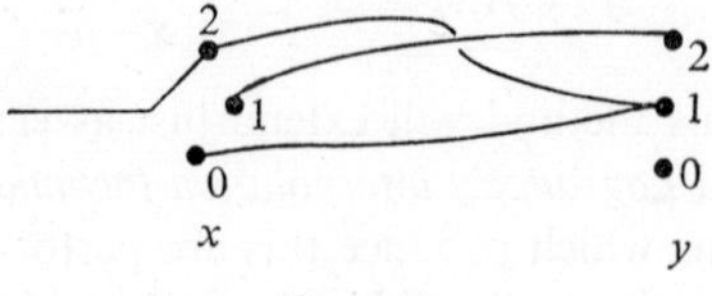

Fig. 3.27

For the circuit drawn

x	0	1	2
y	1	2	1

can we express y as a quadratic function of x? Using the general result above

$$y = 1(2x^2+1)+2(2x^2+2x)+1(2x^2+x)$$
$$= 2x^2+2x+1.$$

Repeat this calculation for other systems of interconnections. The point of this is that it describes a network of wires by an algebraic formula. This may or may not be a practical convenience. At the present time computers are being used to design the layouts for the far more complicated computers which are going to be built in the future, so it is important to investigate various ways of describing circuits by algebraic formulae.

Exercises

(1) A circuit is described by the formula x^2+1. What positions of the switch (if any) activate channels 0, 1, 2?

(2) In a similar way there is a theory of functions of two variables. Find an algebraic expression for the function defined by the table

	0	1	2
0	0	0	0
1	0	1	2
2	0	2	2

Use formula (1) for each of two stages; first find $f(0, y)$, $f(1, y)$, $f(2, y)$ and thence $f(x, y)$, which is $x^2y^2+2xy(x+y+1)$.

Functions of this kind can be used to describe circuits controlled by two three-way switches.

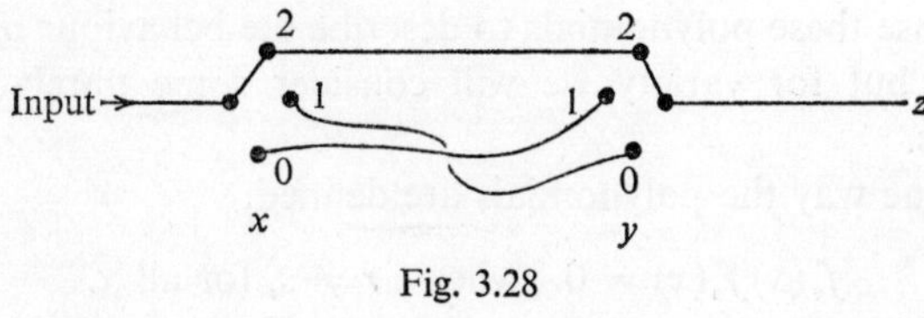

Fig. 3.28

x and y can take three values, but z can take only two, because the current is either switched on or it is not. Adopt the convention

$$z = 1 \text{ means } z \text{ is switched on,}$$

$$z = 0 \text{ means } z \text{ is switched off.}$$

Then the action of this circuit is described by

$$z = 2x^2+2y^2+xy+2x+2y.$$

(3) Construct the operating functions for various circuits, for example the binary adder on p. 164.

(4) (*Harder.*) Construct circuits which realize various operating functions. Can you find a system?

This is a hard question because this algebraic theory has certain unsatisfactory features when compared with, say, the Boolean algebra which described two-way switching circuits (p. 153). In that algebra an expression may be interpreted as a set of instructions for wiring up a circuit that will behave in a particular way. In this present algebra, as far as we have developed it, that is not so; and this is a shortcoming.

Now let us return to GF (5). As before we may find the interpolation polynomials from formula (1), calling them now $f_0(x)$, $f_1(x)$, etc.

We have

$$\begin{aligned} f_0(x) &= 4x^4+1 \\ f_1(x) &= 4(x^4+x^3+x^2+x) \\ f_2(x) &= 4x^4+3x^3+x^2+2x \\ f_3(x) &= 4x^4+2x^3+x^2+3x \\ f_4(x) &= 4x^4+x^3+4x^2+x. \end{aligned}$$

Over this field any function satisfies the identity

$$f(x) = f(0)\,f_0(x)+\ldots+f(4)\,f_4(x).$$

We may use these polynomials to describe the behaviour of five-way switches; but for variety we will consider some purely algebraic properties.

From the way the polynomials are defined,

$$f_r(x)\,f_s(x) = 0 \quad \text{when} \quad r \neq s, \text{ for all } x.$$

This may be verified in a more roundabout way. We have already used the table of powers to demonstrate the relations

$$x^4 = 1 \quad (x \neq 0) \quad (\text{mod } 5)$$

and

$$x^5 = x \quad \text{for all } x$$

$$x^6 = x^2, \text{ etc.}$$

If now the expressions $f_r(x) f_s(x)$ are multiplied out in full, these relations reduce all the products to zero. The polynomials $f_r(x)$ are an example of a set of orthogonal polynomials—an idea of great importance in mathematical physics. We might put the results in the form

$$\sum_{x=0}^{4} f_r(x) f_s(x) = 0, \quad r \neq s,$$
$$= 1, \quad r = s.$$

Similar calculations may be done over any field GF (p^n), and these also have applications in electrical circuitry. The field GF (2^2) may be used to describe the interconnections of two two-way switches, x and y, working two lamps u and v. As an algebraic device write $z = x+my$ and $w = u+mv$, where m is defined on p. 76. This amounts to regarding the two switches as represented by the 'real' and 'imaginary' parts of a complex number z, the lights being represented by the two parts of the complex number w. It may be verified that the interpolation polynomials are

$$f(z) = z^3+1$$
$$f_1(z) = z^3+z^2+z$$
$$f_m(z) = z^3+mz^2+(m+1)\,z$$
$$f_{m+1}(z) = z^3+(m+1)\,z^2+mz.$$

To describe the circuit on p. 167, which was part of a binary adder, replace the c of that page by v. The appropriate calculation shows that the operation is described by

$$w = mz^3+mz^2+z.$$

Extensive calculations with polynomials of this type and their applications to switching problems are given in [84].

4

NUMERICAL METHODS AND FLOW CHARTS

1. YOUNG CHILDREN WITH MACHINES

Here are the machines. See what you can find out. All I ask is that you work only one lever at a time!...You think you can multiply already, Smith? Try checking it on paper....Yes, that lever clears the top window, doesn't it?...Well there must be a reason for the bell ringing, Jones. You can see what happens when you turn the handle the other way....You've got seven times fifty-seven, Williams? Have you checked it? What about thirty-seven times fifty-seven?... Smith's winding away like mad! How can we shorten his work for him?...Check the working in your book and compare it with what you did on the machine....You can add, can you, Perkins? And I see you have found out how to clear the bottom windows as well.... Devlin's got every window full of nines and can't get rid of them.... Yes! of course, just press the lever down. Nobody seems to know much about these nines. You say it's because there's nothing to take away from, Jones? Have you found out about taking away?...

Children, whatever their age and ability, are intrigued by any calculating device however simple. Instructive lessons can be built around the history, manufacture and use of such devices. Some examples are a cardboard addition/subtraction machine, a set of cardboard strips used as 'Napier's Bones', and a simply constructed wooden slide-rule. For details of their construction, consult the references listed at the end of §5 of this chapter, e.g. Birtwistle.

In this discussion we shall assume that a number of standard desk calculating machines are available. Ideally there should be one machine per child (particularly with older children), but group working with up to five or six children per machine is possible, with the

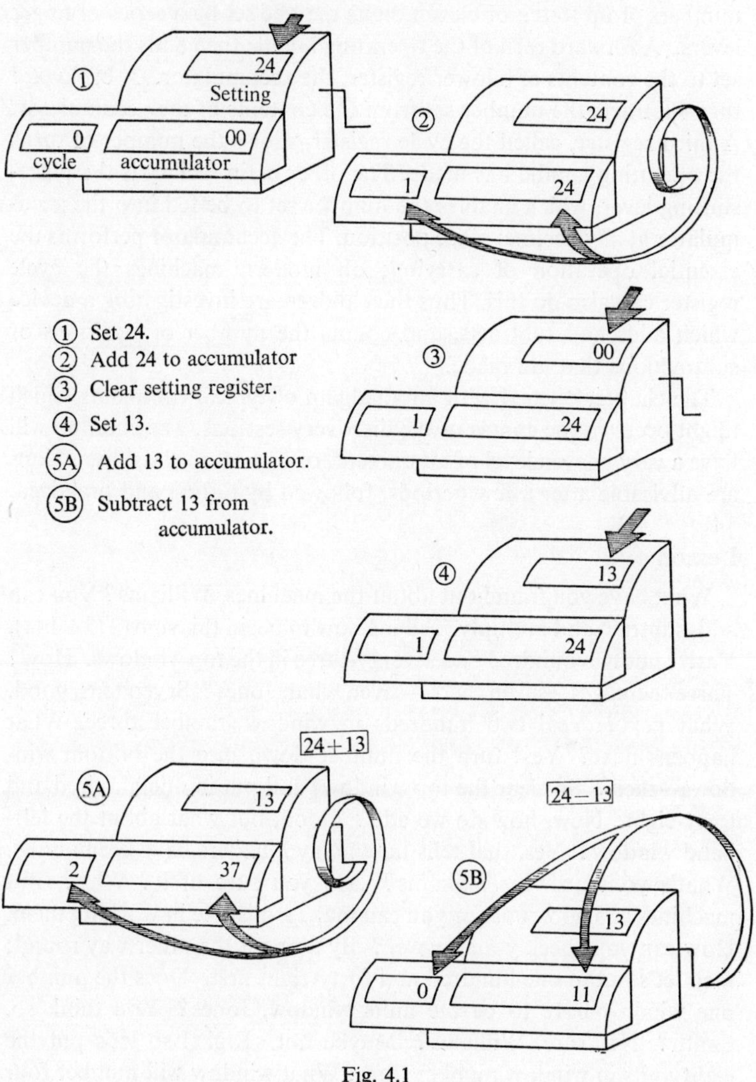

Fig. 4.1

children discussing and writing down what they find out. The first few sessions should be devoted to letting the children discover the various operations themselves.

A typical desk calculating machine has a setting register, in which numbers of up to ten or eleven digits may be set by a series of finger levers. A forward turn of the operating handle then adds the number set to the contents of a lower register, the accumulator. A backward turn subtracts the number set from the contents of the accumulator. A third register, called the cycle register, shows the number of turns the operating handle has made. The other main facility is the place-shifting lever, which enables the number set to be fed into the accumulator at any decimal place position. The accumulator performs the essential operation of carrying; on modern machines the cycle register can also do this. Thus the children are investigating a device which adds and subtracts, and counts the number of additions or subtractions that are made.

The chapter opened with an amalgam of typical comments which might occur in the course of the discovery sessions. The children will have a very uneven level of attainment, of course, so class discussions are advisable after a few periods, followed by testing and practice.

Lesson

What have you found out about the machines, Williams? You can add, subtract and multiply. Tell me how to begin this sum (273 + 148). Yes! I put two hundred and seventy-three in the top windows. How? Three there, yes—seven there—seven what, Jones? Seven tens, good. What next? Yes! two hundreds in window number three. What happens next? Yes! turn the number down into the bottom windows—then? So clear the top windows and put in one hundred and forty-eight. Now, how do we add? Good, but what about the left-hand window? Yes, that tells how many numbers have been added. What's your answer, Williams? Are you sure of it? Why? But machines can go wrong, or you can make a mistake in working them. How can you check your answer? By doing it the other way round; then let's put in one hundred and forty-eight first. Does the number one window have to be the units window, Jones? You think so. Smith? Not sure. Williams? Maybe not. Right! so let's put the eight units in window number three. What window will number four be then? and number five?

Indication of the units window by the sliding pointer can be demonstrated, and the handling of decimal fractions. Discussion of the other operations and facilities can follow similar lines. Use the machines frequently at first, if possible. If they are only available at intervals, there will be a need for revision at the beginning of each session. There is no need to lay emphasis on speed and technique. Inevitably the children will attain widely varying degrees of skill, but it is the understanding that is important. Demonstration examples should be taken at a slow pace, but should give scope for the faster workers to move ahead where they can.

A set of examples, as realistic as possible, must be worked out beforehand, involving fairly complicated numbers. Business calculations, problems on factory production and costings (houses, ships, etc.) finding of total floor areas from plans, straightforward sets of statistics from H.M.S.O. publications are all possible sources. The use of the machines should lead to opportunities for discussion and work on the following ideas:

(1) A comparison of the machine working and human working of sums. Working on paper can be used as a check on machine calculations or interspersed with purely machine workings. The first method could be used to begin with (and continued if the number of machines is small), the second method later on. Attention can be directed to layout, place value and carrying. The latter aspects of the work are particularly fruitful; it can be shown that the machine counts in tens. A simple demonstration model of the carrying operation is helpful, since it is fundamental to all digital machines. Examples in the metric system are good, and can be compared with a variable radix system such as that connected with British coinage. A £ *s. d.* adding machine may be obtainable from the school office for comparison. Point out how complicated a machine would have to be to perform all four arithmetical operations in such a system. It is the place shifting that causes the difficulty. The seeds of later discussion on the binary system used in electronic computers can also be sown.

(2) As is mentioned later, the calculating machine exemplifies the laws of operation between numbers. The desk machine uses decimal numbers of up to about ten significant figures. There is a basic operation recognized as addition, in which it is observed that order of operation does not matter. The inverse operation, subtraction, 'undoes' addition. The machine does not recognize any subtraction as impossible, but order of working clearly matters. Multiplication problems are performed by repeated addition, and the inverse of

multiplication, division, is done by repeated subtractions. When calculating a human uses multiplication as an operation in its own right. Tables must be memorized, or stored as with Napier's Bones. It may be felt worth while to use this opportunity to re-examine the way in which 'times tables' are derived in the first place, from repeated addition.

Such a discussion might form an early part of a development of the number system, from integers to rationals and then to real numbers.

Lesson

Our machine can handle numbers—any number? Will it handle one third? But is ·33333 the same as one third? What decimal is then? What is three times one third? Try it on the machine. What do you get? What is three times point three recurring, then? Can you think of other numbers that the machine won't handle? Why can't the machine handle these when it can three quarters? That's how they differ then! But you called them both numbers, so they must be alike in some ways. If I wrote 'five' in Roman numerals would it still be the same number? So the way we write a number is not the number itself.

What types of number have we got so far? Ending decimals and recurring decimals. What does 'recurring' mean? Is that all, just 'unending'? That's better, the same block of digits is repeated unendingly. How do we get these blocks? By division; from sevenths, thirteenths, etc.

Let us draw a set diagram with all the decimals between nought and one as the members. One subset is of ending decimals. So the complement's members are . . . ? Which of these subsets does the machine use? Now let's concentrate on the unending decimals. What do we know about them? Yes! but are we sure that *all* of them are recurring? How should we draw the diagram then? Good—now this subset had the property? And this one? What can you say about the union of these two subsets? That's it! They all come from fractions. So what about this subset? Yes! It must consist of unending non-recurring decimals.

The sense in which this type of decimal is a number could then be considered, discussing perhaps the familiar $\sqrt{2}$. We certainly regard this as a number because we write $\sqrt{2}\times\sqrt{2} = 2$. Most children get something out of the indirect proof that $\sqrt{2}$ is irrational, and Newton's iterative process (p. 95) provides a method of calculating the decimal expansion as far as is desired. Two important points to emerge are:

(i) that while much can be done with forms such as $\sqrt{2}$, $\frac{1}{3}$, etc., most calculations (machine and human) are with decimal expansions, which must be selected from the subset of terminating decimals;

(ii) the selection is by any method (division, iteration) which gives successively closer approximations.

The desk machine can be used to study the recurrence patterns of sets such as $\frac{1}{7}, \frac{2}{7}, \frac{3}{7}, \ldots, \frac{6}{7}$ or $\frac{1}{13}, \ldots, \frac{12}{13}$. Several of the ideas used in the above lesson should emerge. For example

$$x = 0{\cdot}\dot{1}\dot{2},$$

so

$$100x = 12{\cdot}\dot{1}\dot{2}.$$

Therefore $99x = 12$, and x is the fraction $\frac{12}{99}$. By a similar argument every recurring decimal can be shown to be equal to some fraction.

Two major extensions of the domain of integers are made in the lower school—first to fractions, later to directed numbers. This extension might well be done more plausibly and systematically than at present, particularly where the theory of sets forms the basis of the algebra. The calculating machine has a small but useful part to play, not only as a labour-saving device, but because it exemplifies laws.

(3) 'Rough estimates' used as a check can lead to the discovery of 'short cut multiplication', e.g. in a calculation such as 17×85, the rough estimate is 20×85, which suggests that instead of working the sum as $10\times 85+7\times 85$, using eight handle turns, it is quicker to work $20\times 85-3\times 85$, using five handle turns. The use of this method should not be insisted on, since it can be somewhat confusing for some children.

(4) A calculation using rounded-off values from measurements may involve a build-up of errors. In the example below the errors are in lengths measured to the nearest tenth of an inch, so that the true value of say 7·3 is between 7·25 and 7·35. The machines can be

used to calculate the extreme limits of error that may arise in sums, differences, products and quotients.

$$7{\cdot}3 \times 15{\cdot}7 = 114{\cdot}61$$
$$7{\cdot}25 \times 15{\cdot}65 = 113{\cdot}4625$$
$$7{\cdot}35 \times 15{\cdot}75 = 115{\cdot}7625$$

Thus only the first two significant figures are reliable. There is a tendency when using calculating machines to quote too many significant figures, and demonstrations of this kind are very salutary. Further discussions of errors may be found in the references.

(5) The concept of negative numbers may be developed, since the machines handle such numbers in complement form. Note the comment in the first exploration lesson 'Nobody seems to know much about these nines—you say it's because there's nothing to take away from, Jones?' The machine cannot treat $0-7$ as impossible. Just as $010000-000007$ yields 009993, so $0000-0007$ is begun as if there were a 1 available at the end of the accumulator and ends when the accumulator is full of nines. The class explore this property and find a method of recovering the 7. The property can then be used in a lesson working out bank accounts.

Lesson

So there's the list of cheques paid in and drawn out, entered in correct columns. Now let's use our machines like the bank, to work out Smith's balances. Credit 155 dollars—155 paid into the machine. What is Smith's first balance? Debit 73 dollars—good! Credit 82, debit 142—trouble? I see! Does the bank allow that to happen? Has the machine allowed it to happen? Yes, the bank must trust Smith; but then Smith trusts the bank doesn't he, when he puts in 155 dollars? Who owes whom at first?

Let's write to begin with

BANK $\overrightarrow{\text{OWES}}$ SMITH.

But what is the machine telling us now? Smith owes the Bank. I won't bother to write it; just reverse the arrow.

BANK $\overleftarrow{\text{OWES}}$ SMITH.

Can we do something about these balances? Good, put them in then.

How much do we put under that last arrow? Good, notice how

the machine tells you that Smith owes the bank. Next a credit of 62 dollars; so that's Smith's balance column. Let's study these entries. Would it be enough just to write '32' as a balance entry? Good—this way round it means? And the other way? Then an entry like this tells you more than just '53'?

Now Smith, in fact, had two bank accounts, one at home where the balance was $\overrightarrow{254}$, and one near where he worked, where the balance was $\overrightarrow{580}$. This was on the day he moved, so he asked the two managers to transfer his accounts to a third new bank. What balance would the manager of the new bank send him? How would you write that as an equation? You propose adding these balances? In that case you must be thinking of them as ...? But they're not quite the same as before, are they? '$\overrightarrow{580}$' tells you more than '580'. Would the new manager be able to do anything if he got just '254' and '580' as Smith's balances? Suppose he got $\overleftarrow{254}$ and $\overrightarrow{580}$? You still say he is adding the balances? Suppose he got $\overrightarrow{254}$ and $\overleftarrow{580}$? It looks as if balances can always be added. Is that enough to make them like numbers?

The further discussion would link up in the usual way with other directed number situations, such as velocities and times, where multiplication naturally occurs.

2. THE NUMERICAL APPROACH

'The calculating machine, in view of its importance, should become known in wider circles than is now the case. Above all, every teacher of mathematics should become familiar with it', as Klein pointed out—in 1908! The importance which he attached was fundamentally a mathematical one. 'In the existence of such a machine we see an outright confirmation that the rules of operation alone, not the meaning of the numbers themselves, are of importance in calculating; for it is only these that the machine can follow; it is constructed to do just that.'

Nevertheless, it would be totally misleading to suppose that, at school level, no numerical work can be undertaken without, at the least, a hand calculating machine. What we seek to encourage is a numerical methods philosophy, quite as much as any specific technique. We firmly believe that this can and should be woven into the mathematics course at all levels, without needing expensive equipment and without it being regarded as yet another separate 'topic'.

Computation has been described as obtaining, with a minimum of calculation, a numerical result which is known to be a solution of the given problem. This implies that:

(i) Checks to ensure freedom from blunders are an integral part of any numerical work;

(ii) the precision with which the problem was stated and the degree of accuracy of the suggested solution must be considered carefully;

(iii) the alternative methods available in any particular case must be considered in advance, the most suitable chosen, and then the individual steps of the calculation planned to yield the final optimum result.

Checks

Although we are inclined to give credit for correct reasoning alone, we must face the fact that, in practice, numerical work containing blunders is worse than useless. But the need to design and carry out checks may well double the value of any given problem.

For example, a satisfactory treatment of an early trigonometry problem might be as follows.

To find the distance risen (R yards) in walking 100 yards up a slope inclined at 10° to the horizontal.

(1) Draw a diagram including the given data.

(2) Evaluate $R = 100 \sin 10°$.

(3) Evaluate $H = 100 \sin 80°$. (The horizontal distance, H yards, was not asked for, but it provides further practice and contributes to the check.)

(4) Check that $R^2 + H^2 = 100^2$. (Use of a table of squares.)

Many problems may be checked graphically and a valuable feature of some methods is that they are inherently self-checking.

Degree of accuracy

Most problems have not *a* right answer, or even two, but an infinity of answers consistent with the given data. The alarming range of values open to an innocent expression like $\frac{1\cdot0}{1\cdot7-1\cdot6}$, if the numbers have been rounded to one decimal place, is enough to bring this home! However, frequent discussion of the precision of data used in problems can not only increase the reality of the work but again increase the potential usefulness of each problem. Thus, to take the above trigonometry example further:

(5) Estimate the possible precision of the data (e.g. for a boy pacing out distance and measuring angles with a home-made inclinometer, distance ± 5 yards, angle $\pm 1°$).

(6) Evaluate maximum rise using 105 sin 11° (including check as in (3) and (4), of course).

(7) Evaluate minimum rise using 95 sin 9°.

(8) Problem. If he could halve the error in just *one* of his measurements, which would yield the greater improvement in the accuracy of the result? Does this apply to any slope?

Later, the well-established habit of error analysis will find further stimulation from the work on differential coefficients, and the concept of partial differentiation will seem a natural extension of the familiar ground. Also, at the sixth-form level can come a statistical approach to rounding errors. This simple idea of error analysis is indeed a fertile source of interest.

Methods

Traditionally, a problem was considered solved once a formal solution had been obtained. But the numerical results derived from such a solution are always subject to errors as mentioned above so that the exact formal solution proves to be an illusion. Methods which are intrinsically approximate may well yield results to the required accuracy more economically and in a manner more subject to checks. Thus iterative methods come into their own as powerful and elegant techniques which are applicable to problems ranging from mental estimation of a square root, to solution of systems of differential equations using an electronic digital computer. Indeed, the formal solution may be so uneconomical as to be of very little practical value at all. What navigator would use the cosine rule to obtain a dead-reckoning position when drawing is so much quicker and yields all the precision which the problem justifies?

3. TRIAL AND ERROR

The inclusion of elements of numerical methods in lower school work is essentially a matter of seizing opportunities as they arise. The lesson that follows adopts a heuristic approach to the Babylonian problem of finding the sides of a rectangle with given perimeter and area. The principal object of the lesson is to present an open situation and to take advantage of the different contributions that can be made to evolve a useful numerical technique.

Lesson

If two circles have the same perimeter have they got the same area? What about two squares? Two rectangles? Do you agree? Look at it this way. Do you fix a particular size of circle by quoting its perimeter? A square? A rectangle? How do we distinguish different sizes of dowelling? Of square battening? Of general battening? Yes, we talk of 1 inch round dowelling and 1 inch square battening but in general we have to give two measurements; we ask for 'four by two'.

Suppose the sides of a rectangle are 2 inches and 3 inches. What is the perimeter? The area? Units? Good. Now suppose I gave you the perimeter 14 inches and the area 12 square inches could you tell me the sides? How did you get that? Suppose the perimeter is 20 inches and the area is 1 square inch. What are the sides? It is not so easy to guess this time. Suppose the sides are x inches and y inches. What is their sum? Do you agree? Yes, the perimeter counts each side twice. So ...? $x+y = 10$ and $xy = 1$. Give me any two numbers that add up to 10. What is their product? Some other numbers please?

$x+y = 10$

x	5	8	0·5	1	0	...
y	5	2	9·5	9	10	...
xy	25	16	4·75	9	0	...

What do you notice? Why can't the numbers we want be whole numbers? Is there anything else you notice? Give me two numbers whose product is 1 then. What is their sum? This is another way of looking at the problem. Try some other combinations.

$xy = 1$

x	2	2·5	8	9	...
y	0·5	0·4	$\frac{1}{8}$	$\frac{1}{9}$	...
$x+y$	2·5	2·9	$8\frac{1}{8}$	$9\frac{1}{9}$	...

What do you notice? Can you give a reasonable guess at one of the sides? Yes, one is certainly pretty small. Let's take x to be zero.

If $x = 0$ what is y? Oh dear, what can we do about that? Yes, we could choose y from the first table. That means...? What now? Use the second table for xy? How? We tried $x = 0$ and got into difficulties. Use y did you say? What value of y? What does that give for x? What have we got now? $x = 0{\cdot}1$, $y = 10$, $x+y = 10{\cdot}1$, $xy = 1$. That's getting warm. Now what? Good; let's use $x = 0{\cdot}1$ in the first table then. We seem to be getting warmer each time.

Note that this problem implies a quadratic equation, yet the lesson could easily have been with a first form.

The following is based on a discussion which arose with a second form which had been brought up on the 'Factors and Multiples' chapter to be found in practically any appropriate textbook.

Lesson

So, to get the square root, you divided the indices of the prime factors by two. Yes Jones? Ah! that's a point. Well, in that case, you couldn't write the number as the product of two *equal* factors, could you? But you could probably write several pairs of *unequal* factors. For example, $12 = 1\times 12 = 2\times 6 = 3\times 4$ and of course $= 4\times 3$. All of you; think of a number and write down its pairs of factors. Have you done that? How does your list go, Smith? Oh, I might have known that *you* would choose a prime number! What about yours, Brown? Yes, that's a good choice. So

$$24 = 1\times 24 = 2\times 12 = 3\times 8 = 4\times 6,$$

but where do you think the square root is? Good, in other words, 5 is just bigger than the root we are after. Well, what factor would go with 5? True, so we can't restrict ourselves to whole numbers this time. Yes, $24 = 5\times 4{\cdot}8$. What did you say about the root compared with 5? And compared with 4·8? So the root we are after lies between 5 and 4·8. Any suggestions? Why 4·9? Good, but do you really think that $\sqrt{24} = 4{\cdot}9$? No, of course not; after all $(4{\cdot}9)^2$ must end in 0·01, but you think the average is probably better than the 5 or the 4·8. So where do we go from here? Yes! I'm afraid so, find the factor to go with 4·9 (roughly will do) and then find *their* average.

Now try with my 12, and it wouldn't even matter if we thought that $\sqrt{12}$ might be 6. You can start, $12 = 6\times$ (whatever you think it is!) How is it going, Jones? You got 6 and 2, then 4 and 3, then $3\frac{1}{2}$ and $3\frac{3}{7}$. But what's that in decimals? Good, but we won't need an infinite number of decimal places, will we? 3·50 and 3·43 will do, and so we go on; or at least we *would* go on if we had one of the calculating machines out! Now see how far you can get towards the square root of any number you choose. What is it, Smith? How are you getting on with your prime number? Well, I don't know $\sqrt{13}$ offhand, though that sounds reasonable. But if you square your 3·6 you will see how close you are getting. Better still though, would be to use your 3·6 as the next estimate and go through the process again. If 3·6 is right so far, you will get something like 3·63, or perhaps 3·58, and if it's wrong, you will get an answer which is not so wrong as 3·6 was! You can't lose, can you?

Now for homework! The figures given in the Highway Code show that if the distance a car takes to pull up, after the brakes have been put on hard, is D feet when it was travelling at a speed of S miles per hour, then $D = \frac{1}{20}S^2$. For example, at a speed of 30 m.p.h., $D = 45$ and $45 = \frac{1}{20}(30)^2$. I want you to work out how fast two cars were travelling if one halted in 245 feet and the other in 300 feet. Now I only paced out those two distances roughly, and the Highway Code formula is only approximate, so don't miss your tea to work out the thirteenth decimal place in your answer.

The details of whatever process emerges from such a discussion are not important. What is of value is that (i) the children have a natural question answered on the spot, and (ii) they are introduced to a systematic method of solving a problem by successive approximation. This, like the method evolved in the previous lesson, is an example of iteration.

The scorn poured on solution by 'trial and error' is sadly misplaced if the error is used *to guide the next trial*, and this feed-back loop is the essence of the iterative technique.

There are many topics in a school syllabus which can be illuminated by such an approach. It may also free us from the traditional order of development which so easily becomes regarded as an order of difficulty. The order of treatment of mathematical topics often concerns the other school departments as well, though liaison is so often imperfect. The process which emerged from the second lesson was, in fact, a special case of the Newton–Raphson rule: to find a real root of $f(x) = 0$, use the iteration $x_{n+1} = x_n - \dfrac{f(x_n)}{f'(x_n)}$, where x_1 is an approximation to the root. Although not without its dangers the rule can provide, without waiting for the binomial theorem, the approximations often needed early in the sixth-form physics course; for small x

(i) $$\sqrt{(1+x)} \simeq \frac{1+(1+x)}{2} = 1+\tfrac{1}{2}x,$$

(ii) $$\frac{1}{1+x} \simeq 1[2-1(1+x)] = 1-x.$$

Any method which helps to give an insight into a mathematical situation, as well as yielding a numerical answer, must rank highly in school work. If it also satisfies the condition of practical authenticity one would have thought it likely to prove irresistible! But the method of relaxation seems hardly to have reached the school level at all.

The equations

$$\left.\begin{aligned} 1{\cdot}5x-2{\cdot}1y+6{\cdot}3z &= 2{\cdot}4,\\ 3{\cdot}2x+8{\cdot}0y-1{\cdot}2z &= 3{\cdot}2,\\ x-\quad y+\quad z &= 4{\cdot}7,\end{aligned}\right\}$$

can be rewritten

$$\left.\begin{aligned} 1{\cdot}5x-2{\cdot}1y+6{\cdot}3z-2{\cdot}4 &= R_1,\\ 3{\cdot}2x+8{\cdot}0y-1{\cdot}2z-3{\cdot}2 &= R_2,\\ x-\quad y+\quad z-4{\cdot}7 &= R_3,\end{aligned}\right\}$$

where the R's are called *residuals.*

Clearly, a solution of the system will be a set of values of the variables for which the residuals are all sufficiently small. *What constitutes 'sufficiently small' can involve an extremely valuable discussion of errors and 'ill-conditioned' equations.* A table is then drawn up showing the effect on each residual of a unit increase in each variable. This is the Operations Table.

	Δx	Δy	Δz	ΔR_1	ΔR_2	ΔR_3
(1)	1·0	—	—	1·5	3·2	1·0
(2)	—	1·0	—	−2·1	8·0	−1·0
(3)	—	—	1·0	6·3	−1·2	1·0
(4)	2·0	−1·0	−1·0	−1·2	−0·4	2·0

Operation (4) represents an attempt to construct a composite operation which has a greater effect upon R_3 than on the other residuals just as operations (1) and (2) act more on R_2 and operation (3) on R_1. The attempt has had only limited success but (i) the damage operation (4) may do to R_1 can be rectified by operation (3), and (ii) the combination of changes of variable is a simple one so we may gain in freedom from blunders what we lose in speed of convergence. We start the relaxation proper by substituting arbitrary values of the variables (or any approximate solution available), and applying multiples of operations (2), (3) and (4). These *displacements* are chosen so that, in general, the numerically largest residual is reduced to as small a value as is arithmetically convenient.

The table could be as follows....

Line	Op.	Δx	Δy	Δz	ΔR_1	ΔR_2	ΔR_3
1	1	+1·0	—	—	+1·5	+3·2	+1·0
2	2	—	+1·0	—	−2·1	+8·0	−1·0
3	3	—	—	+1·0	+6·3	−1·2	+1·0
4	4	+2·0	−1·0	−1·0	−1·2	−0·4	+2·0
—	—	—	—	—	—	—	—
		x	y	z	R_1	R_2	R_3
5	Start (Arbitrary)	0	0	0	−2·4	−3·2	−4·7*
6	4	+4·0	−2·0	−2·0	−4·8	−4·0	−0·7†
7	3	—	—	+1·0	1·5	−5·2	0·3
8	2	—	+0·5	—	0·45	−1·2	−0·2
9	*Check*	+4·0	−1·5	−1·0	0·45√	−1·2√	−0·2√ ‡
10	2	—	+0·15	—	0·135	0	−0·35
11	4	+0·4	−0·2	−0·2	−0·105	−0·08	0·05
12	3	—	—	+0·02	0·021	−0·104	0·07
13	2	—	+0·012	—	−0·0042	−0·008	0·058
14	*Check*	+4·4	−1·538	−1·18	−0·0042√	−0·008√	+0·058√
15	4	−0·06	+0·03	+0·03	+0·0318	+0·004	−0·002
16	3	—	—	−0·005	+0·0003	+0·010	−0·007
17	*Check*	+4·34	−1·508	−1·155	+0·0003√	+0·010√	−0·007√

* Now because 4·7 is largest $|R|$, use op. 4 $\times 2$.

† Now because 4·8 is largest $|R|$, use op. 3.

‡ Checks. The current *net* values of the variables are substituted into the expressions defining the residuals to confirm the values obtained in the table.

The above is almost certainly not the most economical relaxation scheme which could have been used to obtain an answer to a comparable accuracy. The speed of convergence depends upon the skill of the computer and there is a powerful incentive, especially when no machine is available, to the exercise of careful thought and ingenuity. If, perhaps, a single machine is being shared by a class, it is convenient to use the check stages as examples of the accumulation of products and to work the rest of the table mentally. Further details of this intriguing method can be found in Fox[50] or the textbooks referred to at the end of the chapter.

Note

During discussions on most iterative processes a graphical commentary on the progress of the work can be of enormous value. The benefit derived is not confined to the understanding of the process but extends to the ideas of convergence and continuity.

4. SIXTH-FORM COMPUTATIONS

The Associated Examining Board 'A' Level paper in 'Computations' (one of a number of options for Paper II of the Mathematics Examination) was introduced to cater for junior staff of establishments, such as the Royal Aircraft Establishment, Farnborough, where there was, and still is, a rapidly increasing demand for trained personnel to operate and programme electronic digital computers. An introductory course is required and it is deliberately based on the use of desk calculating machines. This is because it is felt that not only can the most important principles of numerical methods be illustrated, given a modest degree of expertise, but some desk machine experience will assist a programmer to obtain the best from an electronic machine. In the school context, this background has a great appeal, but the value of the course is rather different and its attendant problems are perhaps greater than is the case in the colleges where the same work is done. These points will be considered again below.

The following is based on a typical two-hour session towards the end of the first year of a two-year grammar-school course. It is designed to bring together a number of techniques which have been studied previously.

Warm-up

About a quarter of an hour is devoted to use of the machines for whatever calculations the students care to carry out—such as working out the current physics practical results and assessing their degree of accuracy. It has been found that with two-hour sessions spaced a week (and sometimes more) apart this warm-up period is very necessary. Also free access to machines during lunch-breaks, etc., is desirable and students are not slow to take advantage of such facilities.

Review of homework

In response to requests from students, arrangements have been made for the loan of machines over weekends, etc., to complete classwork or to do voluntary 'overtime' but, in general, homework has to be set on the assumption that a machine will not be available. One such homework simply challenged the class to find out as much as they could about the tabulated function:

x	1	3	5	7	9	11
$f(x)$	1	32	61	115	185	271

First they explored the function by differencing (see Table 1). The Δ^2 column reads -2, 25, 16, 16 and suggests that all these second differences might have been 16 were it not for some error which has caused the erratic changes in the other values. Such an error, $+e$ say, in a particular tabulated value of $f(x)$ would superimpose on the accurate table a diverging pattern of disturbances as in Table 2.

Considering Table 1, with Table 2 in mind, suggests that the 25 in the Δ^2 column might be made up of $16+9$, which would in turn demand $16-2\times 9$ as the entry above it. Since this entry is indeed -2, the suspicion that $f(3)$ is too great by 9 is strengthened. That this would make the corrected value of $f(3)$ 23 instead of 32 is further confirmation since such a transposition of digits is a slip which commonly occurs in numerical work.

Having corrected the table, the students also deduced, from the fact that the second differences were now constant, that the function tabulated was a quadratic in x; $f(x) = ax^2+bx+c$, say. Thus after solving the equations

$$a(1)^2+b(1)+c = 1$$

$$a(3)^2+b(3)+c = 23$$

$$a(5)^2+b(5)+c = 61$$

and checking that the results were consistent with the other entries in the table, they were able to report their findings as follows:

(1) the tabulated function was $f(x) = 2x^2+3x-4$; $x = 1(2)11$.

(2) there had been a slip in the copying of the table.

Table 1

x	$f(x)$	Δ^1	Δ^2
1	1		
		31	
3	32		−2
		29	
5	61		+25
		54	
7	115		16
		70	
9	185		16
		86	
11	271		

Table 2

x	$e(x)$	Δ^1	Δ^2	Δ^3
x_{-3}	0			
		0		
x_{-2}	0		0	
		0		$+e$
x_{-1}	0		$+e$	
		$+e$		$-3e$
x_0	$+e$		$-2e$	
		$-e$		$+3e$
x_1	0		$+e$	
		0		$-e$
x_2	0		0	
		0		
x_3	0			

etc.

No doubt while working on the above problem, some students felt the need to revise their knowledge on, say, differencing. They would have consulted the flow chart which was evolved during the original lessons on the technique and which is given in Fig. 4.2.

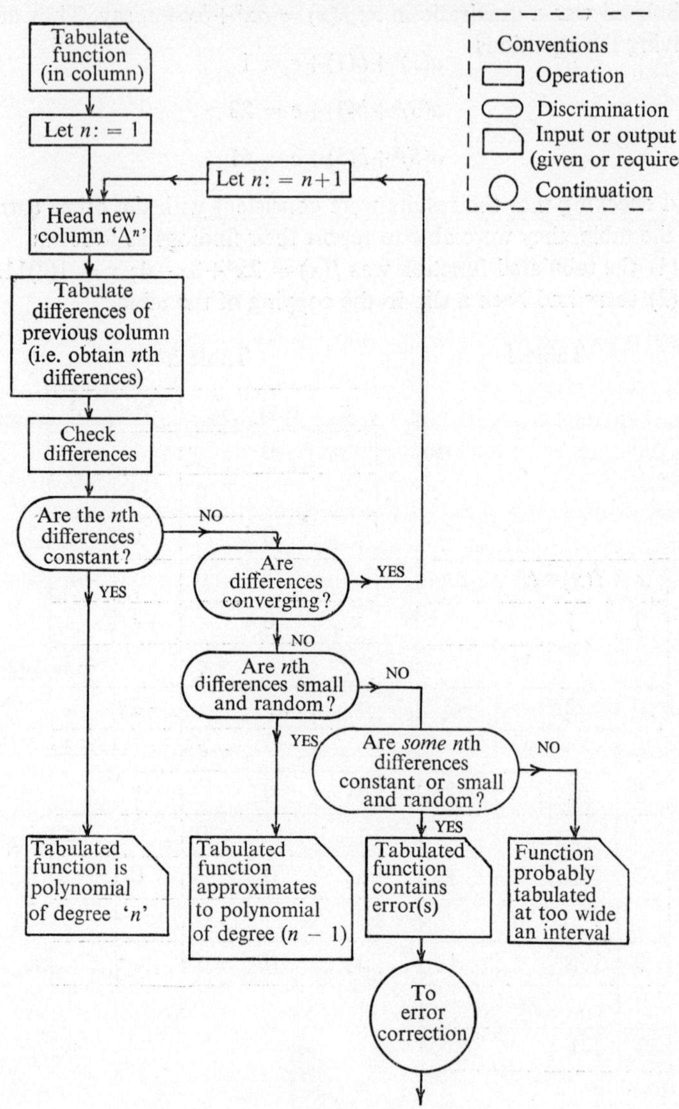
Conventions
Operation
Discrimination
Input or output (given or required)
Continuation
Tabulate function (in column)
Let $n := 1$
Let $n := n+1$
Head new column 'Δ^n'
Tabulate differences of previous column (i.e. obtain nth differences)
Check differences
Are the nth differences constant?
NO
YES
Are differences converging?
YES
NO
Are nth differences small and random?
NO
YES
Are *some* nth differences constant or small and random?
NO
YES
Tabulated function is polynomial of degree 'n'
Tabulated function approximates to polynomial of degree $(n-1)$
Tabulated function contains error(s)
Function probably tabulated at too wide an interval
To error correction

Fig. 4.2

Current lesson material

The following question was then considered:

The given tabular values of $f(x)$ are those of the first derivative of a quartic $F(x)$,

$$\text{i.e. } f(x) = \frac{d}{dx} F(x).$$

Correct any errors in the table and evaluate the first differences of $F(x)$ for the same range.

x	1·2	1·3	1·4	1·5	1·6	1·7	1·8	(*A.E.B.*
$f(x)$	0·288	0·988	2·576	4·500	6·784	9·452	12·528	*1958*)

The first steps were, of course, the analysis of the problem and the sketching of a brief flow chart or block diagram by each member of the class. The criterion of success here is simply whether the student at the next desk can see at a glance how the author of the chart proposes to tackle the problem.

Fig. 4.3 shows a possible chart for this stage.

When the problem was understood clearly enough to complete such a chart, computation began and the student had, perhaps, an hour of steady work ahead of him.

Current homework

Designed to continue the revision process; but again without requiring access to a machine.

By graphical means, find an approximation, x_0, to the positive value of x which satisfies the simultaneous equations $x^2+y^2 = 4$, $x+y^3 = 1$. Hence, using the equations $y_0^3 = 1-x_0$, $x_1^2 = 4-y_0^2$ in sequence, with cyclic repetition if necessary, evaluate x to three decimal places (A.E.B. 1958).

The above problem to be analysed, a flow chart prepared and work carried to the point at which a machine is desirable. It can then be completed at the next session.

Comment

Obviously, students will not all work at the same speed. Some may need to arrange extra machine time, while others might well complete a solution and be able to compare alternative approaches to some sections as indicated in Fig. 4.3.

Once a certain minimum proficiency has been achieved it is quite feasible to provide a modest computing service to other school departments. Surveying is one obvious source of work; biology and economics are others. In addition, the presence of the calculating machine in school removes the only valid objection to a course in statistics, which is of undeniable social relevance today.

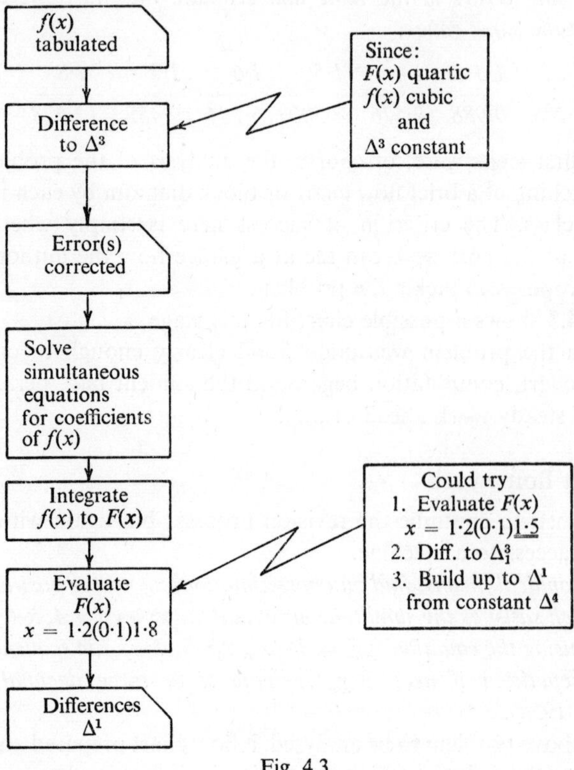

Fig. 4.3

The field is an extremely wide one. Colleges of Advanced Technology offer similar courses, so both staff and students are, in general, actively engaged upon some aspect of practical computing between their course lectures. Without these advantages, the schoolteacher has an even greater obligation to maintain close contact with professional computing practice.

But even if experience is confined to a course based upon the A.E.B. Syllabus, the following facts emerge:

(1) Computation has an appeal to students which persists long after the machines have lost their novelty.

(2) It is evident that the work is relevant to current developments in the world around us.

(3) The subject is by no means mathematically trivial. At all levels of students' ability work in the computations class has shed light upon their more conventional mathematics course.

5. HISTORICAL NOTE

It is curious how ambivalent our attitude to the subject of numerical methods can be. We never know what to call it and as Robert Recorde noted: 'Many praise it, but few dooe greatly practise it: onlesse it bee for the vulgare practice, concernying Merchaundes trade.' The ambivalence goes back to the Greeks. Philolaus could say that it was impossible that anything could be conceived or known without number but he left the 'vulgare practice' of addition and multiplication to his slaves. It is hardly surprising that an alphabetic notation for numbers was retained for so long. Meanwhile the study of number (arithmos) was arithmetic in the sense of Diophantus's *Arithmetica*, that is number theory or what is sometimes called the higher arithmetic to distinguish it, presumably, from 'sums'. Even the Romans, to whom we owe the word 'calculate', made distinctions. The *calculones* were slaves, but teachers of calculation from good families could be called *numerarii*.

It is inevitable that it was the inheritors of a different tradition who first introduced a satisfactory numeral system into Europe and that the first great advances in numerical methods were made in response to the needs of a nascent commerce. The Hanseatic League did more for mathematics than the universities and schools of its time. It was the Rechenmeisters who spread the Arabic notation and the use of decimal fractions.

The importance of developing powerful numerical methods was fully realized by the end of the sixteenth century when navigators and astronomers had amassed an indigestible wealth of observations. Logarithm tables were a welcome aid in the labours of computation and were widely used before the exponential nature of the logarithm was properly understood. Logarithms played a crucial part in the great scientific discoveries of the seventeenth century. Conveniently tamed in the slide-rule they remain a useful calculating tool. The slide-rule is a simple analogue device; it is as well to remember that soon after the publication of the first logarithmic tables Pascal invented a digital calculating machine. The modern digital computer is the twentieth-century equivalent of the logarithm table.

Cajori ascribed the miraculous power of modern calculation to three inventions: Arabic notation, decimal fractions and logarithms. To this list we must add a fourth, the computer. The computer means 'mechanical' arithmetic. It is in the mechanization of arithmetic that the solutions to the difficult logistic problems of a complex industrial society lie. The teaching of numerical methods will be at its liveliest and most relevant when it takes into account that this is the age of the computer. It does not begin to do this if school arithmetic ends with the logarithm table and applied mathematics inhabits a quaint Pythagorean number world of its own. An important change of emphasis is required. This may be summarized briefly by saying that we want algorithms not existence theorems. Iterative techniques may well be to the Common Market what the Arabic notation was to the Hanseatic League.

Further reading

There is nothing new in the main principles of modern numerical methods and they are already applied in many parts of school mathematics. There is, however, no book that gathers these strands into a school context though there is an excellent account of elementary computational procedure by A. Bakst in the 12th year-book (1937) of the American National Council of Teachers of Mathematics. See also ch. 5 in the 24th-year book, 1959.

Some relevant articles from Mathematics Teaching are: R. Fairthorne: Computing wants and computing needs, 2, 1956; A. J. Walker: Introduction, use and future of numerical and practical methods of mathematics, 4, 1957; Editorial: A mechanical analogue computer for educational purposes, 13, 1960; P. French: Calculating machines in a primary school, 18, 1962; C. Birtwistle: Calculating aids in the secondary school, part 1, 19, 1962; part 2, 20, 1962.

The bibliography of the National Physical Laboratory's monograph *Modern Computing Methods* [86], is prefaced by the following important remarks: 'It must be emphasized at the outset that facility in computation can only be gained by practice, reading alone is not sufficient. This is true whether the equipment used is a desk machine or an electronic computer. In the latter case, moreover, although the approach is a little different, it is essential for the programmer to be acquainted with the fundamentals of desk machine practice before he can be efficient.'

It is clear that desk machines will increasingly be used in schools at all levels and some manufacturers are providing courses for teachers throughout the country. Machines are certainly required for the 'A' Level computation paper of the Associated Examining Board. The syllabus for this paper is covered by various textbooks, e.g. Butler and Kerr [21]; Redish [103]; Wooldridge [132].

Numerical tables virtually act as textbooks in a sixth-form course. The following are useful and inexpensive: *Interpolation and Allied Tables* (H.M.S.O. 1956); *Chambers' Six-figure Mathematical Tables* (Chambers, 1950).

The liveliest introduction to the modern electronic computer is still undoubtedly Bowden [17]. A brief selection from an enormous literature is Hollingdale [64]; Hersee [62]; Murphy [85]; Livesley [80].

The importance of numerical methods lies in their applications to industry and research. Although the hand machine implies that the equations we study at school can at least have likely coefficients, it is particularly difficult to find suitable examples of the applications of numerical methods. However, industrial firms and government establishments are usually very co-operative in this respect. Various professional journals can usefully be consulted; the following books indicate briefly the very wide field: Sutton [118]; Crank [31]; D.S.I.R. [37]. The reports of the conferences of teachers and industrialists at Oxford (Technology, 1957); Liverpool (Technology, 1959) and Southampton [123] are also relevant.

Since this book was compiled, Moakes [138] and Thomas and Thomas [139] have produced books devoted to the use of desk machines in schools.

6. FLOW CHARTS

At an elementary level flow charts provide an introduction suitable for juniors to problem analysis, operational methods, and programming. No mathematics is assumed.

Introduction—a lesson

Distribute to each member of the class two pieces of paper. Now make the request, 'Write on one sheet a description of how you came to school this morning'. When this has been completed get them to fold the other piece of paper into eight. 'Fold in half, fold in half and fold in half.' Now they unfold the paper. Ask them to decide which are the main statements in their story. Tell them to write each of these in one of the spaces on the creased paper. If asked, restrict the number to eight. 'If you have finished that—tear your paper straight along the folds.' They have now got eight pieces of paper. 'Pass your pieces to your neighbour. The idea now is that you arrange the pieces you have just got into the order they come in the story.' It is a good idea to ask them to number the pieces of paper in order. When this is finished select a pupil and ask him to read out the statements in the original order. Write these on the board. In doing this, space them out, and enclose each one in a rectangle. Connect the boxes up in order by using arrows. 'We can indicate that this statement is next like this....'

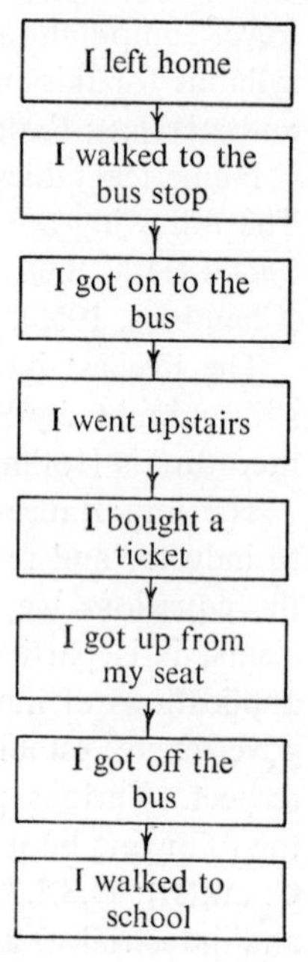

'This charts the story.'

'This is a picture with order in it so that your attention is led from one part to another.'

Ask the pupils to check each others' solutions. Discussion can follow on the importance and nature of the statements in the story (actions, decisions, facts, etc.). Emphasize the importance of selecting the quotations carefully if the story is to be told in only eight statements. Some similar problems follow.

FLOW CHARTS

Insert arrows in the following to suggest a plausible order.

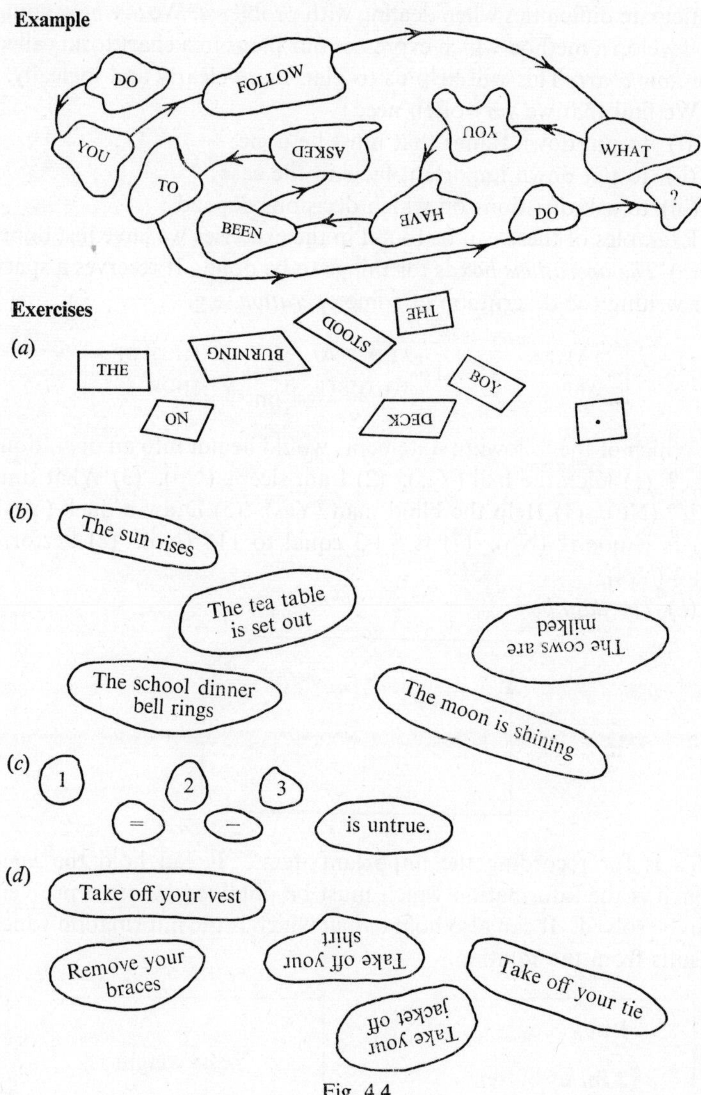

Fig. 4.4

You have all made lists of items required when going on holiday or camping. In the same way, it is often desirable to plan ahead and anticipate difficulties when dealing with problems. We are now going to develop a method which expresses our plans in a chart form called the *flow chart*. This will help us to plan more clearly and logically.

We find that we very often need:

(i) to put down things that must be done;

(ii) to put down important facts in the case;

(iii) to ask questions on which decisions depend.

Examples of these can be found in the exercises we have just done.

(*a*) *The operations box* is for things to be done. It reserves a space for writing the description of some *operation*, e.g.

EAT AN APPLE	ADD TWO TO THREE	SHUT UP SHOP

Which of the following statements would be put into an operations box? (1) Kick the ball (Yes). (2) I am sleepy (No). (3) What time is it? (No). (4) Help the blind man (Yes). (5) Draw a duck (Yes). (6) Is it noon? (No). (7) Is 8+3 equal to 11? (No). (8) Perform 8+3 (Yes).

(*b*) *The data box*

This is for recording the important items. It can hold the *input* which is the information which must be obtained before a problem can be solved. It can also hold *output* which is the information which results from the solution.

I buy 8 eggs,
2 lb. of butter,
1 pint of milk

Your weight is
10 st. 6 lb.

To distinguish information boxes from others the top right-hand corner is cut off.

In the following problems pick out the *input data.*

(1) A pint of milk costs 8*d.* and Mrs Jones has 2*s.* 6*d.* How many pints can she buy? (1 pint costs 8*d.*, Mrs Jones has 2*s.* 6*d.*)

(2) A racing car completes a lap of the $2\frac{1}{2}$ mile race-track in 47 sec. What is its average speed? (Track $2\frac{1}{2}$ miles long, 1 lap takes 47 sec.)

(3) A rugby team wins a match by 10 points to 6 points. What is the highest possible number of tries which could have been scored? (A try is 3 points, a conversion 2 points; one team has 10 points; one team has 6 points.)

(4) A bowler at cricket takes 4 wickets in a match for which he concedes 96 runs. If the bowler has bowled 100 overs, what is the average number of runs scored off his bowling in an over? (96 runs, 100 overs.)

In the problems you have just looked at, what is the *output data*?

(*c*) *The decision box*

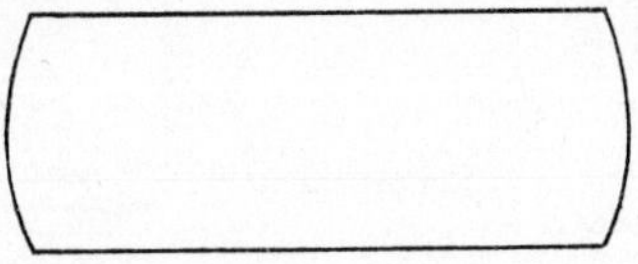

The decision box is used to note the things that must be decided before going on. In most cases the questions will be 'alternatives'. For example, if the answer is 'yes', you will do one thing. If the answer is 'no', you will do something else. These boxes will be a very important part of our chart.

Have we more apples than pears?

Is it one o'clock?

Which of the following statements would be put into decision boxes? (1) Is it cold? (Yes). (2) How much is the orange? (No). (3) Are the oranges too dear? (Yes). (4) Am I overweight? (Yes). (5) Is the price greater than £400? (Yes). (6) I am tired. (No). (7) It is April (No). (8) We all go to the pictures, don't we? (Yes).

Comment

Later you can introduce different charts, like route maps, industrial process diagrams, historical charts, and genealogical trees, which involve branching. This introduces in a simple way the idea of 'decision' in a sequence. Initially restrict these decisions to the 'Yes/No' variety.

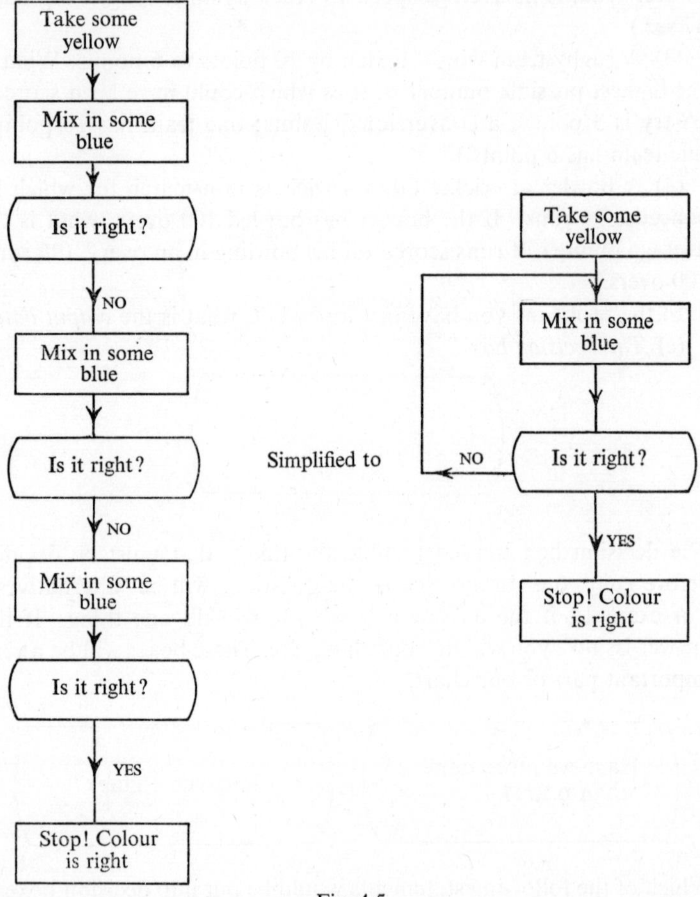

Fig. 4.5

This is followed by the introduction of the idea of *looping*. This can be illustrated by using some repeating process. Try something like mixing blue and yellow paint to get the right shade of green. The sequence of charts in Figs. 4.5 and 4.6 might be evolved. The first two contain features which are obviously unsatisfactory, and these are subsequently corrected.

It can be well worth while to give the children flow charts to interpret. Indeed, this technique can be used as an alternative introduction. Several processes can be put into chart form, such as paper folding, treasure hunts and solving puzzles. Overleaf is an example giving instructions on how to perform a geometrical construction (Fig. 4.7).

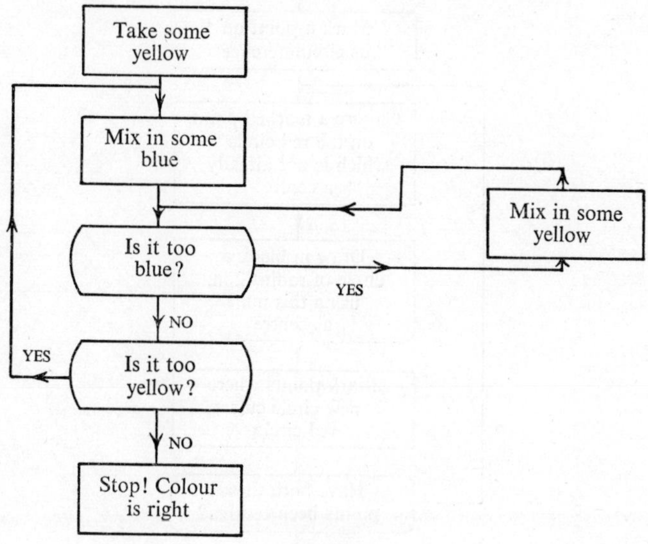

Fig. 4.6

These instruction flow charts can be made for many things, such as using a gramophone, building up a plastic model, or using do-it-yourself kits. The pupil can be given a homework task which involves discovering everyday situations which can be charted.

Eventually flow charts may be used to describe computational problems, such as the iterative methods discussed in a previous section. They are easier to follow if standard symbols are used, but at this stage the pupils may very well develop their own notation as they see the need for it. Flow charts can be used as an aid in teaching a method, as revision notes, and as a way of increasing understanding and diagnosing misunderstanding.

The last two points are perhaps worth enlarging. If a child is asked to draw a flow chart to describe an arithmetical process with which he is familiar, he will find it easy if he fully understands the method. If he does not, he will be forced to think hard about those steps which cause difficulty. Any misconception will appear clearly described for the teacher to see.

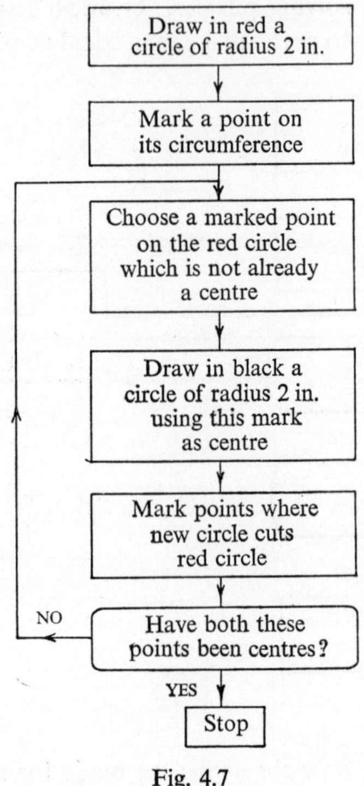

Fig. 4.7

As an example of a flow chart describing an arithmetical problem, consider the task of decomposing a given number into its prime factors. Can we describe this process by a flow chart? A first attempt might result in something like Fig. 4.8. Reconsideration might lead to something more like Fig. 4.9. This might be a better programme to put into a machine. Flow charts are extensively used in computer technology, and more capable pupils might well appreciate the point of wishing to reduce a problem to terms in which it can be handled by a machine.

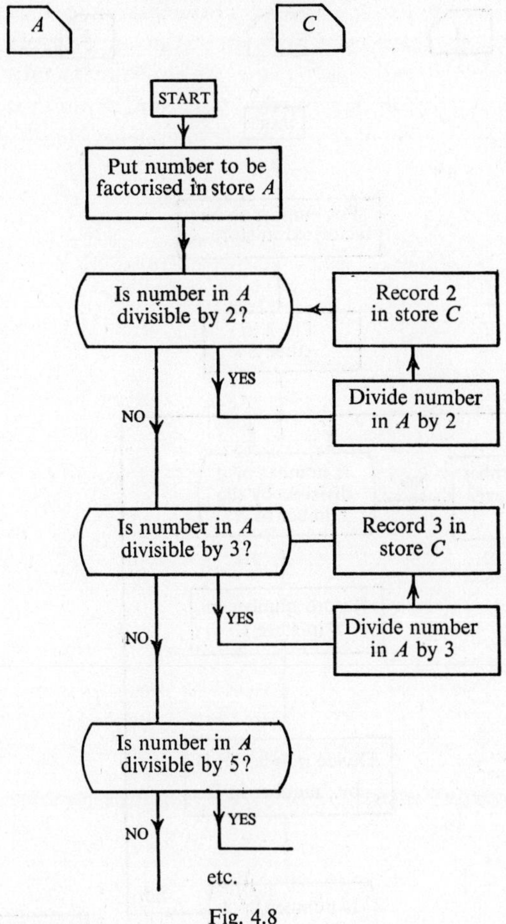

Fig. 4.8

Exercises in problem analysis

(1) Make a knitting pattern for a woollen scarf with a different patterned stripe down the middle. Choose your own stitches and use standard abbreviations.

(2) Construct a flow chart for the method of finding the greatest common divisor of 60 and 132.

(3) Describe clearly the ruler and compass construction of an angle bisector.

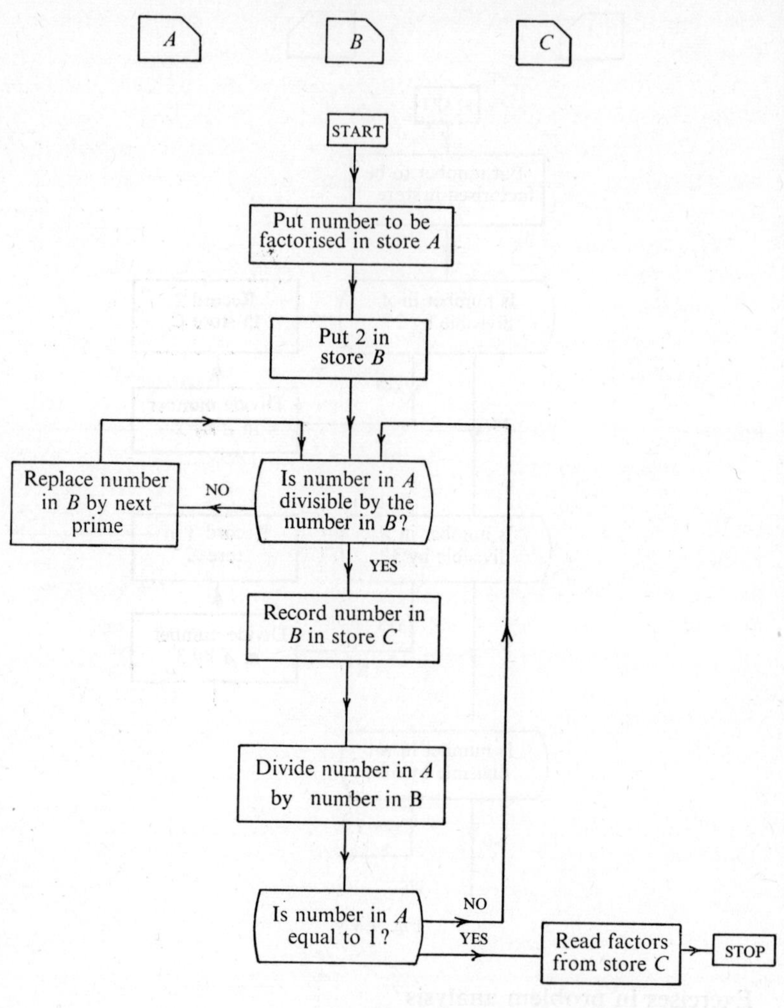

Fig. 4.9

(4) Give precise rules for the use of the sieve of Eratosthenes to find all primes less than a million.

(5) Write out in the form of a flow chart a recipe for making an omelette.

(6) Provide sufficient instructions to enable someone who has not

used logarithm tables to find the logarithm of a number greater than 1.

(7) Describe in detail the operations required to calculate a standard deviation on a hand calculating machine.

(8) Outline the main steps in the solution of a standard problem in mechanics, e.g. the ladder leaning against a wall or the impact of moving spheres.

The first example reminds us that girls are said to make good programmers. A possible knitting pattern in this case is:

Using no. 8 needles, cast on 42 st.

1ST ROW. Sl. 1, k. 15, (p2, k2) twice, p2, k. to end of row.

2ND ROW. Sl. 1, p. 15, (k2, p2) twice, k2, p. to end of row.

Repeat these 2 rows until work measures 60 in. (or required length).

This provides an example of a coded set of instructions.

The flow-chart solution to the second example provides an opportunity to set up a human computing system. A general flow chart of Euclid's algorithm for the determination of a greatest common divisor might be as Fig. 4.10. This chart contains some instructions in a shortened form which is an established convention.

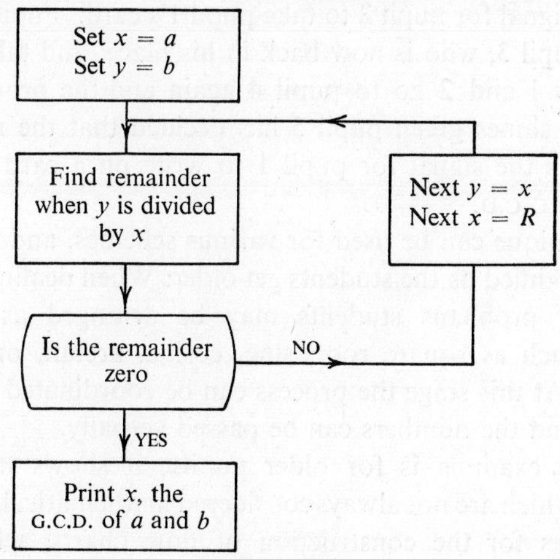

Fig. 4.10

Now select some pupils as follows: pupil 1 as x, pupil 2 as y, pupil 3 as R, pupil 4 as the divider and pupil 5 as the decision operator. Arrange three tables as shown.

Pupil 5 is equipped with an indicator, a box with a red and green lamp. The operation then proceeds as follows. Pupils 1 and 2 go to the teacher and collect a and b, two numbers on cards. They take these to pupil 4 who copies them down on a piece of paper. While he performs the division pupils 1 and 2 return to their seats. Having completed the division pupil 4 gives the remainder, on a card, to pupil 3. (Pupil 4 can be provided with a set of numerals printed on cards as well as paper to write on.) Pupil 3 then takes the card to pupil 5 who

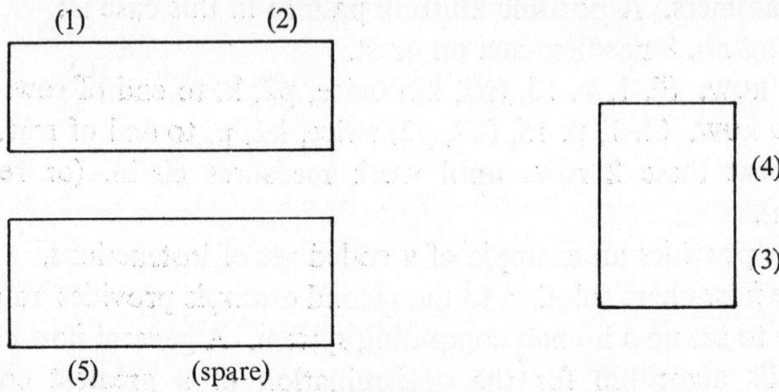

Fig. 4.11

decides whether it is zero or not. If it is not he shines the red lamp. This is the signal for pupil 2 to take pupil 1's card. Pupil 1 then goes across to pupil 3, who is now back in his place, and takes his card. Then pupils 1 and 2 go to pupil 4 again and the process repeats. If the lamp shines green pupil 5 has decided that the remainder is zero. This is the signal for pupil 1 to write on a card his number which is the G.C.D.

This technique can be used for various schemes, and the idea can easily be modified as the students get older. When dealing with more complicated problems students may be arranged as computing elements, such as square root, sine, cosine, arctan, or a function evaluator. At this stage the process can be coordinated by a central controller and the numbers can be passed verbally.

The final example is for older pupils: it shows how familiar situations, which are not always considered mathematical, can provide opportunities for the construction of flow charts, which offer a genuine challenge and present logical problems in programming which are by no means trivial.

Everyone knows the game of noughts and crosses. Construct a programme for the player who plays second, so that he can avoid

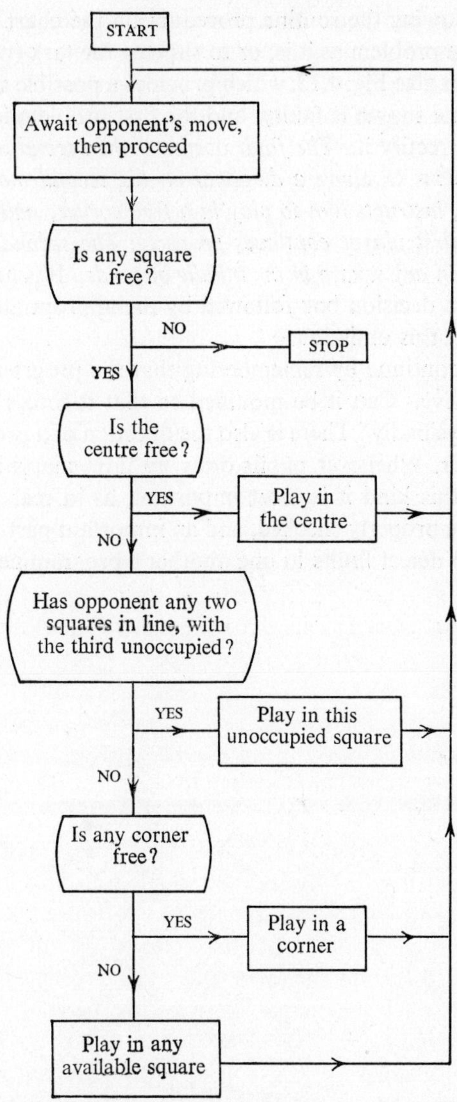

Fig. 4.12. An attempt at a programme for the second player to avoid defeat at noughts and crosses. The programme is faulty. Can you rectify the fault? Can you improve the programme still further?

defeat by following the routine procedure on the chart. This can be presented as a problem as it is, or to shorten the task (which is quite long) we might give Fig. 4.12, which provides a possible solution. But the programme shown is faulty, and the first problem is to discover the fault and rectify it. *The fault occurs if the second player is confronted with* ○ × ○ *along a diagonal on his second move. The flow chart, as it is, instructs him to play in a free corner; and this leads to defeat if the first player continues properly. The second player must play, instead, in any square in the middle of a side.* It is not difficult to insert an extra decision box followed by the appropriate instruction box to correct this omission.

We might continue by remembering that the programme given is merely defensive. Can it be modified so that it forces a win if the opponent plays badly? There is also the question of a programme for the first player. Whenever pupils draw up flow charts in answer to problems of this kind it is most important, as in real life, that the programme be properly checked, and an important part of the lesson is for them to detect faults in one another's programmes.

5

SETS, LOGIC AND BOOLEAN ALGEBRA

1. INTRODUCTION TO SETS

A sequence of lessons

For my birthday I received a tie, a pipe, a pen and a book token. This collection of gifts we call 'my birthday presents for this year'. We shall call such a collection a *set* and each separate gift an *element* of the set; let us write it like this: {tie, pipe, pen, token}. The brackets show that we have grouped these objects together.

How many elements are there in my set of birthday presents? Name one: another.

Jane has had her birthday this year. What presents did you get? Yes, and how many were there? John has not had his birthday this year; what is his set of presents?

Let us think of another set. Indicate the desk, the board rubber, the chalk, yourself and Dick's left shoe. How many elements? Name them. Is Dick's right shoe an element of this set? *Discuss further examples given by the class.*

Sets may be determined by naming each element as we have done. Are there other ways of defining a set? All the desks in this room; all the pupils in this school; a swarm of bees; all the letters of the alphabet; etc.

In some cases, such as {tortoise, desk, pencil, potato} we can only know the set when we have stated each element; in others, such as 'the set of objects in John's pocket', a phrase is sufficient to define the set, but we could, if we wished, name each element in the set.

Consider 'all people who are in their twelfth year now'. Can you name each one? We can name some—those we know—but we realize there are a great many more we do not know. Is this therefore a set?

Yes, because for any particular person we can say, without doubt, whether he is included or not.

Can we always decide without any doubt? Let us think of all the pupils in the school with fair hair. Name some. Do we agree? Would we always agree? No. In this case we say that we do not have a *well-defined* set.

Will all those who have had their twelfth birthday put up a hand? How many are there? Now those in their eleventh year. How many? Now those in their ninth year. How many?

Since, in this last case, we have defined a set by the words 'those in the class in their ninth year', and since we find that it contains no elements, we call it the *empty* (or *null*) set. We write it { } or ϕ.

How many elements are there in the set of 'the months in the year having 31 days'? And in the set of 'all the months in the year having only 26 days'? Give other examples of the empty set.

Think of the odd numbers less than 20. Do they form a set? How many elements are there? Now, the odd numbers less than 20 which are divisible by 3. Do they form a set? What is the number of elements?

We will write:

$$A = \{1, 3, 5, 7, 9, 11, 13, 15, 17, 19\},$$
$$B = \{3, 9, 15\}.$$

The set B is contained in set A, that is, all the elements of B are also elements of A. This can be written in shorthand form as

$$B \subset A \quad \text{or} \quad A \supset B,$$

where $\subset$ is read as 'is contained in', and $\supset$ is read as 'contains'. B is called a *subset* of A. Is {5, 15} a subset of A? {1, 7, 11, 15}, {17}, {1, 3, 5, 21}? Consider some other cases. Which elements of A are *not* also in B? Do they form a set? Is it a subset of A?

Write $$C = \{1, 5, 7, 11, 13, 17, 19\}.$$

If no element of C is in B and no element of B is in C, and together they make up the set A, we say that C is the *complement* of B (written B'), or B is the complement of C (written C').

In set A, what is the complement of {5, 15}? {11}? {1, 3, 5, 7}? Have all subsets of a set a complement?

All the elements of a subset are elements of the original set. Because of this definition we can think of the set {1, 3, 5, 7, 9, 11, 13,

15, 17, 19} as a subset of itself. It is a special subset. There is another set which can be thought of as special; which is it? The empty set. What is the complement of the empty set?

Diagrams can be drawn to illustrate these questions. We can think of all the elements of A as being within a closed curve.

The elements of subset $B = \{3, 9, 15\}$ can then be thought of as inside another region, as shown. Having understood this, we think of the region in which any element belongs without actually showing the individual elements. The diagram then looks as follows, and we call such figures Venn diagrams.

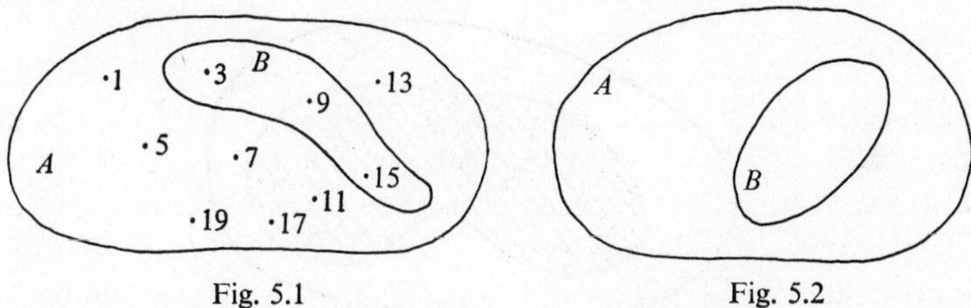

Fig. 5.1 Fig. 5.2

Shade in the region which represents C (the complement of B) in the diagram.

Consider all the pupils in the class. Raise your left hand if you have a brother. Do these pupils form a subset? Raise your right hand if you have a sister. Have we another subset? What about those with two hands up? Are they in both subsets? Draw a diagram to illustrate this.

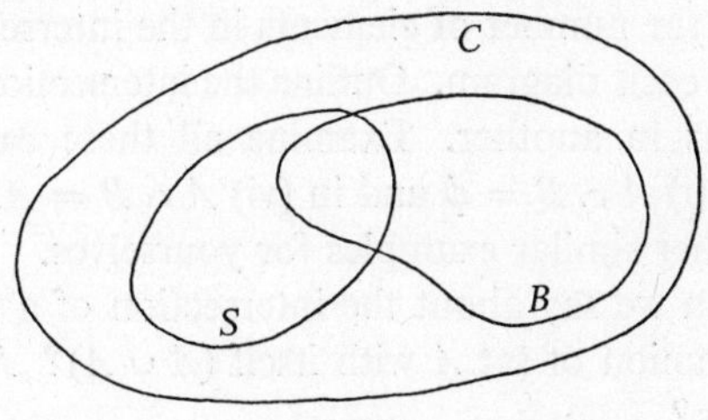

Fig. 5.3

C is the set of all pupils in the class;
B is the set of all those with a brother;
S is the set of all those with a sister.

Find the region to which each of you belongs. Where are those with no brothers or sisters? Where are those with a brother *and* a sister; those with a brother only; anyone with two brothers? etc.

Those who belong to both B and S, that is, those who have a brother and a sister, belong to the *intersection* of the sets B and S. The intersection of B and S is written

$$B \cap S.$$

How many elements are there in this set? Who are the people who are in S *or* in B *or* both (inclusive *or*)? We describe this set as the *union* of S and B and write $S \cup B$.

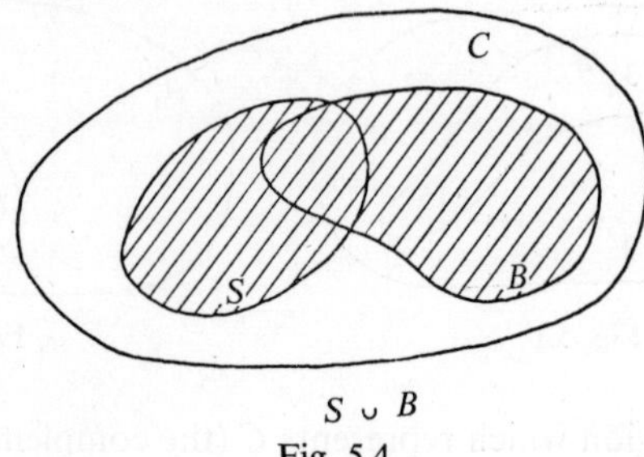

$S \cup B$

Fig. 5.4

Examples

(1) If we have a set A containing 5 elements, and a set B containing 11 elements, how many elements are there altogether? 16? Are we sure? 11? Perhaps. How many different possibilities are there? Illustrate them (Fig. 5.5).

Write down the number of elements in the intersection and union of A and B in each diagram. Outline the intersections in one colour and the unions in another. Examine all these cases thoroughly, noting that in (i) $A \cap B = \phi$ and in (vi) $A \cap B = A$.

Make up other similar examples for yourselves.

(2) What can we say about the intersection of a set A with itself ($A \cap A$)? The union of set A with itself ($A \cup A$)? Are these results true for all sets?

(3) An orchestra O has subsets:

A is the set of all male members;

B is the set of married members;

C is the set of string players.

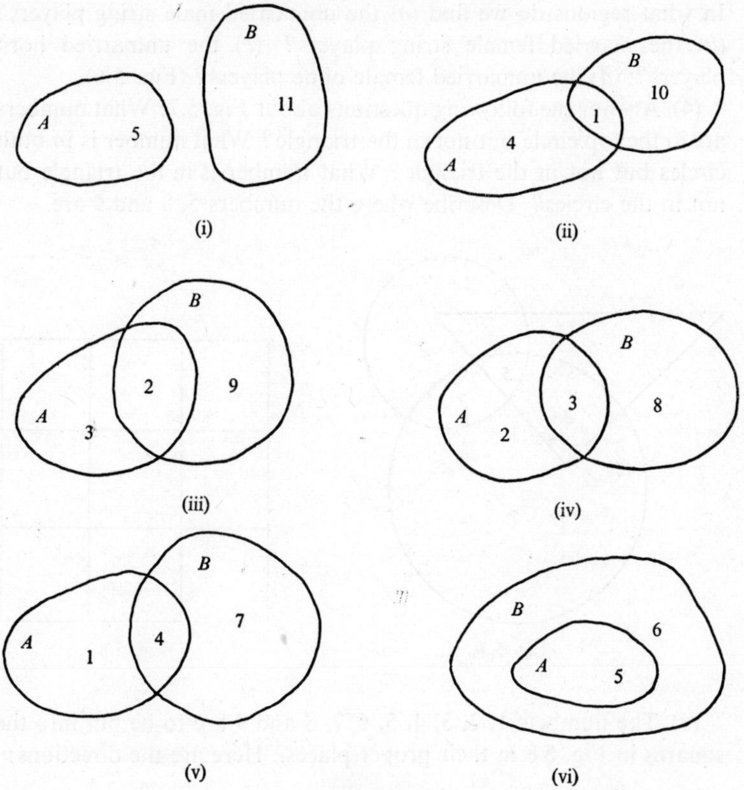

(i) (ii)

(iii) (iv)

(v) (vi)

Fig. 5.5

(The numbers indicate the number of elements in each region.)

Describe in words: A', B', C', $A \cap B$, $A \cap C$, $B' \cap C$, $A' \cap C'$, $A \cup B'$, $B \cup C'$, $B' \cup C'$, $(A \cap B) \cap C$, etc. Now draw the Venn diagram.

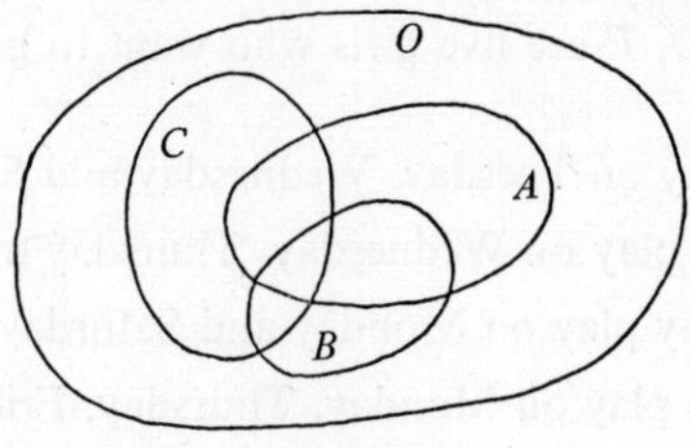

Fig. 5.6

In what regions do we find (*a*) the unmarried male string players? (*b*) the married female string players? (*c*) the unmarried horn players? (*d*) the unmarried female oboe players? (Fig. 5.6).

(4) Answer the following questions about Fig. 5.7: What numbers are in the top circle but not in the triangle? What number is in both circles but not in the triangle? What number is in the triangle but not in the circles? Describe where the numbers 5, 8 and 9 are.

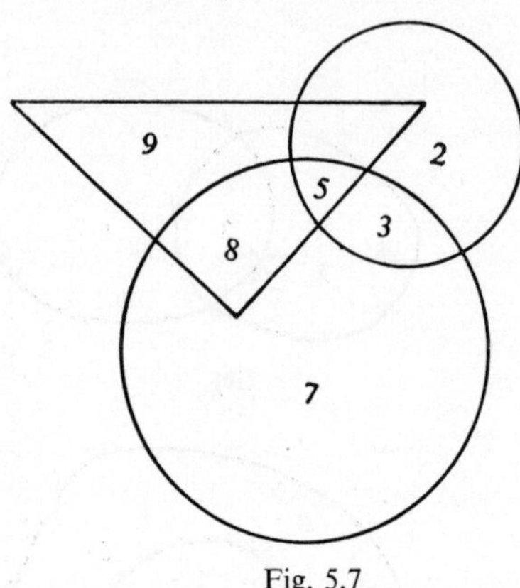

Fig. 5.7

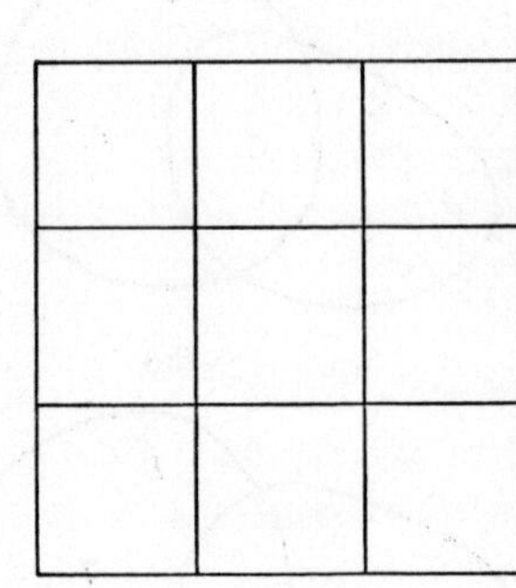
Fig. 5.8

(5) The numbers 1, 2, 3, 4, 5, 6, 7, 8 and 9 are to be put into the squares in Fig. 5.8 in their proper places. Here are the directions:

1, 8, 6 are in the top row;

2, 9, 4 are in the bottom row;

1, 4, 2, 5, 7, 6 are not in the left-hand column;

8, 1, 5, 9, 3, 4 are not in the right-hand column.

Fill in the squares correctly.

(6) *A*, *B*, *C*, *D*, *E* are five girls who want to get together to play tennis.

A can play on Tuesday, Wednesday and Saturday;

B cannot play on Wednesday, Thursday or Sunday;

C can only play on Monday and Saturday;

D cannot play on Monday, Thursday, Friday or Sunday;

E can play any day except Tuesday and Sunday.

On what day can they all play? When can none play? On what day can the fewest play? On what day except Saturday can *A* and *E* play? On what day except Saturday can *B* and *C* play? On what day except Saturday can *B* and *D* play? Can *B*, *C* and *E* play together any day except Saturday? Who can play together on Wednesday? On Thursday? Who can play the least number of days?

(7) *A*, *B*, *C*, *D*, *E* are men of different nationalities. *A*, *C*, *E* speak English; *B* and *D* speak German; *A* and *D* speak French; *B* and *E* speak Spanish; *C* speaks Russian.

Who can speak English and Spanish? Who can speak English and French? Who can speak French and German? Could *A* and *D* make themselves understood to each other? Could *D* and *E*? *A*, *B* and *D*? *A* and *C*?

2. PUNCHED CARDS

Punched cards can be used to provide the same transition from concrete to abstract in the study of sets that structural apparatus supplies in elementary arithmetic. They would seem to need a period of familiarization corresponding to free-play with coloured rods and this might best be achieved by keeping a set of class (or school) record cards. Such instructions as 'Bring me the cards of all children staying to school lunches', 'Find me someone who wears spectacles', 'I want a list of all those in Red House' would quickly reveal the advantages of a punched-card system and a knitting needle (see plate I). These would be followed by increasingly complicated tasks, such as finding all children in Red House who stay to school lunches, all children who wear spectacles *and* are excused games, all children without school uniform who do *not* stay to school lunches, all children who wear spectacles *or* are excused games (or *both*, possibly).

In this way the ideas of union, intersection, complement, empty set and symmetric difference may emerge. *The symmetric difference of two sets A and B is the set of elements of A which do not belong to B, and the elements of B which do not belong to A. It can be described as the 'exclusive or'*. Terminology and notation can be gradually introduced, operations on the cards described verbally, and the sets obtained illustrated by diagrams. The differences between 'or' and 'or but not both' can be clearly shown, and the empty set can be approached in a more concrete way than with other methods.

The complement of a set now becomes 'the rest of the cards', and this considerably simplifies such sets as '*not* those in Red House who do not stay to school lunches', at the same time motivating deep thought about exactly what such a statement means.

The next stage is a generalization. Record cards are determined by facts about the pupils, and some of them are identical as far as slots and holes are concerned, while some possible arrangements of holes and slots are omitted because no children fit the requirements. Suppose the cards are determined by the various possible combinations of properties instead of by the children. Consider a property A, determined by the first positions on the card. If a *hole* means 'has the property A', or 'belongs to the set having the property A', or A for short; then a *slot* means 'has *not* the property A', or 'does *not* belong to the set having the property A', or A' for short. There are two possibilities, A or A', and so two cards are needed. Introduce a second property B. The cards now needed are AB, $A'B$, AB', and $A'B'$. This may be discussed verbally and diagrammatically. Continuing in this way by trial and error, by formulation of rules and/or by commonsense it can be seen that n properties require 2^n cards, and the positioning of holes and slots may be connected with the binary system.

Once a pack of cards has been prepared all the processes carried out with the knitting needle may be described in set notation, and the one pack can be used to solve a variety of puzzles and logical problems [47, 48].

3. DEVELOPING THE ALGEBRA OF SETS

Before attempting formal set algebra the introductory ideas and notations should be revised. It is also helpful to work through some examples with numbers which use the bracket rule (or *distributive law*),

$$a\times(b+c) = a\times b+a\times c$$

We can point out that if we did not have the accepted rule '$\times$ before $+$' more brackets would be needed:

$$a\times(b+c) = (a\times b)+(a\times c).$$

The use of this law in the less common form

$$(b+c)\times a = b\times a+c\times a$$

brings out the otherwise trivial-seeming observation that

$$x \times y = y \times x,$$

when x and y are numbers.

Packs of punched cards provide a stimulating way of investigating set relations. We suppose that each pupil has a pack of these cards and is thoroughly familiar with the *and* operation which sorts out the intersection $A \cap B$ of two sets, and the *or* operation which sorts the union $A \cup B$. At this stage the pupils may think of $\cap$ and $\cup$ only as descriptive shorthands. We wish to make them denote operations; in this case by connecting them with sorting.

The pack of cards should be designed in the first place to contain three sets A, B, C related in the most general way, as in the figure:

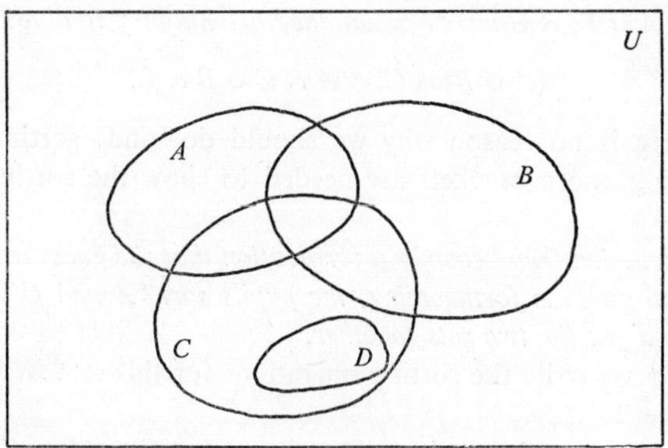

Fig. 5.9

The rectangle represents all the cards in the pack. D can be a fourth set which is included in C. The intersection of A and D contains no members, and so yields the *empty* (*null*) *set*. The individual cards should be numbered or distinguished in some way.

Sort out the set of cards with property A. You may describe this by just writing A. Now sort out the cards with property A and property B. Have you noted which they are? How do we describe this set? *In this informal work it is convenient to use A to stand for a property and for the set of cards having this property.*

Now sort the cards with property B and property A. Did you expect that? How do we describe this set? Look at what is written—

could we put anything that shows what we have just found out about the two sets?

We hope for $A \cap B = B \cap A$, *but accept reasonable alternatives. The pupils replace their cards and are asked to sort the set* $A \cup B$ (*A or B*).

I want to think of these cards very much as one set, and emphasize it in the description. How do I emphasize that I am thinking of '$x+y$' as one number?

Brackets are placed round the $A \cup B$.

Now from set $(A \cup B)$ lift those with property C. Make a note of the cards. What properties has the set lifted out got? Yes; it must have properties A or B, and C. How can that be written? The brackets are telling us that we do 'A or B' first, aren't they? *Then* 'and C'.

When $(A \cup B) \cap C$ *has been obtained, we may try to obtain a guess at expanding*

$$(A \cup B) \cap C = A \cap C \cup B \cap C.$$

Since there is no reason why we should do 'and' sorting before 'or' sorting, more brackets are needed to show the sorting order. Where?

The discussion then becomes a verification that the guess was a good one. If no guess is forthcoming, the pupils sort 'A and C', then 'B and C' and put the two sets together.

How do we write the sorting operations for this set? What cards did you get?

In this way the identity $(A \cup B) \cap C = (A \cap C) \cup (B \cap C)$ *is established. Just as in the algebra of numbers, there is a law for expanding brackets in the algebra of sets.*

The following sorting operations can be investigated by the pupils individually. It is important that they try to cast their results in algebraic form.

(1) Guess an expansion of $(A \cap B) \cup C$, and compare it with $(A \cup C) \cap (B \cup C)$. Note the need for the extra brackets again. Why is there no rule that $\cap$ is performed before $\cup$ (or vice versa)?

(2) Consider $A \cap A$, $A \cup A$. No powers arise in this algebra.

(3) Consider $A \cap D$. The necessity for introducing a symbol for the empty set arises.

(4) Sort the set of cards which have *not* got property A. The symbol A' can be used to denote this set.

(5) $A \cap A'$ is the empty set again. $A \cup A'$ produces all the cards in the pack.

(6) Consider $A' \cap B$ and $A' \cup B$. Sorting these sets calls for some thought over which parts of the pack to put on one side.

(7) Consider A or B, but not both—how is this to be written in terms of $\cap$ and $\cup$?

(8) Consider $A' \cap B'$ and $A' \cup B'$, and compare them with $(A \cup B)'$ and $(A \cap B)'$, respectively.

Alternative methods

(1) *Make three sets of cards labelled A, B, C. Select three sets of children (intersecting in the same way as in Fig. 5.9) and hand them cards.*

Stand up children with card A or card B. *The children write this operation as $A \cup B$ and decide to enclose it in brackets because more operations are to come.*

If you have card C as well, remain standing. Otherwise sit down. *The class decides that $(A \cup B) \cap C$ has been selected.* Now all sit down, and start again.

Stand up if you have cards A and C. Now stand up if you have cards B and C. *$(A \cup B) \cap C$ is seen to be the same set as $(A \cap C) \cup (B \cap C)$. Investigation of other laws can follow, but for individual working the next two methods may be used.*

(2) Regions of the Venn diagram (Colour-fig. 1) are enclosed, using coloured chalks and pencils. (Colours are more intriguing than the cross-hatching usually suggested.) A colour is attached to each binary operation. Thus in $(A \cup B) \cap C$, $(A \cup B)$ is underlined in red, and the intersection of this set with C in blue. In the diagram we enclose the union of A and B with a red curve; then consider the intersection of the red region with C and enclose it with blue. If only a few colours are available, the other half of the identity is considered on a different diagram.

In $(A \cap C) \cup (B \cap C)$, underline $(A \cap C)$ in red, $(B \cap C)$ in green, and the union of these two sets in blue. In the diagram enclose region $A \cap C$ with red, region $B \cap C$ with green. Finally, enclose the red region and the green region with blue, and the identity is shown (Colour-fig. 2). To perform the whole demonstration in one diagram requires five colours, since five distinct binary operations appear.

The colouring method has the important advantage that it can show clearly the order in which the binary operations were performed. Be careful to enclose each coloured line with the next colour, and not draw over it. The downwards order in underlining the equation corresponds to the outwards order in enclosing the diagram.

(3) Nailed geometry boards and coloured elastic bands can be used in the same manner as Venn diagrams and coloured pencils. The nails inside a particular board can represent a finite set (Colour-fig. 3).

4. BASIC FORMULAE IN THE ALGEBRA OF SETS

Notation

In the preceding sections of this chapter, the signs $\cup$ ('cup') and $\cap$ ('cap') have been used to symbolize the operations of union and intersection. This is not the only notation in common use, and readers of other texts will meet several alternatives.

Union	Intersection
$\cup$	$\cap$
$\cup$	Juxtaposition
$+$	$\cdot$
$+$	Juxtaposition

There are arguments in favour of each of the alternatives, and each has certain disadvantages. The drawbacks to the cup and cap notation are that: (i) the similar shapes of the symbols are easily confused, and (ii) formulae involving more than one or two operations are cumbersome to write and difficult to read. For example,

$$A \cap (B \cup C) = (A \cap B) \cup (A \cap C)$$

appears as $A(B \cup C) = AB \cup AC$ when written with the second alternative notation, and gains immediately in clarity.

Nevertheless, we believe that this is the best notation to use when set language is first introduced. The symbols are quite different from those used in 'ordinary' algebra and carry no overtones of meaning to short-circuit the discovery of the basic properties of the operations symbolized. An eleven-year-old will not see any point or purpose in discussing, say, the commutativity or associativity of union if this is given away in advance by the use of a plus sign.

The pupils can discover that union and intersection behave in ways which resemble the ways in which addition and multiplication of numbers behave. In this approach they will not suppose that the analogy is complete, as they may do if the plus and dot (or juxtaposition) notation is used from the beginning. Using $+$ and . at a later stage for its brevity and clarity will seem quite acceptable, and will have been suggested by the correspondences discovered in the preliminary exploration.

The choice between using $\cup$ and $\cap$, or $\cup$ and juxtaposition, is resolved for us by the parallelism that we wish to observe between set notation and the usual symbols for disjunction and conjunction—viz. $\vee$ and $\wedge$—in the sentence calculus. This decision is part of a wider policy. We believe that the pupil should be allowed to discover the isomorphism that exists between set algebra, sentence calculus and the behaviour of simple electrical circuits. So we suggest that the existence of different notations should be put to good use by being, in the early stages, associated specifically with one of these three areas. A summary of the correspondences is given on p. 180 in the section on Axiomatics.

However, when the interchangeability of $\cup$ and $\cap$, and $+$ and juxtaposition has been established, the latter form is better for work involving the manipulation of formulae and is used here to summarize the main statements of set algebra.

The formulae

The following list of statements contains the basic information about the behaviour of sets under the operations of union, intersection and complementation. These formulae can be thought of as rules for manipulation. An algebra (in a non-technical sense) is a set of calculation procedures, and these rules form such a set. Examples of their use in calculation can be found on pp. 152, 160 and in many texts, for example [32, 47, 63, 129].

We do not suggest that these rules should be developed at once. They can be acquired quite informally from Venn diagrams or simple verbal arguments at suitable stages in the school treatment, and the full range is not at all necessary until a complete 'calculus' is required. In fact, by the time a serious study of these rules is undertaken the pupils will be ready to look at the interdependence of rules and, perhaps, the need for a strong axiomatic foundation.

A, B, C are subsets of a set E; ϕ is the null set.

ASSOCIATIVE LAWS

(I. 1) $A+(B+C) = (A+B)+C$ (II. 1) $A(BC) = (AB)C$

COMMUTATIVE LAWS

(I. 2) $A+B = B+A$ (II. 2) $AB = BA$

DISTRIBUTIVE LAWS

(I. 3) $A+BC = (A+B)(A+C)$ (II. 3) $A(B+C) = AB+AC$

IDEMPOTENT LAWS

(I. 4) $A+A = A$ (II. 4) $AA = A$

ABSORPTION LAWS

(I. 5) $A+AB = A$ (II. 5) $A(A+B) = A$

COMPLEMENTATION LAWS

(I. 6) $A+A' = E$ (II. 6) $AA' = \phi$

(7) $(A')' = A$

DE MORGAN'S LAWS

(I. 8) $(A+B)' = A'B'$ (II. 8) $(AB)' = A'+B'$

LAWS OF ϕ AND E

(I. 9) $E+A = E$ (II. 9) $\phi A = \phi$

(I. 10) $\phi+A = A$ (II. 10) $EA = A$

(I. 11) $E' = \phi$ (II. 11) $\phi' = E$

The duality of the above pairs of statements is noteworthy. It suggests that the interchanging of union with intersection, and E with ϕ, does not affect the truth of an algebraic identity.

The informal derivation of these results is illustrated in the following two examples.

Example 1. Let A be the set of children in the class who have at least one brother, and B the set of children with at least one sister. The set $A+B$ (the union of sets A and B) is the set of those children with either a brother or a sister. The children without a brother or a sister (only-children) make up the set $(A+B)'$.

But since A' is the set of those without brothers and B' the set of those without sisters, the set of only-children is the set of children common to sets A' and B', which is $A'B'$. So it follows that $(A+B)' = A'B'$ (I. 8).

Example 2. If A, B are two subsets represented in a Venn diagram, the set formed from the intersection of A with the union set of A and B is seen to be the set A itself (Colour-fig. 4).

5. APPLICATIONS OF THE LANGUAGE OF SETS

The terminology which we have been describing provides a convenient language with which to discuss many items in different parts of the traditional syllabus. This is one of the main reasons for wishing to introduce the language, but it is not without its difficulties and dangers: and this section is more in the nature of a warning of a difficulty, and an indication of how it may be overcome, than a positive recommendation to study a topic in a particular way.

The operations of finding the highest common factor and the least common multiple might, at first sight, appear to be very simple applications of the ideas of intersection and union. Thus, faced with the problem of finding the highest common factor and the least common multiple of 24 and 30 one might, naïvely, suppose that one could proceed as follows:

$$24 = 2\times 2\times 2\times 3 \qquad 30 = 2\times 3\times 5$$

The set of prime factors of 24 is {2, 2, 2, 3}.

The set of prime factors of 30 is {2, 3, 5}.

The intersection of these two sets is {2, 3}, so the highest common factor is 6.

The union of these two sets is {2, 2, 2, 3, 5}, so the least common multiple is 120.

Unfortunately this simple process encounters a technical objection. It is established usage to regard a set as being made up of its elements, with each element, inevitably, counted once only. Thus '{Nelson, Drake}', '{Nelson, Nelson, Drake}', and '{Nelson, Drake, Nelson, Nelson}' all denote the same set—a set with two members, Nelson and Drake. This convention appears to be necessary in order to have the law $A \cup A = A$, but on this convention '{2, 2, 2, 3}' denotes the same set as '{2, 3}'. This invalidates the calculation above; but it can be saved if we introduce some label to distinguish the twos from one another. One way of doing this is to use the idea of *primary divisors*. This point is elaborated in the Notes following the next lesson.

Factors and divisors

'Which whole numbers divide exactly into 24?' '1, 2, 3, 4, 6, 8, 12, 24.' Each of these numbers is called a *factor* or a *divisor* of 24. Taken all together we can call them *the set of divisors of 24* and write the set {1, 2, 3, 4, 6, 8, 12, 24}.

We can call the set $D(24)$.

$$D(24) = \{1, 2, 3, 4, 6, 8, 12, 24\}.$$

The set can also be illustrated by representing each divisor by a point on the number line.

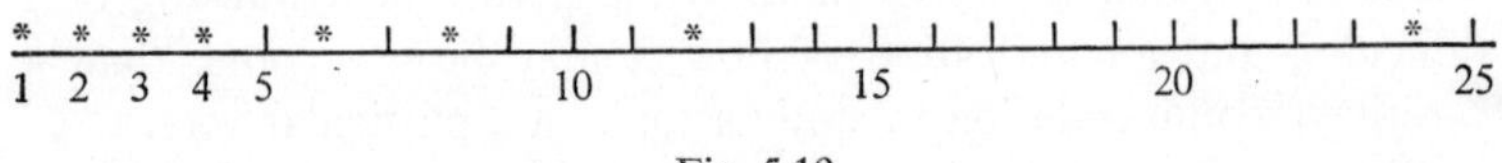

Fig. 5.10

Examples

Find the set of divisors of each of the following numbers:

16; 21; 36; 20; 72.

Illustrate each answer by marking the points on a number line.

Common factors or common divisors

These are the divisors which are *common* to two or more numbers, that is, the elements which are contained in the set of divisors of each of the numbers. Or, in set theory terminology, the intersection (set) of the given sets of divisors. If $D(A)$ and $D(B)$ are sets of divisors the set of common divisors is $D(A) \cap D(B)$.

Examples

Find the common divisors of 24 and 18.

$$D(24) = \{1, 2, 3, 4, 6, 8, 12, 24\},$$

$$D(18) = \{1, 2, 3, 6, 9, 18\},$$

$$D(24) \cap D(18) = \{1, 2, 3, 6\} = D(6).$$

The largest number in this set of common divisors, $D(24) \cap D(18)$, is called the Greatest Common Measure (G.C.M.) or, more commonly, the Highest Common Factor. In this example the H.C.F. is 6. We can also illustrate this with a number line, using either different colours or different kinds of marks to distinguish between the two sets $D(24)$ and $D(18)$. The intersection set, that is, the set of common factors, is distinguished by being marked by both signs.

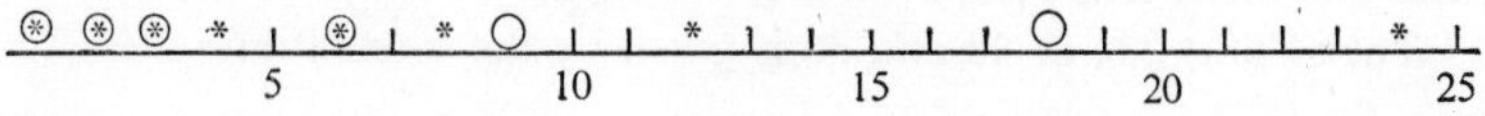

5 10 15 20 25

Fig. 5.11

Set $D(24)$ is marked $*$; set $D(18)$ is marked $\circ$; $D(24) \cap D(18)$ is marked $\circledast$.

Exercises

In each of the following questions find the set of common divisors and in each case state the H.C.F. (1) 28, 42; (2) 25, 40; (3) 15, 22; (4) 105, 245, 175; (5) 36, 48, 60, 96.

Notes

In the above, the word 'divisor' is used rather than the word 'factor' because, as normally used, the word 'factor' implies an association with, and an operation uniting, one or more other factors. Hence, although the elements can be called either factors or divisors, when we speak of *a set of factors* we shall mean that their product is equal to the number in question. A set of factors of 24 is $\{2, 12\}$ or $\{3, 8\}$ or $\{2, 3, 4\}$ and so on. Each of these can also be called *a* set of divisors. As above, we will speak of $\{1, 2, 3, 4, 6, 8, 12, 24\}$ as *the* set of divisors of 24. This, however, under the convention used is not *a* set of factors of 24; it is *a* set of factors of the number $1.2.3.4.6.8.12.24$.

A further difficulty arises when we consider sets of prime factors. The distinct prime factors of 24 are 2 and 3 (1 not being a prime). If 24 is factorized into its prime factors we have $24 = 2.2.2.3$. The confusion can be avoided by introducing the notion of *primary numbers* and *primary divisors*, where we define a primary number as a power of a prime, the indices being natural numbers or zero. The sets of primary divisors of 48 and 36 are

$$P(48) = \{1, 2, 2^2, 2^3, 2^4, 3\},$$

and

$$P(36) = \{1, 2, 2^2, 3, 3^2\}.$$

As we explained above, there is a technical objection to writing:

the set of prime factors of 24 *is* {2, 2, 2, 3},

or *the set of prime factors of* 25 *is* {5, 5},

and this form of set notation is not open to us.

Then the H.C.F. of 48 and 36 is given by the calculation

$$P(48) \cap P(36) = \{1, 2, 2^2, 3\} = P(12),$$

and the L.C.M. is given by the calculation

$$P(48) \cup P(36) = \{1, 2, 2^2, 2^3, 2^4, 3, 3^2\} = P(144).$$

Euclid's algorithm

Let us now consider the numbers 40 and 68. Using sets of divisors we have

$$D(40) = \{1, 2, 4, 5, 8, 10, 20, 40\}$$

$$D(68) = \{1, 2, 4, 17, 34, 68\}$$

$$D(40) \cap D(68) = \{1, 2, 4\} = D(4)$$

The H.C.F. *of 40 and 68 is 4.*

Suppose we add 40 to 68, obtaining 108, and find $D(108)$. Now consider the difference between 40 and 68, that is, 28, and find

$$D(28),\ D(28) \cap D(40),\ D(28) \cap D(68),\ D(28) \cap D(40) \cap D(68).$$

$$D(28) = \{1, 2, 4, 7, 14, 28\},$$

$$D(40) = \{1, 2, 4, 5, 8, 10, 20, 40\},$$

$$D(68) = \{1, 2, 4, 17, 34, 68\}.$$

So $D(28) \cap D(40) = \{1, 2, 4\} = D(4)$,

$$D(28) \cap D(68) = \{1, 2, 4\} = D(4),$$

$$D(28) \cap D(40) \cap D(68) = \{1, 2, 4\} = D(4).$$

Now find the difference between 28 and 40, that is, 12.

$$D(12) = \{1, 2, 3, 4, 6, 12\},$$

$$D(12) \cap D(28) \cap D(40) \cap D(68) = \{1, 2, 4\} = D(4).$$

Find the difference between 12 and 28, that is, 16.

$$D(16) = \{1, 2, 4, 8, 16\},$$

and $D(16) \cap D(12) \cap D(28) \cap D(40) \cap D(68) = D(4)$.

Repeat; that is, $16-12 = 4$.

$$D(4) \cap D(12) \cap \ldots, \text{etc.} = D(4).$$

Thus we have found that $D(4)$ is the set of divisors of the original two numbers and all the differences as found above. It follows from this that the H.C.F. of all these differences and of the two original numbers is 4. This suggests the method of finding the H.C.F. by differences (shortened where appropriate by division) called *Euclid's algorithm*—which is useful for large numbers and later for polynomials.

Another approach to Euclid's algorithm

Consider a shopkeeper who cannot find his weights. He uses his ingenuity and finds goods in 4- and 6-ounce packets. What can he weigh with them?

4, 6 oz., 8, 12, 16, 20 oz., 12, 18, 24, 30 oz.

He can weigh any multiple of 4 ounces and any multiple of 6 ounces. Is there anything else he can do? He can weigh 2 ounces, obtained from the difference, and hence any multiple of 2 ounces. This is more general than the first discoveries, but his problem is not completely solved. Why? Because he cannot weigh any odd number of ounces. Can we think of a situation where he could weigh the odd as well as the even weights, apart from the trivial case of 1-ounce packets?

Suppose he can weigh multiples of 3, multiples of 5, and, by subtraction, multiples of 2. He has all the even numbers and in addition all the odd numbers that are multiples of 3 or 5. But is this all? We have:

3, 6, 9, 12, 15, 18, ...,

5, 10, 15, 20, 25, 30, ...

From this we can see that there are multiples of 3 and multiples of 5 which differ by 1, for example, 6 and 5, 9 and 10, etc. Hence he can weigh 1 ounce and therefore any number of ounces.

What pairs of weights will give differences of 1 ounce in this way? 7 and 11, 4 and 7, 5 and 9, etc. Have these pairs of numbers anything in common? They are co-prime but need not be pairs of primes. Co-prime pairs enable all weights to be obtained. Can we analyse the situation when other pairs are taken?

We already know that 4- and 6-ounce packets give the set of even weights. What about 6 and 15? *We will use {6} to represent the set of all positive integral multiples of 6.* We have:

$$\{6\} = 6, 12, 18, 24, 30, \ldots,$$

and $$\{15\} = 15, 30, 45, 60, \ldots.$$

What is the smallest weight obtainable by consideration of the differences in this case?

$$1 \times 15 - 2 \times 6 = 3, \quad 3 \times 6 - 1 \times 15 = 3, \text{ etc.}$$

Therefore he can weigh multiples of 3 ounces.

Give other pairs which give multiples of 3 ounces. 21 and 27, 33 and 42, etc. How can we describe 3 with respect to 6 and 15? Was 2 the H.C.F. of 4 and 6? What can we say in the case of co-prime pairs?

What numbers can be generated from the pairs: (1) 9 and 24; (2) 10 and 16; (3) 16 and 36; (4) 8 and 15; (5) 20 and 60?

We are hoping to establish the idea that there is an h which divides a and b, that there is no larger number with this property, and that this h is expressible as the difference between a multiple of a and a multiple of b.

As far as our shopkeeper is concerned, his problem is solved if he chooses co-primes, but the H.C.F. of any pair will tell him what is the best he can do with any particular pair.

Apart from obvious topics, such as further work on H.C.F., this could be followed by lessons on simple Diophantine equations; many classical problems about these are firmly established as puzzles. See also the note on pp. 144–5.

Sets of multiples

This lesson is on classification by Venn diagrams. It involves an investigation of many relations between numbers and introduces the notion of the L.C.M. of a set of integers.

It is suggested that different coloured chalk be used to denote integers and classes.

Draw the outlines in Fig. 5.12.

In {2} we classify all the multiples of 2; and in {3} we classify all the multiples of 3.

Where shall we put 4; 9; 12; 120; 67; 1? Can you give any number which will go into this region? And this? Can you place any number you think of? What can we say about the shaded region? It contains all multiples of 6, and only these numbers. How can we describe the region in terms of intersections? The set of multiples of 6 is the intersection of the sets of multiples of 2 and the set of multiples of 3; that is, $\{6\} = \{2\} \cap \{3\}$.

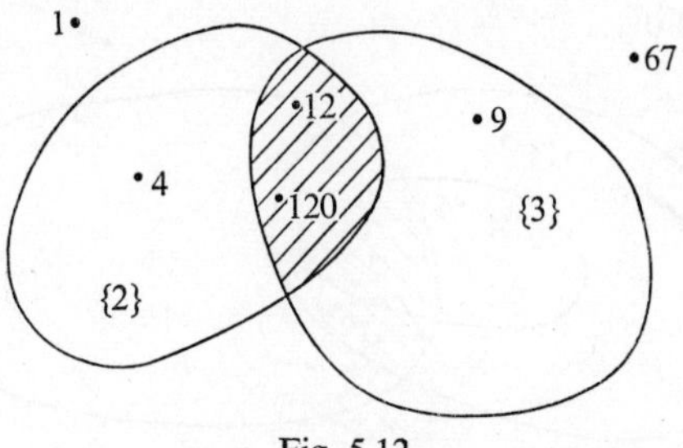

Fig. 5.12

Now let us make the diagram more complicated. *Draw outlines of Fig. 5.13.*

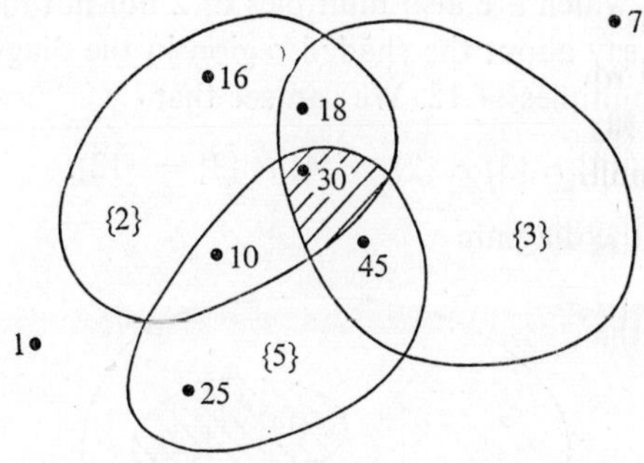

Fig. 5.13

What are the elements of {5}? Where shall we put 25; 10; 30; 45; 16; 7; 60; 1? Can you give any number which will go into this region? And another? And this region? What can we say about the shaded region? It contains all the multiples of 30 and only these numbers.

How can we describe this region in terms of intersections? Could you discuss {2}, {5}, {7} and {2}, {5}, {9} in the same way? *In the first example, the sets of multiples were generated from prime numbers. That is the case in the second example, but the third example has a further complication.*

Now consider {2}, {4}, {3}. *Draw outlines of Fig. 5.14.*

The pupils should draw the diagram, since they may not all realize that {4} is contained in {2}. Psychologically, there may be a barrier because of the 'opposing' expressions $4 > 2$ and $\{2\} \supset \{4\}$.

Is every multiple of 4 also a multiple of 2? Is every multiple of 3 also a multiple of 6?

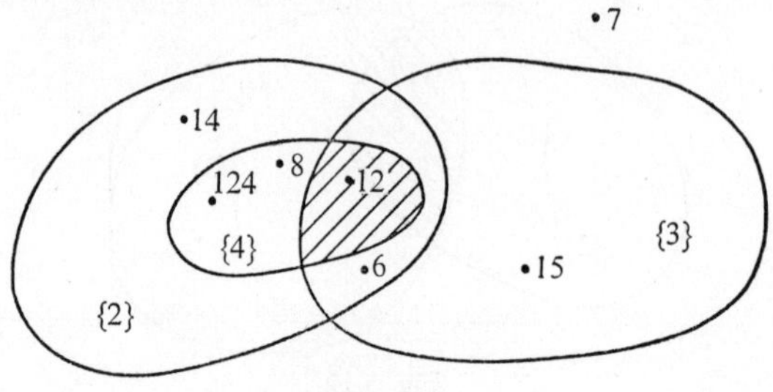

Fig. 5.14

Where will the numbers 15, 7, 6, 12, 8, 14, 124 be? Where are the multiples of 3 which are also multiples of 2 but not multiples of 4? What can we say about the shaded region in the diagram? It contains all the multiples of 12. We can see that

$$\{4\} \cap \{3\} \cap \{2\} = \{4\} \cap \{3\} = \{12\}.$$

If we look at the diagram

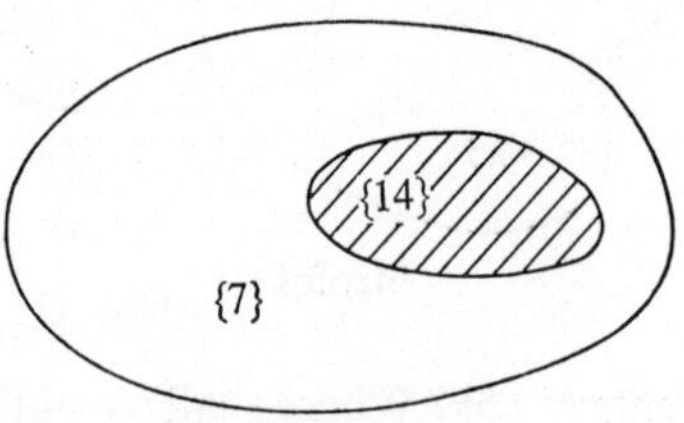

Fig. 5.15

we can see that the shaded part can be described as the intersection of the multiples of 7 and the multiples of 14, which is just the multiples of 14.

Can you think of any example which will have a diagram like {2}, {4}, {3}? Is there still a possibility of a different diagram? What about {5}, {10}, {20} and {4}, {8}, {12}? (Fig. 5.16).

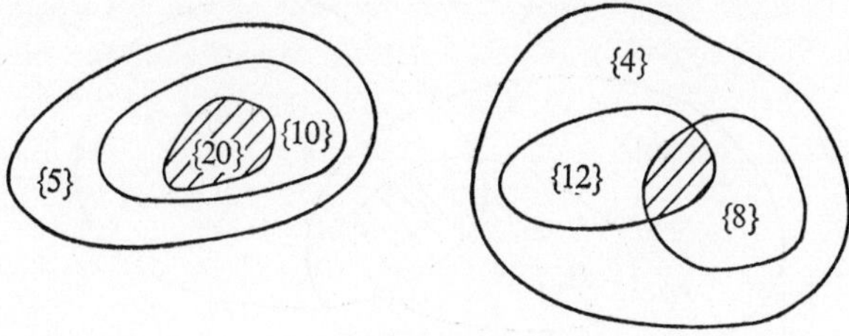

Fig. 5.16

What can you say about the shaded region in each case? Think of many examples of the type $\{a\}$, $\{b\}$, $\{c\}$. Can they all be illustrated by one or other of the diagrams already shown? What about {6}, {18}, {24}; {35}, {24}, {33}?

Given an arbitrary set of three numbers, how do we find the number associated with the intersection of the sets of multiples? Let us take the triple 4, 9, 15 and consider the sets of multiples {4}, {9}, {15}. Which type of diagram is required? (Fig. 5.17).

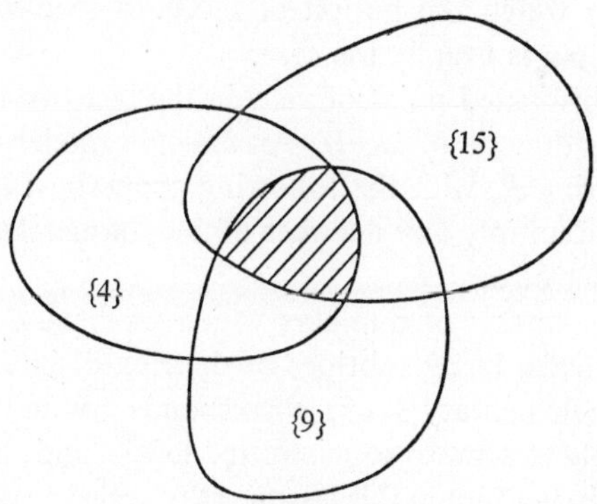

Fig. 5.17

The shaded region is the set {180}, which is the intersection of the multiples of 4 with the multiples of 15, with the multiples of 9. These numbers are called the *common* multiples of 4, 5 and 9, and the smallest element in the set, viz. 180, is called the *least common* multiple or L.C.M.

Now consider the three numbers 6, 12, 16. Can we find the common multiples and in particular the L.C.M.?

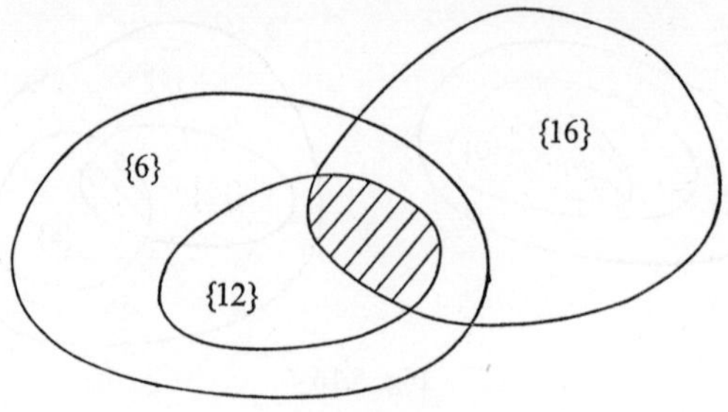

Fig. 5.18

Since $\{12\} \subset \{6\}$, we need only consider $\{12\} \cap \{16\}$. Here

$$\{12\} \cap \{16\} = \{48\}$$

and the L.C.M. = 48. Similarly with the three numbers 12, 18, 30.

Note

This lesson is not providing us with an algorithm for the determination of L.C.M. but is rather presenting a situation, rich with possibilities, in which the notion of L.C.M. is seen within a much wider setting than is usually the case.

From a sophisticated point of view in the last two lessons we are handling the arithmetic of *ideals*. An *ideal*, in modern algebra, is a subset S of a ring R, with the following properties: (i) it is itself a ring; (ii) if we multiply any member of S by a member of R we get another member of S.

In this case the sets of numbers which we have denoted as {5}, {3}, etc., are ideals, being subrings of the ring of integers. Factorization, at an elementary level, corresponds to finding the intersection of ideals at a more sophisticated level—and that is the level which the children adopt in this lesson. (Properly speaking, rings and ideals involve negative quantities as well as positive, but this complication need not concern us here.)

The introduction of the notion of ideals into the theory of numbers was important because number systems had been discovered within which the property analogous to the familiar property of integers of being uniquely factorizable into primes broke down.

However, in these systems, an ideal is uniquely expressible as the intersection of prime ideals, and this restores the analogy with the integers [79].

Likewise, the earlier lesson on H.C.F. was also an exploratory lesson on ideals, centred around the theorem that *in the ring of integers every ideal is a principal ideal*[14].

Definitions

The following lesson is suitable for children of twelve and upwards.

We are going to discuss different kinds of plane quadrilaterals. *Are there quadrilaterals that are not plane?* Tell me some special sorts of quadrilaterals: ... parallelograms, squares, rectangles, trapezia.... Let us omit some of these for simplicity and concentrate on trapezia, rectangles and cyclic quadrilaterals. *What do we mean by* cyclic quadrilaterals?

We must also make clear what we mean by a *trapezium*. Some people (mainly in the U.S.A.) use 'trapezium' for any irregular quadrilateral, but we limit it to 'a quadrilateral having two sides parallel'. With this definition, is a parallelogram a trapezium? None of the books tells us very clearly. A parallelogram certainly has two sides parallel, but also more than this. Similarly, a square possesses all the characteristic properties of rectangles, but also some additional ones.

Everyday language usually prefers definitions to be *exclusive*, so that a square is not a rectangle, and a parallelogram is not a trapezium. Mathematics, on the other hand, always adopts an *inclusive* definition, so we shall regard a parallelogram as a special kind of trapezium. A rectangle, of course, is likewise a special kind of parallelogram, so it is a still more special kind of trapezium.

Let us return now to trapezia, rectangles and cyclic quadrilaterals. Are all rectangles cyclic? How shall we show this on a Venn diagram? Are *all* trapezia cyclic? Are *any* trapezia cyclic? Are *all* cyclic quadrilaterals trapezia?

Rectangles are both cyclic and trapezia. How will this show on the diagram? Are there any cyclic trapezia that are not rectangles? Draw a Venn diagram displaying these results (Fig. 5.19).

Other similar exercises

(1) Draw a Venn diagram for the set E of all triangles, and the subsets H = {scalene}, I = {isosceles}, K = {equilateral}, L = {acute-angled}, M = {right-angled}, N = {obtuse-angled triangles}. Discuss any uncertainties of definition, for example, is an equilateral triangle isosceles? can a scalene triangle be right-angled?

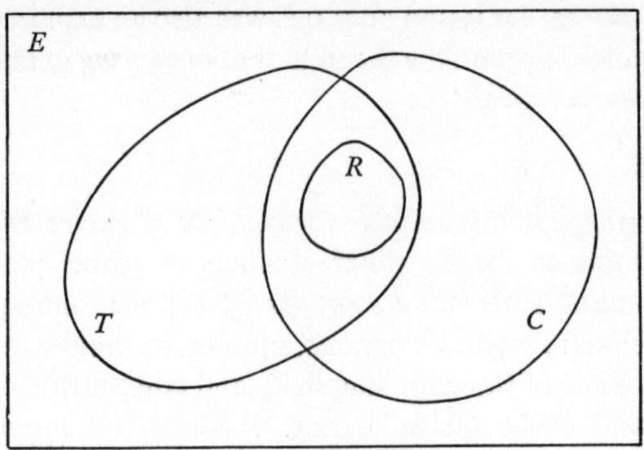

Fig. 5.19

E = {quadrilaterals}; T = {trapezia}; R = {rectangles}; C = {cyclic quadrilaterals}.

(2) Draw a Venn diagram for P = {all parallelograms}, Q = {rhombi}, R = {rectangles}, S = {squares}.

(3) Incorporate the diagram of exercise 2 in a wider one with E = {all plane polygons (including triangles and quadrilaterals)}, T = {triangles}, C = {cyclic figures}.

(4) Represent the natural numbers from 1 to 16 on a Venn diagram showing the subsets A = {perfect squares}, B = {even numbers}, C = {primes}. (*Note*: 1 is not regarded as prime.)

Discuss various intersections and unions. Which two of these subsets are disjoint (and so their intersection is the empty set)?

(5) For E = {natural numbers}, A = {multiples of 2}, B = {multiples of 3}, C = {multiples of 4}, D = {multiples of 6}, F = {multiples of 12}, display the relations between these subsets on a Venn diagram. (See the lesson on p. 140.)

Syllogistic reasoning

The following lesson might be tried in the sixth form, or earlier.

Traditional logic (stemming from Aristotle) discussed relationships of the following type between three categories (or sets):

If some men are good
and all saints are good,
then some men are saints.

In this, a relationship between two sets (men and saints) is sought through the intermediary of another set (that of good people). Is this particular example a valid argument? No, even if the final statement should happen to be true.

Age-long discussion eventually reached the conclusion that there were nineteen valid forms of *syllogism*, as these were called. A discussion of them all was complicated, but set notation and Venn diagrams enable the essentials to be seen more easily. We will discuss three of them here.

(1) All pilots (P) are brave men (B).
All astronauts (A) are pilots.

What conclusion can be drawn?

The first statement or premiss can be expressed as 'the set P of pilots is included in the set B of brave men' or $P \subset B$, and the second premiss is $A \subset P$. Here P is the intermediary or *middle term*. If these two premisses are represented on a Venn diagram as in Fig. 5.20 we see that $A \subset B$, or 'All astronauts are brave men'. This is the valid conclusion.

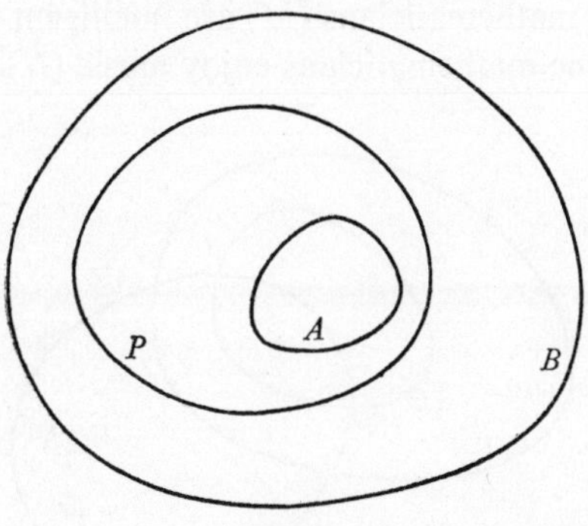

Fig. 5.20

An alternative form of the diagram is shown in Fig. 5.21, where three closed curves are drawn in general position to give the eight possible subsets of the three sets, namely

$$A \cap B \cap P, A \cap B \cap P', A' \cap B \cap P', A' \cap B' \cap P', \text{etc.}$$

The first premiss indicates that 'the set of pilots who are not brave men' is empty, or $P \cap B' = \phi$, which covers two of the eight funda-

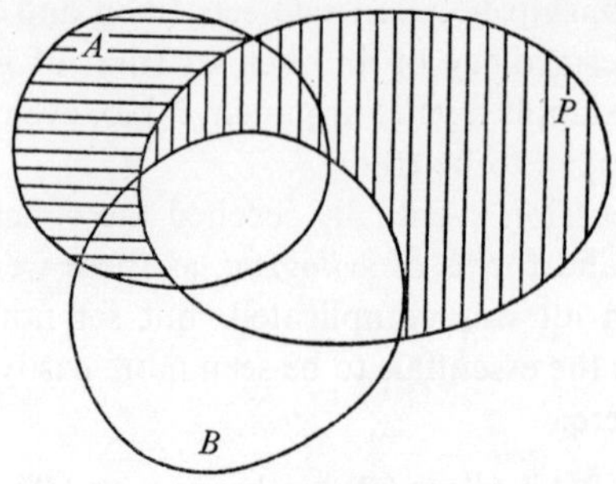

Fig. 5.21

mental subsets, $A \cap P \cap B'$ and $A' \cap P \cap B'$. To show this, these subsets are excluded by vertical hatching.

Similarly the second premiss indicates that $A \cap P' = \phi$ and the corresponding region of the diagram is excluded by horizontal hatching. From the resulting diagram we see that $A \subset B$, which is the same conclusion as was reached from Fig. 5.20.

(2) All mathematicians (M) are intelligent (I).
Some mathematicians enjoy music (N).

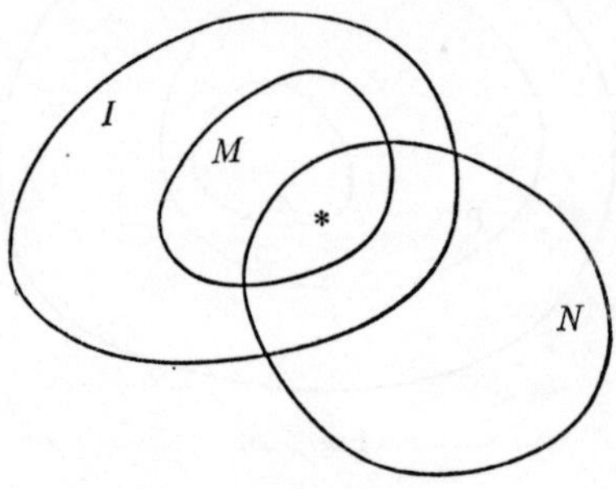

Fig. 5.22

The first premiss asserts $M \subset I$, and this is indicated in Fig. 5.22 by set inclusion and in Fig. 5.23 by horizontal hatching to mark an empty subset.

The second premiss is shown in both figures by an asterisk which denotes that the corresponding subset $M \cap N$ is not empty. We have no certain knowledge whether any of the remaining subsets are empty or not.

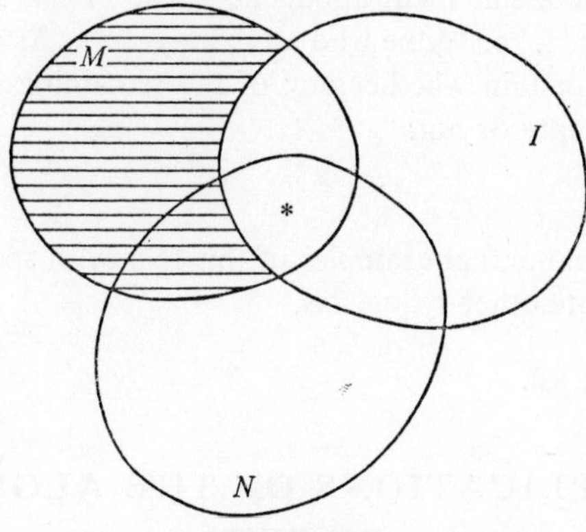

Fig. 5.23

The valid conclusion is either: some intelligent people enjoy music, or: some who enjoy music are intelligent.

(3) No murderer (M) is happy (H).
No selfish person (S) is happy.

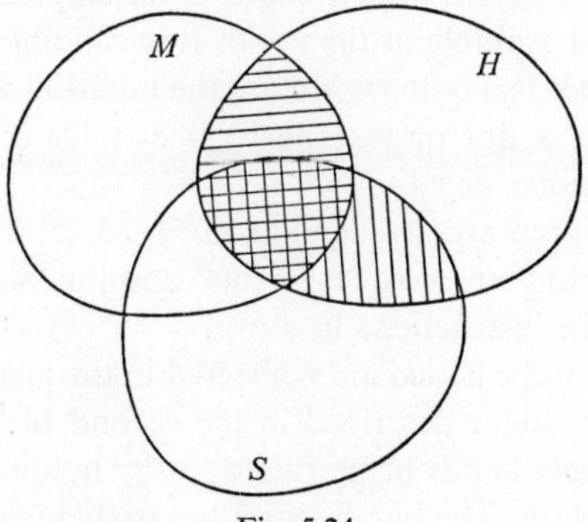

Fig. 5.24

The premisses are represented in Fig. 5.24. Can any valid conclusion be drawn relating the set of murderers to the set of selfish people?

No. To conclude 'All murderers are selfish' or even 'Some murderers are selfish' (or with 'murderers' and 'selfish' interchanged) is unwarranted. It is an example of the commonly observed fallacy in argument called by logicians the fallacy of the *undistributed middle term.* Here the middle term is the set of happy people, and in neither premiss is there a statement about the whole of this set (either 'all who are happy...' or 'none who are happy...'). As a result we do not know for certain whether any of the remaining subsets in the diagram are empty or not.

Exercises

(1) Try to find actual examples of this fallacy in speech or print.
(2) Investigate other syllogisms.

References [23, 83].

6. APPLICATIONS OF THE ALGEBRA OF SETS

An analysis of the Engelbart human binary adder

Here we describe another of the binary games invented by Engelbart [39], in which the ideas of §2.3 are extended further and the children in the class function as a binary adder. It amuses children to work as part of the machine without understanding fully how it works, but the analysis, which could be attempted with advanced pupils, is a good example of the use of Boolean algebra in a modest piece of computer technology. We use the notation where $+$ denotes disjunction (or), a dot or juxtaposition denotes conjunction (and) and a prime denotes negation.

Nineteen children are arranged as in Fig. 5.25. A to K represent binary digits, 1 to 9 are 'logic elements' and the 'signal source', the teacher, gives the instructions to move.

The numbers to be added are registered in the augend and addend as in the binary adder described in the second binary lesson (p. 9) by the appropriate hands being raised, or by holding cards with the correct numbers on. The hands of all the participants may be UP or DOWN. The person acting as the signal source gives two distinct signals, repeated five times. Engelbart suggests the words 'bonk-bleep', but any suitable noises would do. On 'bonk' the *lettered* elements watch certain other participants; on 'bleep' they copy the state observed at 'bonk'. The *numbered* elements move immediately, regardless of signals. Their instructions are given in set notation.

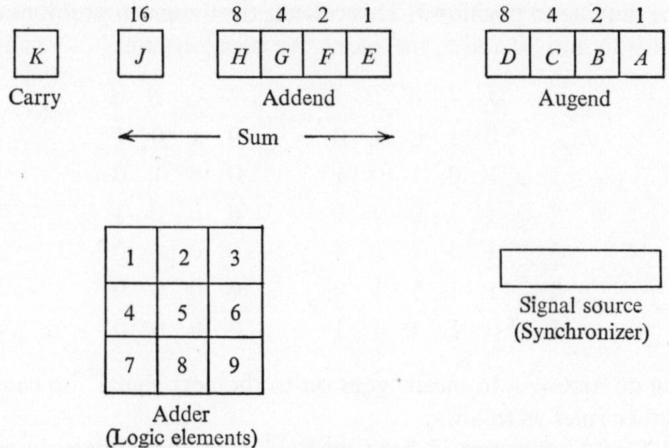

Fig. 5.25

'$1 = A.E$' means '1 puts his hand UP if *both A and E* have their hands UP'; '$3 = A+E$' means '3 puts his hand UP if *either A or E or both* have hands UP'; '$7 = 5'$' means '7 puts his hand UP if 5 has his hand DOWN'. D has a particularly simple task as he keeps his hand DOWN all the time.

Element	Watches	Logic elements
A	B	
B	C	$1 = A.E$
C	D	$2 = 1.K$
D	'Down element'	$3 = A+E$
E	F	$4 = 1+K$
F	G	$5 = 3.4$
G	H	$6 = 3+K$
H	J	$7 = 5'$
J	9	$8 = 6.7$
K	5	$9 = 2+8$

Analysis

Without complications about 'carrying' in binary notation, the system is simply that the figures in the addend and the augend slide five places to the right (hence five signals), with figures in positions E

and A moving to position J. This collects the figures in positions J to E. Thus to add 10 and 5, the successive positions are:

	J	H	G	F	E	D	C	B	A
	0	1	0	1	0	0	1	0	1
1	1	0	1	0	1	0	0	1	0
2	1	1	0	1	0	0	0	0	1
3	1	1	1	0	1	0	0	0	0
4	1	1	1	1	0	0	0	0	0
5	0	1	1	1	1	0	0	0	0

Using an arrow $\rightarrow$ to mean 'goes on to the next signal' we can list the other rules as follows:

(*a*) Digits appearing in both unit columns simultaneously must be carried as one digit one place to the left, and hence transfer to K not J; so $AE \rightarrow K$. But a digit in A or E, but not both, will transfer to J; so $AE' + A'E \rightarrow J$. Of course, $K \rightarrow J$.

(*b*) When one unit place is occupied and K is also occupied, on the signal both digits will want to move to J. Since they cannot do this, K is occupied instead. Hence

$$(AE' + A'E)\,K \rightarrow K.$$

Combining these two requirements we need to have:

$$AE + (AE' + A'E)\,K \rightarrow K \qquad (1)$$

and

$$(AE' + A'E)\,K' + (AE' + A'E)'K \rightarrow J. \qquad (2)$$

The logic elements arrange this as follows:

(1) $AE + (AE' + A'E)K = (AE + AEK) + (AE' + A'E)K$

$= AE + (AE + AE' + A'E)K = AE + (A+E)K + AE + AK + EK$

$= (A+E)(AE+K) = (A+E)(1+K) = 3.4 = 5 \rightarrow K.$

(2) $(AE' + A'E)K' + (AE' + A'E)'K = AE'K' + A'EK' + A'E'K + AEK$

$= AEK + A'E'K + (AE' + A'E)K'$

$= AEK + (A+E+K)\,[A'E' + (A'+E')\,K']$

$= 1K + (3+K)\,(A'E' + 1'K')$

$= 1K + (3+K)\,(3' + 4') = 2 + 6.5' = 2 + 6.7 = 2 + 8 = 9 \rightarrow J.$

The tasks of the logic elements are simple, but they could be equally simple and the number of elements reduced to five. Note the 'one or the other but not both' feature of (1) and (2), that is, $AE'+A'E$. Using five logic elements, L, M, N, P and Q, let

$$AE = L$$
$$AE'+A'E = M$$
$$MK = N$$
$$L+N = P \rightarrow K$$
$$MK'+M'K = Q \rightarrow J,$$

and (1) and (2) are satisfied. (This would look even simpler using the notation $A \triangle E = AE'+A'E$.)

The circuit board

The circuit board is an apparatus that provides a convenient and quick method of wiring together a battery, switches and lights. The lights can be made to go on or off for various positions of the switches (see plate III).

When a switch in a simple electrical circuit is closed, the current is allowed to flow through. It is possible to arrange two switches so that when one is open the other is closed. More simply, a single switch with two poles will send current through one branch of a circuit but *not* the other. Such a switch is called a *changeover*.

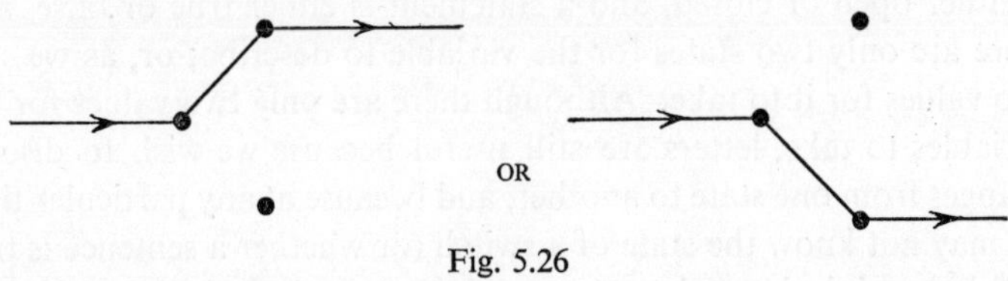

Fig. 5.26

If two switches are wired in series into a circuit supplied with a source of current, the current can flow through only if both *A and B* are closed.

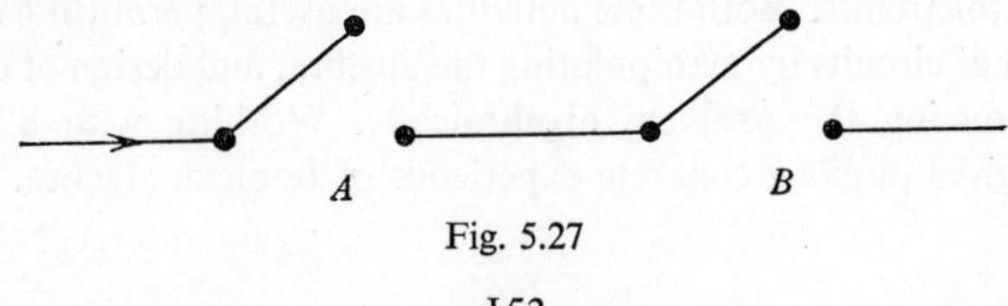

Fig. 5.27

On the other hand, two switches arranged in parallel will let the current flow through if *A or B or* both are closed.

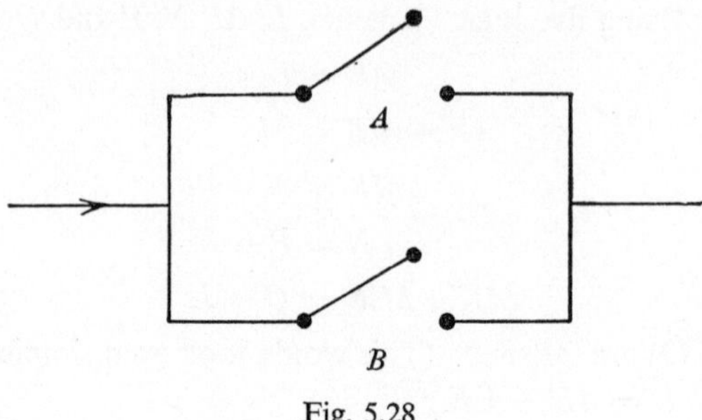

Fig. 5.28

We can say that these two ways of connecting two switches give an *and* and an *or* (inclusive *or*) circuit respectively.

These three operations of *not*, *and* and *or* are analogous to the complementation, intersection and union operations of set algebra and constitute a two-valued Boolean algebra (see p. 180).

The term *two-valued* may require explanation. In the algebra of sets a letter may denote the universal set (everything that is under discussion at the time), the null set, or anything in between. When the algebra is used to describe electrical switching circuits, or statements, as in the sections which follow, there is a difference. A switch is either open or closed, and a statement is either true or false, and there are only two states for the variable to describe; or, as we say, two values for it to take. Although there are only two values for the variables to take, letters are still useful because we wish to discuss changes from one state to another, and because at any particular time we may not know the state of a switch (or whether a sentence is true or false) and the letter denotes an unknown state. Letters may denote either variable or unknown states, just as letters in ordinary algebra denote variables or unknown numbers.

Circuits may be translated into this algebra and this algebra translated into circuits. Both these activities are useful, permitting simplification of circuits by manipulating the algebra, and design of circuits by expressing the problem algebraically. Working with a circuit board gives pupils a concrete experience of Boolean algebra.

Construction of a circuit board

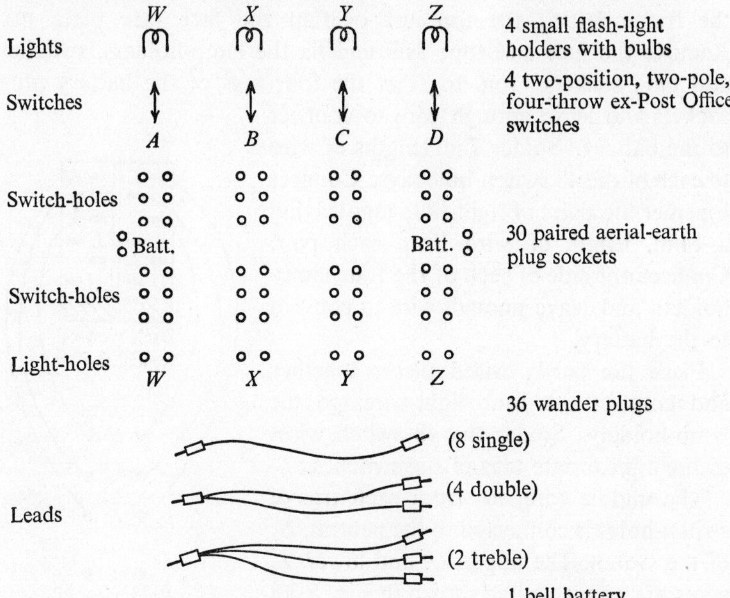

A supply of single core plastic covered wire, different colours preferable, for internal wiring and construction of leads.

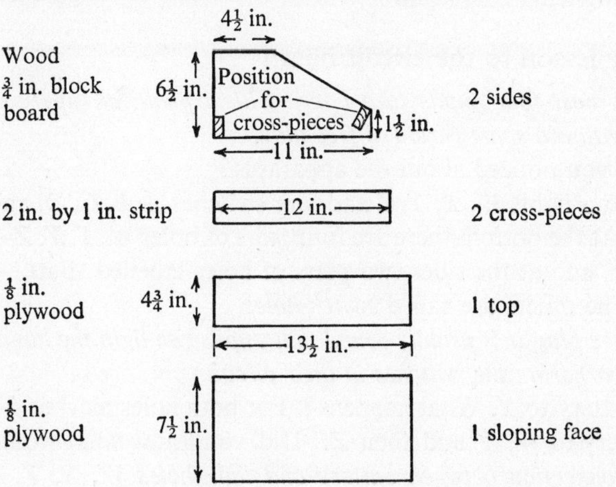

Fig. 5.29. Construction of a circuit board.

Assembly. Fix the sides and cross-pieces together with screws to form a rigid frame. Fit and screw the sloping face and top to the frame letting the top just overlap the face (see plate III). Remove the face and top; drill and fix the lamp-holders, switches and plug sockets. Join together the four tags of the battery plug sockets and leave enough wire to connect to the battery. Solder 7 in. lengths of wire to each of the 48 switch-hole tags. Connect together the pairs of light-hole tags leaving a 14 in. length of wire from each pair. Connect one side of each of the four lamp-holders and leave enough wire to connect to the battery.

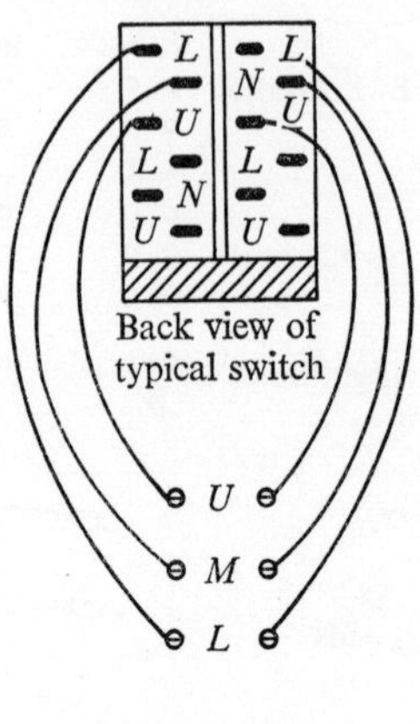

Fig. 5.30

Place the partly wired pieces together and connect the four light-wires to the lamp-holders. Solder the 48 switch wires to the appropriate tags of the switches.

The middle wire, M, from each trio of switch-holes is connected to the neutral, N, of the switch. The upper, U, and lower, L, wires are connected as shown in Fig. 5.30.

Check wiring and replace face and top. Fit battery and bulbs. Label. Make up leads of varying lengths 6–12 in.

Introductory lesson to the circuit board

Arrange at most four pupils to a board with a plentiful supply of single leads. Allow a short period of free play.

What have you noticed about the apparatus?

There are four lights W, X, Y, Z and four switches A, B, C, D, and many holes. At the bottom there are four pairs of holes W, X, Y, Z—the *light-holes*, and at the sides two pairs of holes labelled 'Batt.'—the battery. The others are called *switch-holes.*

After the free play it is usually found that pupils can light the lamps but cannot incorporate the switches in their circuits.

Connect battery to X. What happens? For both holes marked X? Connect battery to W, Y and then Z. Did you know what would happen? *A connection between battery and light-holes W, X, Y, Z lights lamps W, X, Y, Z respectively.*

We shall denote *light Y on* by y, *light Z off* by z'. What do x, w', y' denote? Can you produce w and x' and y and z' simultaneously?

Let us now consider the switches. These have two positions: *up* or *near to light*, and *down* or *away from light*. We shall denote by a, b, c, d switch position *away from light* and by a', b', c', d' switch position *near to light*.

Under switch B we have

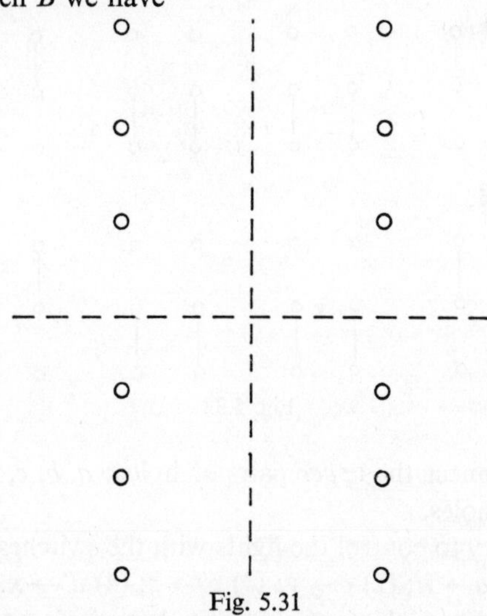

Fig. 5.31

Think of these as four sets of three holes. There are *no* connections across the dotted lines and we must think of the four sets as having nothing to do with each other.

Consider one set of three holes (Fig. 5.32). For switch position b, the middle and lower holes are connected internally. For b' the middle and upper holes are connected internally. Connect battery to the middle hole and also connect the lower hole to X. Move switch B. What do we find? B lights X, that is, $b \to x$. Anything else? Yes, $b' \to x'$.

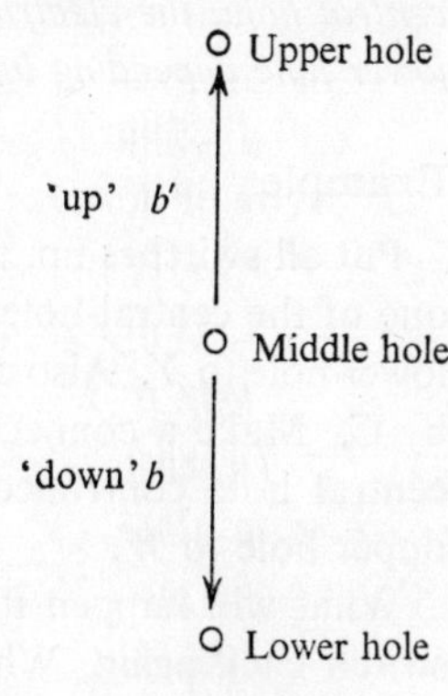

Fig. 5.32

Suppose we now connect the upper hole to Z. What happens?

$$b \to x \quad \text{and} \quad b' \to z.$$

Now try a different set of three holes under *B*. We notice the same result.

Remember

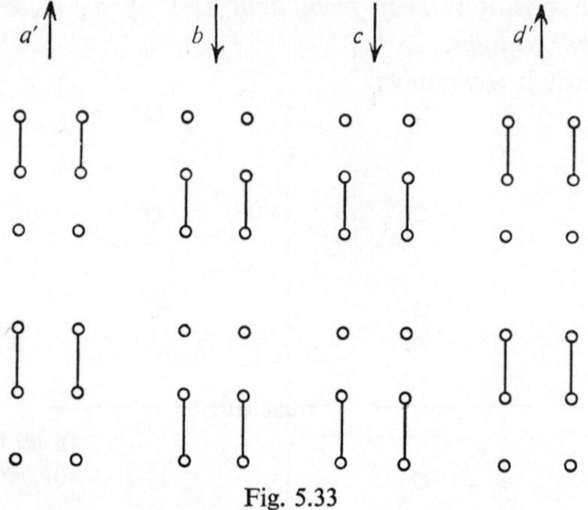

Fig. 5.33

a', b', c', d' connect the *upper* pairs of holes; a, b, c, d connect the *lower* pairs of holes.

Now let us try to control the lights with the switches. Set up these situations: (1) $a \rightarrow w$; (2) $c \rightarrow y$; (3) $c' \rightarrow z$; (4) $d' \rightarrow x$.

It may help if you think of electricity 'pouring' out of the battery holes, and when a plug is put in one of these holes it now 'pours' off the plug at the other end of the lead. If this plug is now pushed into a central hole, the electricity 'pours' out of the corresponding upper or lower hole depending upon the switch position.

Example

Put all switches up, that is, a', b', c', d'. Connect a battery hole to one of the central holes underneath *A*. Connect the corresponding lower hole to *Y*. Also connect the battery to a central hole controlled by *C*. Make a connection from the corresponding lower hole to a central hole controlled by *B*. Finally, connect the corresponding upper hole to *W*.

What will happen if switch *A* is moved? Try it, and change the switch back again. What will happen if switch *C* is moved? Try it and leave it. Now change *B*. Why does *W* go off?

The class must now be given plenty of practice to enable them to become familiar with the boards and the notation. In discussion bring out the use of the words 'and' and 'or'.

Exercises

(1) Two men each have a switch. A warning light is to come on if: (i) either man puts his switch on; (ii) both men put their switches on. Set up these circuits using switches A and B and light W for (i); and C and D and light Z for (ii). This illustrates an 'or' and an 'and' circuit. *They are, of course, circuits in parallel and in series.*

(2) There are four men in a rocket and each has a switch. A 'safe light' is to light up if all are ready to proceed.

(3) There are four men in a rocket and each has a switch. A 'danger light' is to light up if anyone is signalling danger.

(4) A quiz machine. Each switch is associated with two alternative answers to a set of four questions. A bulb is to light if the person being questioned sets each switch to the correct position. Wire up the circuit for the case when the answers are 'Yes', 'No', 'No', 'Yes'.

(5) How may a light on a staircase be put on or off by changing the position of the switches at the top or bottom of the stairs?

(6) The four lights are labelled '7', '2', '9' and '6'. The appropriate light is to light up when two switches are set in response to the two questions: 'Is it an odd number?' 'Is it a prime number?'

During this preliminary work with the circuit boards the pupils may discover various rules of the algebra, such as [(a and b) or (a and c)] $\rightarrow w$ is the same as [a and (b or c)] $\rightarrow w$ and [a or (a' and b)] is the same as [a or b].

Lesson

This is a more advanced lesson which can be given to sixteen-year-olds who have some formal knowledge of Boolean algebra. It is suggested that they use juxtaposition and plus instead of the words 'and' and 'or'.

How can we use our circuit board as a vote controller where four voters have voting powers of 3, 2, 1 and 1?

Class discussion decides that a, b, c and d indicate that the four men have voted and a′, b′, c′ and d′ indicate they have not voted.

Light W is to come on if a majority vote is recorded. What is the total vote? A majority vote? Good, give me some examples of a particular majority voting. 3 votes and 2 votes; 2 and 1 and 1; 3 and 2 and 1.

Let us put these in our notation. *A*'s 3 votes and *B*'s 2 votes correspond to (*a* and *b*). What about voters *C* and *D*? Yes, the fact that only *A* and *B* have voted is shown by (*a* and *b* and c' and d'). In a shorter form? $abc'd'$. List all the possible majorities.

$$abc'd'; \; ab'cd'; \; ab'c'd; \; ab'cd;$$

$$abc'd; \; abcd'; \; abcd; \; a'bcd.$$

What does this tell us about our voting machine? Light *W* is to come on for any of these switch positions. Another way of putting it? *W* must light for $abc'd'$ or $ab'cd'$ or $ab'c'd$ etc. In symbols?

$$abc'd' + ab'cd' + ab'c'd + ab'cd + abc'd + abcd' + abcd + a'bcd \rightarrow w.$$

Are we ready to wire this on our circuit boards? Why not? The expression is too long and can be simplified. This is suggested by the fact that *ab* always implies a majority whatever the positions of *C* and *D*. How can we make our eight-termed expression shorter? Consider

$$abc'd' + abc'd = abc'(d'+d)$$

$$= abc',$$

since $d'+d = 1$.

Also $$abcd' + abcd = abc.$$

These two results give $abc' + abc = ab$. So

$$ab\underline{c'd'} + ab\underline{c'd} + ab\underline{cd'} + ab\underline{cd} = ab.$$

This typical simplification comes from class discussion and all the steps can be explained practically in terms of the circuit board. Similar simplifications will produce

$$ab + ac + ad + bcd \rightarrow w$$

and finally

$$a(b+c+d) + bcd \rightarrow w.$$

Now the wiring of the machine can begin. Let us listen to a pupil arguing to himself.

Connect battery to middle *A*. Now I want [*a* and (*b* or *c* or *d*)]. Yes; a triple lead from lower *A* to middle *B*, *C* and *D*; then another triple from lower B, C and D to W. Checking, $ab \to w$ also $ac \to w$ and $ad \to w$.

Next [or (b and c and d)]. Battery again; but this time to middle B. Then lower B to middle C, lower C to middle D and lower D to W. Check; $bcd \to w$. Now try all possible cases. Good, only the original eight positions light the light.

Note

This lesson shows that the simpler the expression the simpler the circuit. A further economy may sometimes be made by wiring the 'other way'. For example, examine the two circuits for $ab' + a'b$.

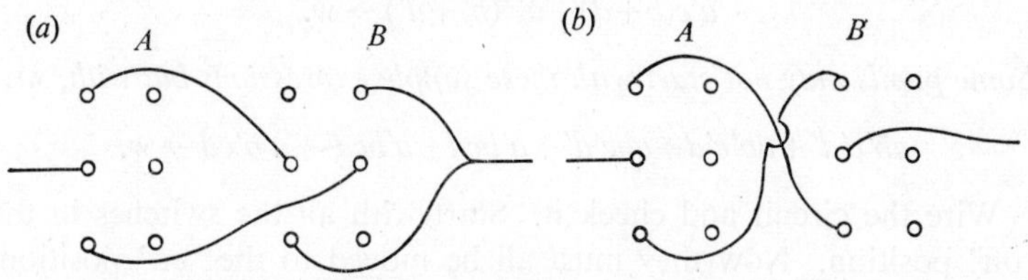

Fig. 5.34

Solution (*b*) saves a wire and a set of contacts. The best circuit, judged by the least number of connecting wires and switches, requires the simplest expression and the simplest method of wiring, a problem which has not yet been solved generally.

Lesson

A lesson for more experienced pupils based on the classic problem of the man, the wolf, the goat and the cabbages.

A man wishes to cross a river with his possessions: a wolf, a goat and a cabbage. The only boat available is small and can only carry the man and one of his possessions. Unfortunately, if left unguarded the goat will eat the cabbage and the wolf will eat the goat. Wire a circuit board showing how the man and his possessions are able to cross the river.

As before, class discussion decides upon the following notation:

a, b, c and d refer to the man, wolf, goat and cabbage on one side of the river and a', b', c' and d' to the other side. A warning light W is to light if a false move is made which allows the wolf to eat the goat or the goat to eat the cabbage.

What is the danger? The goat must not be left with the wolf or the cabbage if the man is not there. In symbols?

$$a'bc + a'cd \rightarrow w.$$

Yes, is that all? All could change sides. Good, and in this case?

$$ab'c' + ac'd' \rightarrow w.$$

So the expression becomes

$$a'bc + a'cd + ab'c' + ac'd' \rightarrow w,$$

or after simplification

$$a'c(b+d) + ac'(b'+d') \rightarrow w.$$

Some pupils may not start with these simple expressions but with, say,

$$ab'c'd' + ab'c'd + abc'd' + a'bcd + a'bcd' + a'b'cd \rightarrow w.$$

Wire the circuit and check it. Start with all the switches in the 'off' position. Now they must all be moved to the 'on' position, making sure that the man is always in the boat, and that he does not carry more than one of his possessions. How many solutions are there? Examine your circuit. Can you use fewer wires and switches? Here is a solution using eleven wires. Can you do better? Nine wires and five switches, good. *The photograph of the circuit board on plate* III *shows the 'best' solution.*

Exercises

(1) A light at the foot of a staircase is to be operated from three switches on different floors of a house.

(2) Binary numbers 0–15 are set up on the four switches. A bulb is to light if the number is prime.

(3) I want a lift to a football match next Saturday. Bill says—'If I am back from Brighton, and my wife does not want to go shopping I will take you.' Jack says—'If my car is repaired in time, or if John is not using his car I can take you.' Set up four switches and two lights to show what happens in the various cases.

(4) We have four switches. A bulb is to light if (*a*) two or more switches are on, (*b*) two or less are on, (*c*) exactly two are on.

(5) Four children are discussing a picnic. Alfred will not go if Doreen goes, and Brenda will only go if Charles is there. Wire up

four switches and four lights so that if the switches denoting a possible combination of children are switched on the corresponding lights light up.

(6) Set up binary numbers 0–3 on switches A and B and the same on C and D. The product is to appear in lights.

(7) A is a control key. A binary number is set up on B, C and D. When A is off, the number is duplicated in lights. When A is on, the square of the number appears in lights.

(8) Wire the circuit such that light W is on if any one switch is on; light X is on if any two switches are on; light Y is on if any three are on; the light Z is on if all four are on.

(9) In a party of explorers A and B are doctors; A and C are surveyors; and B and D can cook. Two are required to go on a journey, and there must be a doctor, someone who can survey and someone who can cook. Wire up a circuit with four switches A, B, C and D and a light which lights up if two switches denoting a possible combination are put on.

(10) In the University at Aarhus, in Denmark, the lights in the students' bedrooms work according to the following pattern: there are three switches A, B and C; and two lights X and Y.

A	B	C	X	Y
0	0	0	0	0
0	0	1	0	0
0	1	0	0	0
0	1	1	0	1
1	0	0	1	0
1	0	1	1	0
1	1	0	0	0
1	1	1	0	1

(We adopt a convention which is common in circuitry. 0 and 1 denote *off* and *on*—they have no numerical significance.)

How are the lights wired, and why are they wired in this way?

(11) Wire up circuits to illustrate the addition and multiplication tables of GF(2^2) (p. 76).

References [4, 32, 47, 129].

Designing and constructing a binary adder

Boolean Algebra can usefully be applied to this particular problem. The basic ideas which are necessary are those of Boolean addition and multiplication and the representation of functions by means of electrical circuits. The circuit board just described may be used to illustrate the work at any stage.

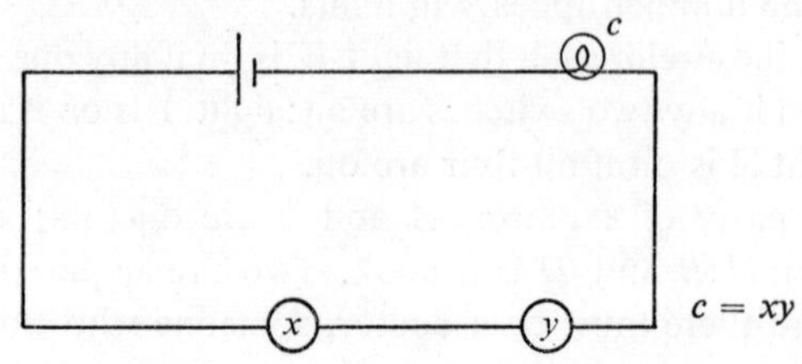

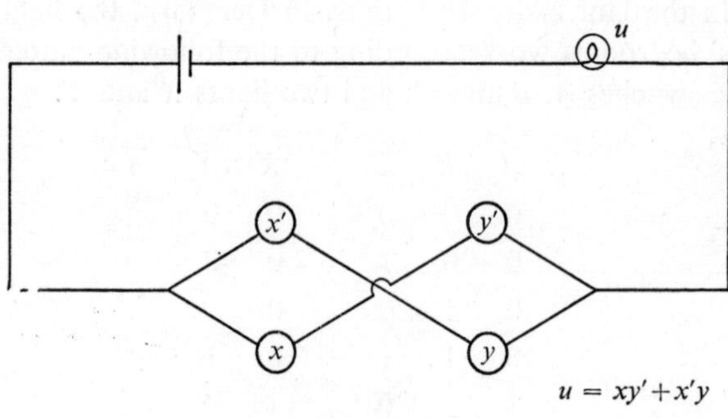

Fig. 5.35

The problem is to design a circuit which will enable one to perform additions in the binary system. For example,

$$\begin{array}{r} 10110 \\ 11111 \\ \hline 110101 \\ \hline \end{array}$$

Q. In any one column there are two digits to be added and possibly a carry figure. Consider the simpler case of just adding two digits x and y, where x and y can take the values 0 and 1. The answers might be 0, 1 or 10. If we call the unit digit u and the second place digit c, what is the table of values for x, y, u and c?

A.

x	y	c	u
0	0	0	0
0	1	0	1
1	0	0	1
1	1	1	0

Q. What formulae involving x and y correspond to u and c?

A. c is simply xy, but u is more difficult. When x and y are either (0, 1) or (1, 0) u is 1. That is to say $u = 1$ when x or y is 1, but not when both are.

Q. What circuits correspond to these expressions?

A. See Fig. 5.35.

(Now, 0 and 1 may be regarded either as digits, or as *off* and *on*.)

This circuit can be constructed and checked on the circuit board. It is the circuit for addition in the right-hand column in a binary adder, but in the remaining columns there is the additional complication of carrying-over from the previous column.

Q. In the further columns there are two digits x and y, and also a carry digit z from the previous column. What is the appropriate table of values now?

x	y	z	u	c
0	0	0	0	0
0	1	0	1	0
1	0	0	1	0
1	1	0	0	1
0	0	1	1	0
0	1	1	0	1
1	0	1	0	1
1	1	1	1	1

Q. What formulae give u and c in terms of x, y and z?

A.

$$u = x'yz' + xy'z' + x'y'z + xyz,$$

$$c = xyz' + x'yz + xy'z + xyz.$$

Q. What is the system here?

A. It is necessary to decide from the table which combinations give u or c, and to add the alternatives together. It appears that two wires will have to be used to carry over from one column circuit to the next, because both z and z' occur in these formulae.

Q. What is the formula for c'?

A. $$c' = xy'z'+x'yz'+x'y'z+x'y'z'.$$

Q. Can we manipulate the formulae so as to economize on switches? Remember that only the letters x and y involve fresh switches, as z and z' are carry-over lines from the previous circuit.

A.
$$\begin{aligned} u &= x(yz+y'z')+x'(yz'+y'z). \\ c &= xy(z+z')+x'yz+xy'z \\ &= xy+x'yz+xy'z \\ &= y(x+x'z)+xy'z. \\ c' &= y'x'(z+z')+x'yz'+xy'z' \\ &= y'x'+x'yz'+xy'z' \\ &= y'(x'+xz')+x'yz'. \end{aligned}$$

The total number of appearances of x and x' is eight and the total number of appearances of y and y' is also eight, so the circuit needs eight simple switches for each. *Other suggestions may arise, and should be discussed if they do.*

Q. What is the circuit diagram?

A. This is given in Fig. 5.36. Every unprimed letter denotes a contact which is made when the corresponding switch is pressed, and primed letters denote contacts which are broken when the switch is pressed.

Note

(See plate II.)

(1) The c and the c' leads from one column circuit are fed into the z and z' positions on the next column circuit.

(2) The z' lead on the first column circuit must be taken from the battery. (Since $z = 0$, z' must be 1.) The z lead is not required.

(3) The c lead from the final column circuit must be taken to a bulb to give the final digit in the answer. The lead c' is not required.

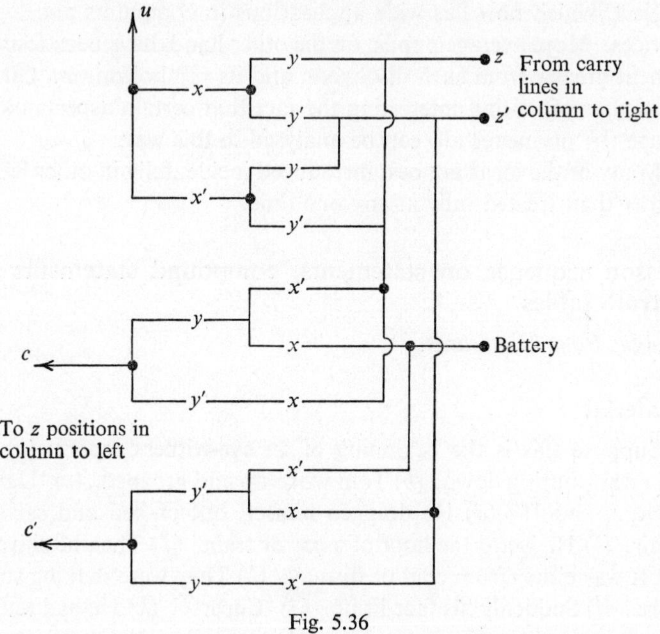

Fig. 5.36

(4) It would be an economy to build the circuit with change-over switches of the type used in the circuit board. These enable a pair of primed and unprimed letters to be realized by one switch, with three contacts, if they occur at a suitable place in the circuit. Four change-over switches are needed for x and four for y. They can be placed at the points in the circuit diagram indicated by dots.

7. THE LOGIC OF STATEMENTS

Introduction

The purpose of these lessons is to show something of the process of formalization of plain language and to lead towards Boolean algebra, but the full symbolism has not been introduced. Some of the ideas symbolized are used to throw light on the method of proof in mathematics and to consolidate standard work.

Probably the ablest pupils will find the treatment given here unnecessarily lengthy, and will grasp the essentials more directly. For them, interest would be aroused mainly by the further algebra of the subject, which now has wide applications in computers and logical devices. More average pupils, on the other hand, have been found to benefit greatly from such discussion and its symbolization. Girls in particular often find interesting the fact that certain aspects of language (by no means all) can be analysed in this way.

Many of the ideas are best introduced incidentally in other lessons rather than treated fully at any one time.

Lesson sequence on statements, compound statements and truth tables

Age: from 13 onwards.

Material

Suppose this is the beginning of an eye-witness report: (*a*) The rain was pouring down. (*b*) Tom woke up and groaned. (*c*) 'Ugh... Ugh...school!' (*d*) He dragged himself out of bed and sat on a chair. (*e*) He heard the hoot of a car or train. (*f*) Then he shivered. (*g*) It was either from cold or dismay. (*h*) They were making such a noise. (*i*) Suddenly his face lit up. (*j*) 'Cheers!' (*k*) He had remembered something. (*l*) 'Term does not start today, but tomorrow.'

When we read a report of something we believe that everything said is true. In a story, of course, everything is imagined, but in suitable circumstances much of what is said could be true. We may not know whether it is true or not, but it could be. We often find individual sentences, however, for which it doesn't make sense to say that they are true or not. Let us look at the individual sentences in the report I have just read out and have written on the board, with letters to help us refer to the sentences.

Statements

First let us pick out the simplest statements, about each of which (if we knew enough about the circumstances) we could say whether it was true or false.

After discussion we arrive at: (*a*) The rain was pouring down. (*f*) Then he shivered. (*i*) Suddenly his face lit up. (*k*) He had remembered something.

Are there any sentences which are not statements in this sense of possibly being true or false? (*c*) 'Ugh...Ugh...school!'

(*j*) 'Cheers!' These are both ejaculations or expressions of feeling. The words themselves state nothing about school or anything else. They could perhaps be taken to state (c^*) 'Tom felt annoyed at the thought of school', and (j^*) 'Tom felt pleased', but he might have been pretending in uttering these sounds (though that is no worse than making a statement without meaning it). The speech marks certainly indicate 'Tom said: Ugh...Ugh...school!' and 'Tom said: Cheers!', and these are statements, but the sentences in their original form are not.

So we see that sets of words forming a sentence do not always make a statement that can be either true or false. Can any of you invent other sentences that are not statements?

Negation

What is the *negation* of each of the simple statements we found above? (If a stands for the first statement we shall use $\sim a$ to stand for its negation. Compare -5 for the negative of 5.)

$\sim a$. The rain was not pouring down.

$\sim f$. Then he did not shiver.

$\sim i$. His face did not suddenly light up.

$\sim k$. He had not remembered something.

It is not always easy to express the negation correctly in words. With (*i*), *possible alternatives for consideration are* '*his face did not light up*' *and* '*his face lit up, but not suddenly*'. *It is always possible, of course, to negate a statement by prefixing* '*It is not true that...*' *to it.*

If a is true, $\sim a$ is false.

If a is false, $\sim a$ is true.

These facts may be represented conveniently in a table, using T for true and F for false.

a	$\sim a$
T	F
F	T

Will you each write down a simple statement and then negate it, using also a letter of the alphabet to denote it and the sign $\sim$?

Conjunction

Now look for any sentences that, as a whole, are statements but which are made up of parts that are also statements.

(*b*) Tom woke up and groaned.

This can be expressed as

Tom woke up, and Tom groaned;

or as $m \mathbin{\&} n$

(*Notation is discussed at the end of the section, p.* 179), where m stands for the statement 'Tom woke up' and n for 'Tom groaned'. Are there any more like this? Yes:

(*d*) He dragged himself out of bed and sat on a chair,

which may be expressed as $p \mathbin{\&} q$,

where p is 'He dragged himself out of bed' and q is 'He sat on a chair'.

Also (*l*) 'Term does not start today, but tomorrow',

which may be expressed as $x \mathbin{\&} y$,

where x is 'Term does not start today', and y is 'Term starts tomorrow'. The 'but' is a conjunction just as 'and' is, but expressing a shade of contrast between the two statements.

Each of these compound statements b, d, l contains two subordinate statements, each of which could be true or false.

Let us consider $b = m \mathbin{\&} n$. If m is true and n is true, is b true? If m is true and n is false, is b true? How many combinations of truth and falsehood must we try? Collect all the results into a table (often called a truth-table):

m	n	$b = m \mathbin{\&} n$
T	T	T
T	F	F
F	T	F
F	F	F

Now make up some compound statements of this sort for yourself, and consider in what circumstances the statement as a whole is true.

Such a compound statement formed with 'and' is called a *conjunction* of the subordinate statements.

Alternation

Now let us look for any statements consisting of two subordinate statements connected by 'or'. We find

(*e*) He heard the hoot of a car or train,

which may be expressed as

$$r \text{ or } s,$$

where r is 'He heard the hoot of a car' and s is 'He heard the hoot of a train'.

Also (*g*) It was either from cold or dismay,

which is

$$t \text{ or } w,$$

where t is 'It was from cold' and w is 'It was from dismay'.

First considering $e = (r \text{ or } s)$, with r, s each being either true or false, let us try to decide when e will be true and when false.

If r is true and s false, e will be true.

If r is false and s true, e will be true.

If r is false and s false, e will be false.

If r is true and s true, ... ?

We may feel some uncertainty about e in this case. The most likely interpretation is that Tom did not hear the hoot of *both* a car and a train, so in this case e will be false. This interpretation takes 'r or s' to mean 'r or s but not both'.

With the statement g, which we have expressed as (t or w), the truth values in the first three cases are the same as for $e = (r \text{ or } s)$ above. The fourth case is then t and w are both true, that is, the reason for Tom's shivering was cold and also dismay. The most likely interpretation here is that the compound statement is true, that is, here we are taking '(t or w)' to mean 't or w or both'.

It is not always clear from the language or the context whether 'or' is meant to include or exclude the possibility of both constituents being true at the same time. 'Will you have tea or coffee?', while not a statement that can be true or false, indicates the use of 'or' that excludes the possibility of both simultaneously. On the other hand, 'He is too old or too weak for the army' uses 'or' to include the possibility of both simultaneously.

In legal or other documents where precision is needed some variation in wording is used to make it clear; for example, 'Will you have tea, *or else* coffee?' and 'He is too old *and/or* too weak for the army'. (We shall use '& or' instead of 'and/or'.)

With this terminology the truth-tables are

r	s	$e =$ (r or else s)
T	T	F
T	F	T
F	T	T
F	F	F

and

t	w	$g =$ (t & or w)
T	T	T
T	F	T
F	T	T
F	F	F

Can you make up compound statements containing 'or', some of which follow the first truth-table and some the second? It is not always easy to decide.

Open statements

Are there any sentences that we haven't dealt with yet? Yes:

(h) They were making such a noise.

It is not clear what the pronoun *they* refers to (raindrops? the traffic? people downstairs?). Until the meaning of *they* is made perfectly clear this is not a statement that can be said to be true or false. Such expressions are called *open statements* or *propositional functions.*

More complex statements

It is possible to construct more complicated statements. For example, if f and g are run into one sentence we have

f^*: Then he shivered, either from cold or dismay.

After discussion, write this as

$$f^* = f \mathbin{\&} g$$
$$= f \mathbin{\&} (t \mathbin{\&} \text{or } w).$$

In what circumstances will f^* be true? How many combinations of truth and falsehood are there for f, t and w? We obtain the truth-table:

f	t	w	t & or w	f & (t & or w)
T	T	T	T	T
T	T	F	T	T
T	F	T	T	T
T	F	F	F	F
F	T	T	T	F
F	T	F	T	F
F	F	T	T	F
F	F	F	F	F

Exercises

(1) Combine d and e into a statement containing four subordinate statements, and draw up its truth-table.

(2) Make up other compound statements and draw up their truth-tables.

Lesson on implication in algebra and trigonometry

Age: from 13.

(i) If $x = 10$ does $x^2 = 100$? Yes.

If $x^2 = 100$ does $x = 10$? It could be, but it depends on what kind of numbers we allow. If we allow negative numbers the answer is 'not necessarily', for x could be -10.

(ii) If $\sin y = 1$ does $\cos y = 0$? Yes.

If $\cos y = 0$ does $\sin y = 1$? It could be, but not necessarily, if y is not restricted, since $\cos 270° = 0$ but $\sin 270° = -1$.

(iii) If $3x+1 = 7$ does $x = 2$? Yes.

If $x = 2$ does $3x+1 = 7$? Yes.

(iv) If $x = 3$ does $x^2 = 4x-3$? Yes

If $x^2 = 4x-3$ does $x = 3$? Not necessarily, since x could equal 1.

We call this relationship *if...then* between two statements *implication*. The word means *wrapped up in*, for the meaning of the second statement is 'wrapped up in' the meaning of the first.

It is convenient to use the symbol $\Rightarrow$ to represent this, so if p and q stand for statements $p \Rightarrow q$ means *if p, then q* or *p implies q*.

The true relations among those just discussed can be written

$$(x = 10) \Rightarrow (x^2 = 100),$$
$$(\sin y = 1) \Rightarrow (\cos y = 0),$$
$$(3x+1 = 7) \Rightarrow (x = 2),$$
$$(x = 2) \Rightarrow (3x+1 = 7), \text{ etc.}$$

The direction of the arrow is important. Sometimes, however, we have $p \Rightarrow q$ and $q \Rightarrow p$ both true, and then we write it more compactly as

$$p \Leftrightarrow q.$$

From our earlier results, then, we can write

$$(3x+1 = 7) \Leftrightarrow (x = 2).$$

The sign $\Rightarrow$ plays much the same part as the 'therefore' sign $\therefore$: $p \Rightarrow q$ is much the same as p, $\therefore q$. The difference is that with $\Rightarrow$ we are not actually asserting 'p is true' but only saying 'if p is true, then q is true'. In the reverse direction compare $p \Leftarrow q$ with p *because* q.

Exercises

Check whether the following are true, and whether the arrow can truthfully be reversed:

$$(2y-7 = 5) \Rightarrow (y = 6),$$
$$(a = 6 \text{ and } b = 4) \Rightarrow (ab = 24),$$
$$(x = 2) \Rightarrow (x^2+3x = 10),$$
$$(x^2 = 4x+5) \Rightarrow (x = -1 \text{ or } x = 5),$$
$$(\tan u = 0) \Rightarrow (\sin u = 0),$$
$$(z = 45^\circ) \Rightarrow (\tan z = 1),$$
$$(\cos 2x = 1) \Rightarrow (\sin x = 0),$$
$$(\sin x = 0) \Rightarrow (\sin 3x = 0),$$
$$(ABCD \text{ is a rectangle}) \Rightarrow (ABCD \text{ has equal diagonals}),$$
$$(m \text{ and } n \text{ are even numbers}) \Rightarrow [(m+n) \text{ is an even number}].$$

Lesson on implication in the solution of equations

Age: from 14.

Solve the equation $x = \sqrt{x+6}$ in your own books. What do you get? 9 or 4? Try $x = 9$. Yes, it checks. Try $x = 4$; No. $4 = 2+6$

is not true. ($\sqrt{x}$ *always means the* positive *root.*) What has gone wrong? I expect you did something like this:

$$(x-6) = \sqrt{x}.$$

Square both sides: $\quad \therefore\ (x^2-12x+36) = x.$

$$\therefore\ x^2-13x+36 = 0.$$

$$\therefore\ (x-9)(x-4) = 0.$$

$$\therefore\ x = 9 \text{ or } x = 4.$$

Each statement here is perfectly correct, yet $x = 4$ has appeared and it is not a root of the original equation. Let us follow the implications through.

$$(x = \sqrt{x}+6) \Rightarrow (x-6 = \sqrt{x}).$$

Can we reverse the direction to $\Leftarrow$? Yes. So put $\Leftrightarrow$. Now square. $[(x-6) = \sqrt{x}] \Rightarrow [(x-6)^2 = (\sqrt{x})^2]$. Can we reverse this arrow? Compare $(x = 10) \Rightarrow (x^2 = 100)$—but the reverse implication does not hold. So our answer is No.

The next implications, from $(x-6)^2 = (\sqrt{x})^2$ to $(x^2-13x+36) = 0$ and then to $(x-9)(x-4) = 0$ and $(x = 9$ or $x = 4)$ are all reversible.

Now if $p \Rightarrow q$ and $q \Rightarrow r$ we can conclude that $p \Rightarrow r$. (*Discuss.*) Such a property is called *transitivity* of the relation of implication. (*See p. 201.*)

In solving this equation, then, we have found a chain of implications from $(x = \sqrt{x}+6)$ to $(x = 9$ or $x = 4)$. Can it be reversed? Every implication was reversible except the one when we squared the two sides of the equation. This one link is missing in the reverse implication, so we cannot assert

$$(x = 9 \text{ or } x = 4) \Rightarrow (x = \sqrt{x}+6).$$

If this were true it would follow that both 9 and 4 would satisfy the original, but without that link we are not sure. The reason is that a different equation, $x = -\sqrt{x}+6$, when squared would have given all the same implications as before, and finally

$$(x = -\sqrt{x}+6) \Rightarrow (x = 9 \text{ or } x = 4).$$

In fact, this is the equation that $x = 4$ satisfies.

The correct reversed implication would be

$$(x = 9 \text{ or } x = 4) \Rightarrow (x = \sqrt{x}+6) \text{ or } (x = -\sqrt{x}+6).$$

More precise results are

$$(x = \sqrt{x+6}) \Leftrightarrow (x = 9) \text{ and } (x = -\sqrt{x+6}) \Leftrightarrow (x = 4).$$

Further exercises

(1) Insert $\Rightarrow$, $\Leftarrow$ or $\Leftrightarrow$ as appropriate between the pairs of statements on each line:

$$x+\sqrt{(2x)} = 4 \qquad (x = 2) \text{ or } (x = 8).$$
$$x+\sqrt{(2x)} = 4 \qquad (x = 8)$$
$$x+\sqrt{(2x)} = 4 \qquad (x = 2).$$

(2) Relate all five of these statements by $\Rightarrow$, $\Leftarrow$ or $\Leftrightarrow$ as appropriate:

$8 \sin x + \cos x = 4$; $\quad 64 \sin^2 x = 16 - 8 \cos x + \cos^2 x$;
$65 \cos^2 x - 8 \cos x - 48 = 0$; $\quad (13 \cos x - 12)(5 \cos x + 4) = 0$;
$(\cos x = 12/13)$ or $(\cos x = -4/5)$.

(3) Many well-known paradoxes in mathematics, 'proofs' that $1 = 2$ and the like, depend upon incorrectly reversed implications. Study some of these and find the false steps in the proofs.

Lesson on logical consistency and proof by reductio ad absurdum

Age: *Sixth form.*

Have you heard of the barber in a certain village (a man) who regularly shaves all (and only those) men in the village who do not shave themselves? Does the barber shave himself or not?

The proposition p: 'the barber shaves himself' leads by the data to the conclusion that he is in the category of those whom the barber does not shave, i.e. $p \Rightarrow \sim p$. The conclusion would seem to be that $\sim p$ is true. Equally, however, the proposition $\sim p$ puts the barber in the category of those who do not shave themselves and therefore are shaved by the barber, i.e. $\sim p \Rightarrow p$, so it would seem that p is true. But p is either true or false, so which are we to believe?

To find the right answer we must examine the implication $p \Rightarrow q$ a bit more. If $p \Rightarrow q$ then if q is not true, p cannot be true. For example, if $(x = 10) \Rightarrow (x^2 = 100)$, (as it does), then if $x^2 \neq 100$, $x \neq 10$. *Use also examples from everyday life, or from geometry.* Further discussion leads to the fact that

$$p \Rightarrow q \quad \text{and} \quad \sim q \Rightarrow \sim p \tag{1}$$

are equivalent in their meaning. *Each is called the contrapositive of the other.*

Now, if $p \Rightarrow q$, either p is false or (if p is true), then q is true. So let us put

$$(p \Rightarrow q) = (\sim p \text{ or } q). \tag{2}$$

See the note on implication on p. 179.

We also need the relations

$$\sim (a \;\&\; b) = \;\sim a \text{ or } \sim b, \tag{3}$$

where 'or' allows the possibility of *both a* and *b* (see Lesson on Statements)

and
$$\sim (a \text{ or } b) = \;\sim a \;\&\; \sim b. \tag{4}$$

(*Discuss.*)

With these results, we can return to the 'barber' paradox.

The relations $p \Rightarrow \sim p$ and $\sim p \Rightarrow p$ do not express the whole truth, for these implications occur only within the framework of assumptions A of the problem. The true implications are therefore

$$(A \;\&\; p) \Rightarrow \sim p \tag{5}$$

and
$$(A \;\&\; \sim p) \Rightarrow p, \tag{6}$$

or, in terms of the definition (2), (6) takes the form

$$\sim (A \;\&\; \sim p) \text{ or } p$$

or
$$(\sim A \text{ or } \sim \sim p) \text{ or } p,$$

from (3), substituting A for a and $\sim p$ for b,

or
$$(\sim A \text{ or } p) \text{ or } p$$

or
$$(\sim A \text{ or } p). \tag{7}$$

Similarly (5) takes the form

$$(\sim A \text{ or } \sim p). \tag{8}$$

So our 'barber' arguments lead to the assertion that both ($\sim A$ or p) and ($\sim A$ or $\sim p$) are true. It can be seen from this (or obtained symbolically) that therefore $\sim A$ *must be true.*

This is the correct solution to the paradox, namely, that the original assumptions as a whole were untrue because they were inconsistent. In other words, there cannot be a barber fulfilling the conditions specified.

This result is highly important in any logical system. An inconsistent set of assumptions can lead to *any* statement whatever within the field considered.

Exercise

With the addition of $2+2 = 7$ to the ordinary laws of arithmetic, prove (i) that $100 = 8\frac{1}{2}$; (ii) that any natural number is less than 10.

The above method of argument is called 'reductio ad absurdum'. Examples can be given from all branches of mathematics. One of the best known is the proof that $\sqrt{2}$ cannot be expressed as the ratio of two integers a/b. If we assume that $\sqrt{2} = a/b$, where a, b are integers with no common factor, we can easily prove that a, b must have a common factor 2, and this is absurd. Therefore our assumption was wrong.

An interesting geometrical proof by this method is that if the segments of two angle bisectors of a triangle, cut off by the opposite sides, are equal in length, then these two angles are equal. This is the converse of a simple isosceles triangle property, but though innocent-looking it is notoriously difficult to prove.

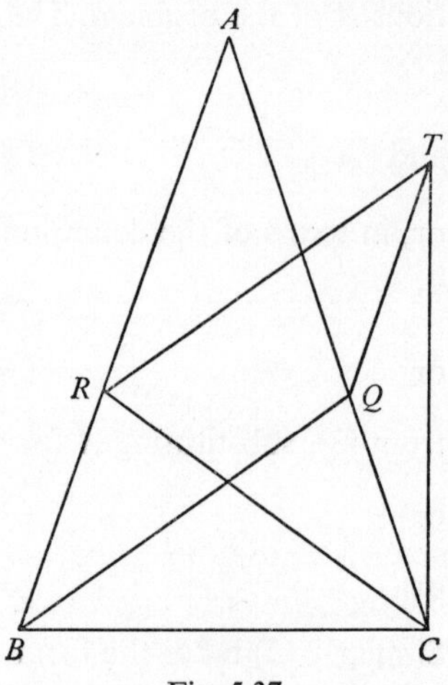

Fig. 5.37

In the figure, it is given that BQ bisects angle ABC and CR bisects angle ACB, and $BQ = CR$. It is required to prove that angles ABC, ACB are equal.

Complete the parallelogram $QBRT$, join CT. If angles ABC, ACB are not equal, assume (for definiteness) that $ABC > ACB$. Hence... $CQ > BR$, and hence, in triangle CQT,

$$CQ > QT,$$

so angle $CTQ > QCT$. It follows that angle $CTR > TCR$, but $CR = TR$, which gives a contradiction in triangle RCT.

Note on implication. The most important difficulty in this symbolic treatment is that 'implication' as used normally is not quite the same as '$\sim p$ or q'. A full discussion of this is essential for those interested in the translation of everyday ideas into a formalized system. The logic of electrical circuits (see p. 153) does conform to

$$(p \Rightarrow q) = (\sim p \text{ or } q),$$

but to apply this to the normal use of implication involves some distortion of meaning. Some books slur over this difficulty. A clear discussion is given in Kemeny *et al.* [71], chap. 1. Refer also to Exner and Rosskopf: *Logic in Elementary Mathematics*. For a good discussion of this and other paradoxes in non-symbolic terms see Quine [101].

Notation for the sentence calculus

The preceding lessons have been deliberately kept within the limitations of an informal approach and have introduced few new symbols. If the work is to be developed beyond this elementary stage, it is advisable to employ the conventional symbols for the conjunction and alternation of statements.

The conjunction of the statements p and q is written $p \wedge q$.

The two meanings of alternation are shown by $p \vee q$, which is the inclusive ('and/or') form, and $p \veebar q$, which is the exclusive ('or else') form.

It can be shown that the inclusive form of alternation is a more mathematically convenient partner for conjunction than exclusive alternation, and that a restriction to one kind does not limit the expression of logical connectives in symbolic terms. In fact, it is always possible to construct a compound statement having a prescribed truth-table, using only the connectives $\wedge$, $\vee$ and $\sim$.

For example, $p \veebar q$ can be written

$$(p \wedge \sim q) \vee (\sim p \wedge q).$$

(Further, the use of de Morgan's laws enables either of $\wedge$ and $\vee$ to be expressed in terms of the other and $\sim$.)

Reference

J. G. Kemeny *et al.* [70].

A NOTE ON AXIOMATICS

When we compare the algebra of sets that we have developed with the logic of statements and the behaviour of simple switching circuits we find that their algebraic formulations correspond.

Sets (A, B)	Statements (p, q)	Switches (X, Y)
A'	$\sim p$	X'
$A \cup B$	$p \vee q$	$X+Y$
$A \cap B$	$p \wedge q$	XY
$A \cup A = A$	$p \vee p = p$	$X+X = X$
$A \cap A = A$	$p \wedge p = p$	$XX = X$
$A \cup A' = E$	$p \vee \sim p$ logically true	$X+X' = 1$
$A \cap A' = \phi$	$p \wedge \sim p$ contradiction	$XX' = 0$

This identity of structure is known as *isomorphism*. The generic name for this particular structure is a *Boolean algebra* (and we can think of set algebra, the statement calculus and switching circuitry as particular cases of it). We believe that this isomorphism will reveal itself to the pupils after they have met the applications informally, and that, at the sixth-form stage, they will be in a position to study the algebra systematically.

We have already given, in the section on the basic formulae of set algebra (p. 134), a list of the main results which can be established. These can be derived, as indicated there, by informal arguments about sets or statements; but with some pupils we can go further and endeavour to develop this Boolean algebra from a set of initial postulates.

It should be clear from the lessons that appear in this book that we do not confuse what is often called 'modern' mathematics with mathematics treated axiomatically. For most purposes axiomatics are not important—most of the mathematics we use (even if we use a good deal of it) is connected only very distantly with a set of axioms—and for most children the appeal of axiomatics is slight. But we should fail in our duty to some of our pupils if we did not let them become aware of the peculiar compulsion exerted by mathematics, when it becomes self-conscious, to start the search for the *minimum* assumptions that are needed to launch it. Some pupils will enjoy the strict deduction of a sequence of results, played as a kind of game, and a few may see more deeply and realize that it is more than a game.

The attempt to base a part (or the whole) of mathematics on a set of axioms from which everything else in that part can be derived has always appeared historically at the end of a long period in which the key theorems and processes were known and extensively used. The desire to axiomatize seems to come when an effort is made to co-ordinate and codify what is known already, to tidy up the loose ends and to discard what is unimportant or of dubious value. The accounts are squared and the achievement passed on with a clear conscience.

Perhaps the only motivation that can be used in teaching some part of mathematics from the beginning as an axiomatic system is that of game-playing. Pedagogical experience about this is slight and it seems likely that this motivation will rapidly peter out, long before any considerable development from the axioms has taken place. Games are soon tired of if they are too complex or too easily mastered.

But if we are not concerned to *teach* from axioms we can still find a motivation, a kind of intellectual pride, in the desire to tidy up. And if this motivation does fail nothing much has been lost, since the material which can be deduced is already common property.

However, there are circumstances in which there is a strong case for returning to material which is already known and reconsidering it from an axiomatic point of view. A very pragmatic case can be made for the axiomatic approach in the case of systems like Boolean algebra, where the pupil has already seen that the abstract mathematical structure has many, entirely different, material counterparts. The case is then as follows. We have seen that precisely the same formulae describe such apparently unconnected things as card indexes, electrical circuits, logical puzzles, areas in a plane, to name only some, so if we find what seems to be a new application of the algebra do we need to verify every one of the formulae, or can we merely verify a few basic ones and rest safe in the knowledge that the others are bound to follow? This is a genuine question because such further applications of Boolean algebra do exist. Sometimes there is not, as yet, any technical literature utilizing them; this may develop in the future—a case in point being the production of trick effects in colour cinematography.

This makes a practical case for the understanding of the axioms of a subject, but it is considerably more difficult to apply this argument to geometry because unless the pupil has some appreciation of non-Euclidean geometries he tends to feel that there can only be one geometry anyway, and the case for axiomatizing it then appears to be entirely theoretical. When the axiomatic method is first encountered it is most desirable that it should be in some part of mathematics which is obviously polyvalent (that is, corresponding to many clearly distinct, concrete situations) and which can be axiomatized without too much difficulty. On both these counts Euclidean geometry is a bad place at which to begin; but if the case for a stage C treatment of some branch of mathematics is accepted then Boolean algebra comes high on the list. It can be axiomatized without too much difficulty, and moderately capable sixth-formers can help to select the axioms and develop the most important theorems for themselves.

As a preliminary to this work, and assuming that all the main equations of Boolean algebra are already well known and have been proved or demonstrated informally, the pupils should have their attention drawn to the fact that some of these equations can be established by starting with some of the others. For example, to prove $AB+AB' = A$, one proof might run:

$$\begin{aligned} AB+AB' &= A(B+B') \\ &= A.1 \\ &= A. \end{aligned}$$

What has been assumed? That . is distributive over + (line 1); that the sum of an element and its complement is 1 (line 2); and that an element is unchanged by multiplication by 1 (line 3).

The pupils now choose other formulae and derive them from whichever of their store of equations they like to use. As a result of this experience some equations will seem more powerful, or more easily manipulated, than the others and will be selected for the axiom set. With some help, the pupils can now make a complete choice of a set of axioms and try to derive many of the other results from them.

It is not likely to be advisable to discuss the *consistency* of the axioms with most pupils. The idea of the *independence* of the axioms will, though, almost certainly arise. But there is no need, at least in the first place, to make superhuman efforts to ensure that the axiom set does not contain any redundancies. There are, after all, other criteria than economy for a set and in the early stages of work of this sort these will override the search for an absolute minimum.

A good set of axioms is: In a set E provided with two binary operations . and +; for all $A, B, C \in E$,

$$\left.\begin{array}{l}(1)\ A+(B+C) = (A+B)+C.\\(2)\ A(BC) = (AB)C.\end{array}\right\}\text{ associativity}$$

$$\left.\begin{array}{l}(3)\ A+B = B+A.\\(4)\ AB = BA.\end{array}\right\}\text{ commutativity}$$

$$\left.\begin{array}{l}(5)\ A+BC = (A+B)(A+C).\\(6)\ A(B+C) = AB+AC.\end{array}\right\}\text{ distributivity}$$

There exist elements $0, 1 \in E$ such that for all $A \in E$

$$\left.\begin{array}{l}(7)\ A+0 = A.\\(8)\ A.1 = A.\end{array}\right\}\text{ identity elements}$$

For each $A \in E$ there exists an element $A' \in E$ such that

$$\left.\begin{array}{l}(9)\ A+A' = 1.\\(10)\ A.A' = 0.\end{array}\right\}\text{ complements}$$

(The symbol $\in$ indicates that A, B and C are elements of the set E.)

This set is easy to work with and reasonably easy (partly because of the obvious duality) to remember. It is not, however, an independent set as the first pair (associativity) can be deduced from the others. If this pair is removed, the independence of the others can be demonstrated by constructing operation tables which obey seven of the axioms but not the eighth (see Stoll [115], p. 170).

Shorter sets can be used, but they are not of such simplicity and will not lead to such easy development as that suggested. One example of an alternative set which will tax the very ablest pupils is:

(1) $A+B = B+A$.
(2) $A+(B+C) = (A+B)+C$.
(3) $(A'+B')'+(A'+B)' = A$.
(4) $A+A = A$.

References [115, 130].

6

RELATIONS AND GRAPHS

1. RELATIONS

Binary relations

Elementary mathematics makes frequent use of phrases such as

is a factor of	is less than	is a multiple of
is congruent to	is the square of	is greater than or equal to
is parallel to	is the supplement of	is perpendicular to,

and so on.

Each of these phrases needs a first and last word for completion, and the sentence then says that a certain relation exists between the two things named. The phrases impose a certain order on the pair of things inserted, so that to say '24 is a multiple of 6' is quite different from saying '6 is a multiple of 24'. It is also clear that the things selected for insertion in the sentences have to be of a particular kind if the statement is to be meaningful. It will not do to say '3 is parallel to 7' or '*ABC* is the square of *XYZ*'.

The relations we have mentioned are *binary*—that is, they relate two things. *It is possible to have relations connecting more than two things, but binary relations are of so much greater importance that they will be the only variety dealt with here.* The relations are directional, so that any pair of related things is *ordered*, and the relations are unambiguous in the sense that, given two elements a and b drawn from their respective sets, we can immediately say whether a relates in this specified way to b or not. Now, this same notion of relation is intimately linked with experience and is available to us through the medium of ordinary language. Phrases like

went for a holiday to	has the same colour socks as
is in love with	is in...House
was born in	is on the same main road as
is in front of	had...to drink yesterday,

and so on, define binary relations in the same way as the more obviously mathematical phrases already quoted.

Lesson A

This is suitable for any age from 11 upwards; it needs no prior mathematical experience.

Can anyone tell me the meaning of the word 'beverage'? Give me some examples, please. Any more suggestions? I will write these in a list on the board (*at the right*).

Did any of you drink any of this beverage yesterday? Or of this one? Let us do this in a little more detail; suppose we investigate your 'drinking habits'! I think that the whole class is too big for this experiment, so we will just take all of you in these two rows.

Now, David, will you come out to the board and write down your name somewhere about here (*on the left*). Did *you* have any of these beverages to drink yesterday? Tea and lemonade? Stay where you are and point to them in the list. Before you sit down again, can you draw something on the board to show which two in the list you pointed at?

A chalk line from the child's name to each drink is a natural expectation. The teacher can easily suggest this if the child does not think of it; or, of course, he can follow up any other suggestion that is made.

Will the next person in the row come and show us what he had to drink yesterday? Yes, put your name near David's. What did Trevor drink yesterday? Now the next.

As more children draw lines on the board it will become increasingly difficult for them to draw lines which do not cross some already there. Confusion of lines should be avoided as far as possible.

Now let us look at the diagram we have made. Can anyone tell us anything that he sees from the drawing? The beverage that most children drank; the beverages that nobody drank; the children who drank the largest number of different beverages; the children who drank least, etc.

If the teacher draws a closed curve or a box around the names of the children and the names of the drinks, the diagram will show two sets—children on the left, drinks on the right—with a number of lines linking elements in the two sets. Similar examples can now be constructed by choosing two sets and some criterion—which we shall call a relation—*that pairs up some of the elements.*

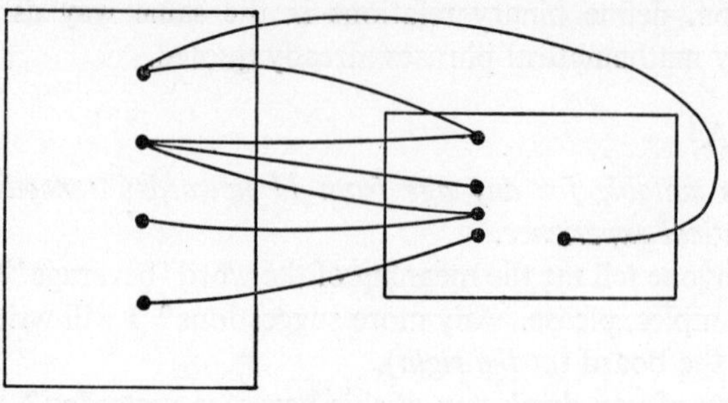
Fig. 6.1

Examples

(1) Draw the relation 'belongs to' between a set of children in the class and the set of School Houses.

(2) Draw the relation 'is disliked by' between the set of subjects in the curriculum and a set of pupils.

(3) Draw the relation 'was a pupil at' between a subset of the class and the set of neighbouring Junior Schools.

(4) Draw the relation 'is a multiple of' between the set {2, 15, 17, 8, 50} and the set {10, 4, 1, 5, 20, 3}.

The diagrams that are produced give a picture of the way in which elements of the second set are linked to those in the first. It will probably be noticed that examples (1) and (3) above give situations in which no element of the first set is linked to more than one element in the second. If John (see example (1)) belongs to Red House, he cannot at the same time belong to Green House or Blue House. In the diagram, this is seen as one line (or none) leaving each member of the set of children. Relations of this type are called *functions*. These assume considerable mathematical importance later, but need not be stressed when they are first encountered.

It should be realized by the children that each of the relations is represented by lines going in a particular direction. This can be brought out by putting arrow-heads on the lines (*and this device follows naturally from the pointing suggested in the lesson*). Then a reversal of the arrow-heads implies a change in the nature of the relation. For instance, changing the directions of all the arrows in the diagram for example (4) changes the relation from 'is a multiple of' to 'is a divisor of'. This latter relation is the *reciprocal* or *inverse* of the first. It can easily be seen that a reciprocal relation does not necessarily share any of the properties of the original relation. In particular, the reciprocal of a function is not necessarily another function. The relation 'is the son of' in a set of males is a function, but the reciprocal relation 'is the father of' is not (why not?).

The kind of diagram that we have so far used is not the only device we can use for illustrating a relation. Any method of showing how the relation pairs off an element of the first set with a member of the second will serve. For example, suppose X is the set $\{a, b, c, d, e, f\}$ and Y is the set $\{p, q, r, s\}$ and that the relation R yields the pairs $\{(a, p), (b, p), (d, q), (d, r), (e, r), (f, r), (f, s)\}$; the situation can be represented in the following ways:

(i)

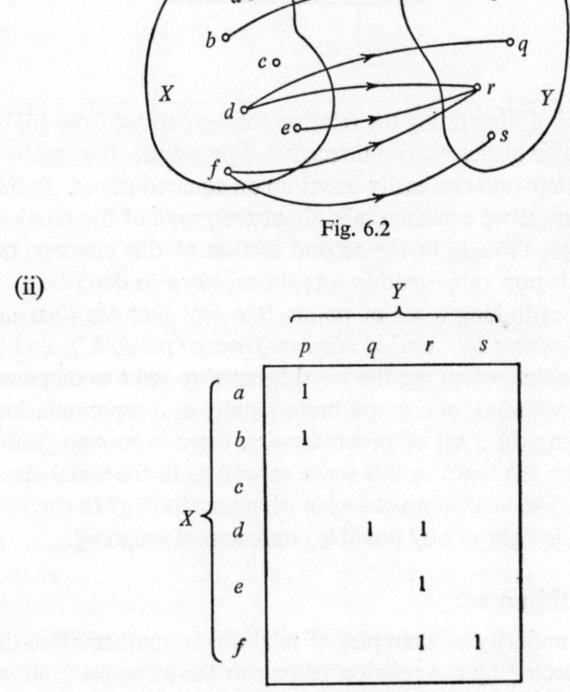

Fig. 6.2

(ii)

X \ Y	p	q	r	s
a	1			
b	1			
c				
d		1	1	
e			1	
f			1	1

Fig. 6.3

In the second representation, we enter the table by the row indicated by the first member of the pair, and the column indicated by the second. *This kind of table can later be shown as a matrix, by omitting the labels of the rows and columns and by filling the blank entries with 0's. It is interesting to investigate the connection between forming the 'product' of two relations and performing matrix multiplication.*

(iii)

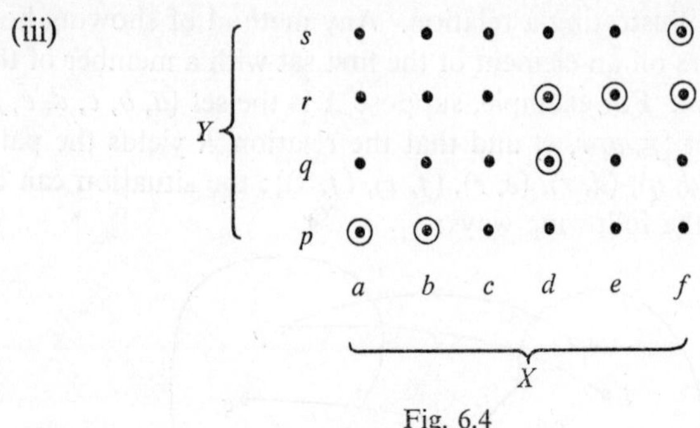

Fig. 6.4

This method of illustrating the relation can be derived from (ii) by rotating the table anticlockwise through a right angle. It suggests a coordinate system and can easily be set up on squared paper. In line with normal usage we can then talk about the *graph* of the relation. As we shall see, though, in the second section of this chapter, the word 'graph' is now often used in a technical sense to describe a set of *directed* lines linking a set of points (*the fact that the lines are directed distinguishes this kind of diagram from a 'network'*), and in this section we shall often use the word 'graph' to refer to diagrams of type (i). If we think of a graph more loosely as a representation, through drawing, of a set of ordered pairs, there is enough justification for using the word in this sense as well as in the usual classroom context. The teacher must decide what terminology to use with his pupils in the light of any possible confusion of meaning.

Relations within a set

In the vast majority of examples of relations in mathematics, the elements connected by the relation belong to the same set (and we say that we have a relation *in* the set, say, X). It is in this context that properties like symmetry and transitivity begin to appear (see p. 201), but many examples drawn from non-mathematical sources can also produce the properties that are later to be mathematically significant, and the process of refining the definitions of these properties, and expressing them in mathematical terms, can be an end-product rather than a starting-point.

Lesson B

Based on a demonstration by Georges Papy in London, Easter, 1961.

Do you all know each others' surnames and Christian names? I want you to point to any of the other boys and girls whose surname begins with the same letter as the initial letter of your Christian name.

How can we show this on the board? We need a plan of the class. Would somebody come out and do that as simply as possible? Would you now show me where you are on the plan? Is he right? Anne, would you come out and show me where you are on this plan? Is she right?

There is now a series of dots on the board indicating the positions of the children.

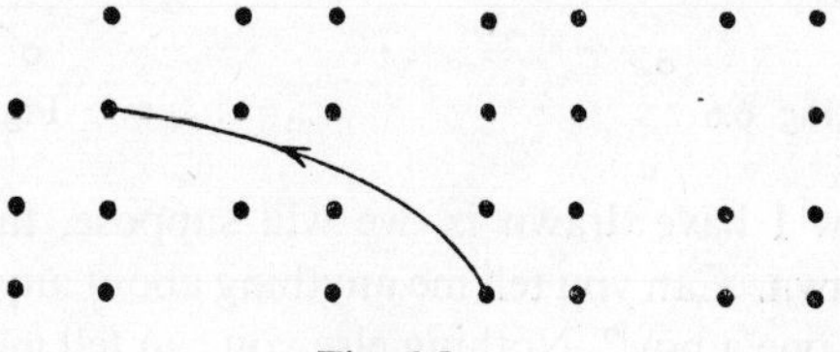

Fig. 6.5

Now will anyone who was pointing come and show this on the board? Are those the right positions? How are we going to know that you were pointing *at* him? We must be clear about the direction in which we are pointing.

An arrow is a good idea. *The arrow is drawn in.*

Now will someone else come out and show who he was pointing at? Is that right? How do you know? Now someone else, please. Can any of you point to more than one other person? If you can, then put in *all* the arrows that are needed. All right; come and show us. We have several lines on the drawing now and we do not want to get them mixed up. So be careful where you draw your line. Is that arrow going the right way? Come and put it right.

This situation can be developed as far as seems necessary. We may have an interesting discussion if it occurs to someone to point to himself. If the class does not provide a suitable example the teacher can fabricate a class member, such as Charlie Chaplin or Brigitte Bardot, who will serve the purpose. How is this kind of pointing shown on the graph? An obvious way is to put a little ringlet joining the point to itself. Does a ringlet need an arrow-head? There may be other important ideas, or germs of ideas, such as two people pointing at each other. Why does this happen? Suppose George Grant and Grace Garvie join the class, how shall the lines be drawn?

Now I want you to imagine that this (*a random set of points on the board*) is a group of children in the playground. They are, as you see, higgledy-piggledy, not like the neat rows in the classroom! Suppose one of them points to another child and calls out: 'He is my brother!' We could show it as Fig. 6.6.

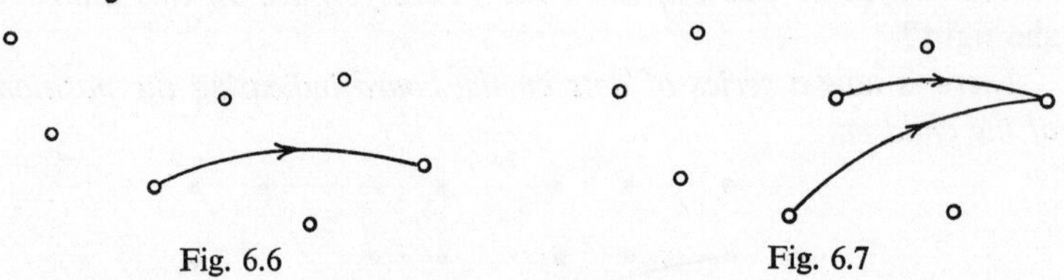

Fig. 6.6 Fig. 6.7

The arrow I have drawn is, we will suppose, the only one that could be drawn. Can you tell me anything about any of the children? Why is that one a boy? Nothing else you can tell me? The other is a girl, is it? How do you know?

Now let us imagine that we can draw Fig. 6.7 instead, and that these are the *only* two arrows that we can put in. What can we say now?

A boy and two girls? They're sisters, are they? Well, what about this one?

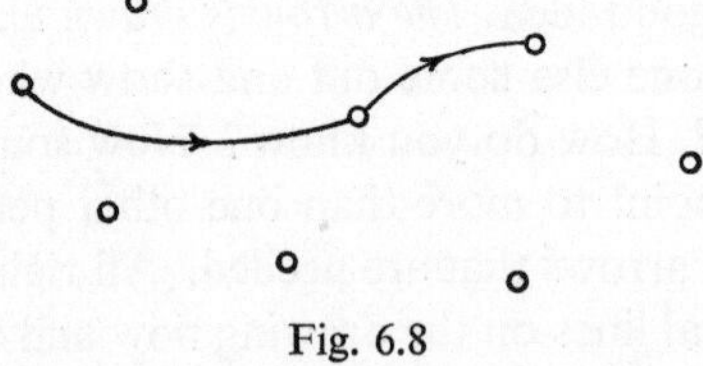

Fig. 6.8

It needs another arrow does it? Why is that? Oh? *Two* arrows? Any more? Four arrows? Are you sure? *There is some disagreement and a decision will have to be made to put in only the certainties and not the possibilities.*

The children can now be asked to make up graphs for themselves based on this relationship, and they can pass them to their neighbours for deductions to be made, missing arrows to be inserted, and so on.

In this situation, can we have cases in which a boy points to himself? That is, have we got any ringlets in the graph? Can two children point to each other? Does this always happen? If child A points to child B, and child B points to child C, does A always have to point to C? Or C point to A?

Examples

(1) *The relation 'is the father of' in a set of people.*

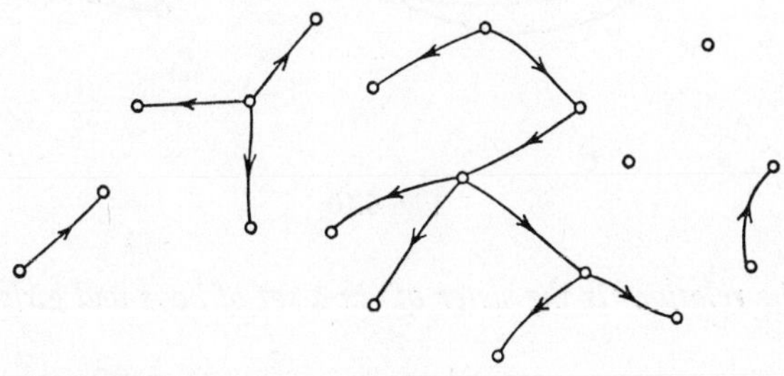

Fig. 6.9

In such a graph we see

(i) At most, one arrow goes to each point.

(ii) If a pair belongs to the relation, the reverse pair does not.

(iii) There is no ringlet (self-connection) in this graph!

(iv) If an arrow goes from a to b and another from b to c, there will not be one from a to c.

Justify the above remarks. Make some observations about the graph. Can you point out those who must be males? Those who are brothers or sisters? Can you pick out any grandfathers?

(2) *The relation 'is the sister of' in a set of girls.*

We see in Fig. 6.10 that

(i) The graph has no ringlets.

(ii) If the graph shows a pair, the reverse pair is also shown.

(iii) If an arrow goes from a to b and one from b to c (a, b, c being three distinct points of the graph), there is sure to be one from a to c.

Explain the significance of these observations.

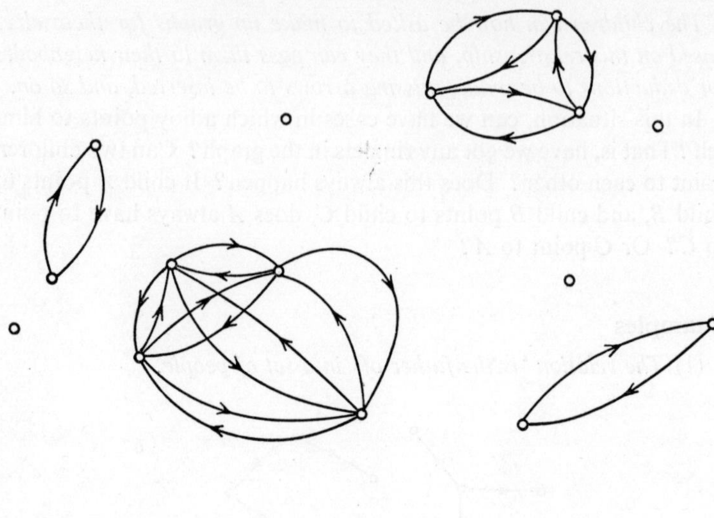

Fig. 6.10

(3) *The relation 'is the sister of' in a set of boys and girls.*

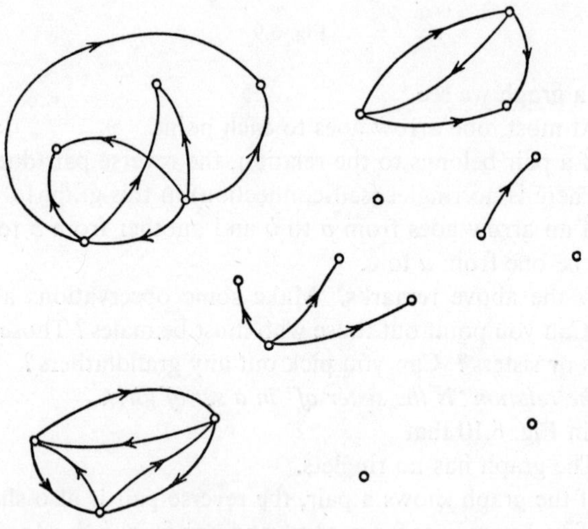

Fig. 6.11

Why aren't there any ringlets?

Both (a, b) and (b, a) can belong to this relation; in other cases (a, b) belongs but (b, a) does not. When does the first case occur? And the second? You can tell the sex of each member of the set with an arrow to it or from it. How? Point out some other things about the graph. Can it be complete as it is?

(4) Draw the graph of the relation 'is the brother of' in a set of boys.

(5) *The relation 'is the brother of' and 'is the sister of' in a set of girls and boys.*

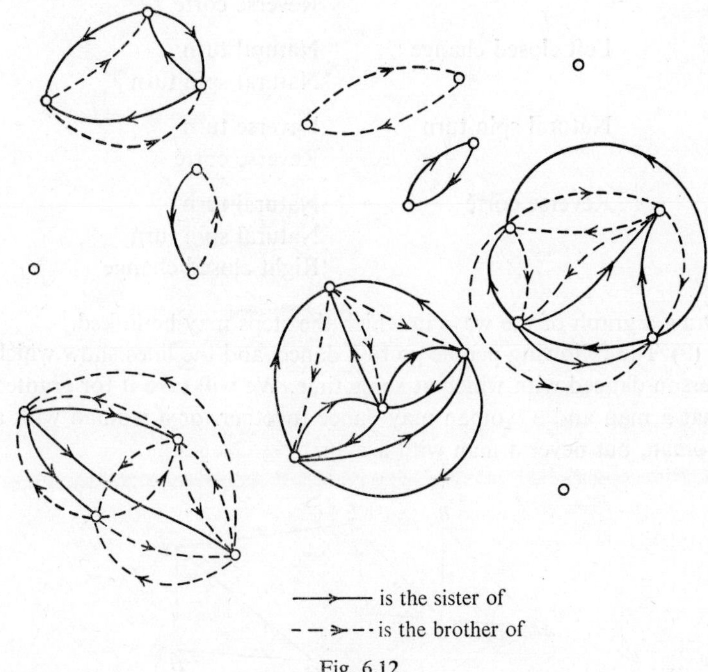

Fig. 6.12

Why are there no ringlets in the graph?

If an arrow, dashed or solid, goes from a to b, then there is one, dashed or solid, from b to a. Why? All the arrows from the same point are of the same kind. Why? Make some more observations.

(6) For points in the plane study the relation 'is symmetrical to' with respect to a fixed point in the plane, using a graph.

(7) For lines in the plane study the relation 'rotation through an angle of 60°' in one direction round a point, using a graph.

(8) Silvester[112] gives the following information about the waltz:

Step	*May be followed by*
Natural turn	Right closed change
Right closed change	Reverse turn Reverse corté
Reverse turn	Left closed change Reverse corté
Left closed change	Natural turn Natural spin turn
Natural spin turn	Reverse turn Reverse corté
Reverse corté	Natural turn Natural spin turn Right closed change

Draw a graph of the ways in which the steps may be linked.

(9) The following people go to a dance, and the lines show which person danced with which at some time. We will take it for granted that a man and a woman may dance together, or a woman with a woman, but never a man with a man.

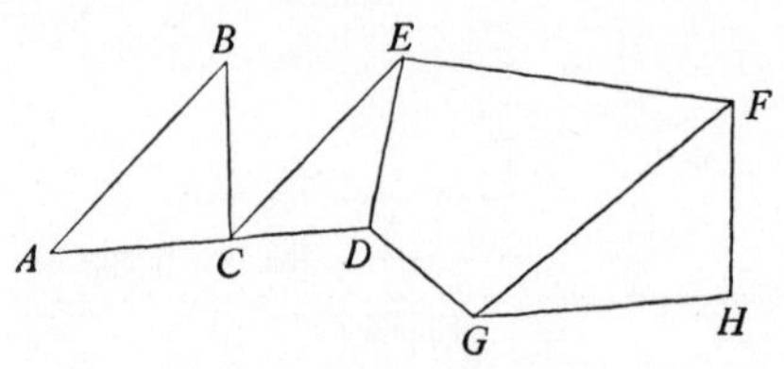

Fig. 6.13

If C is a man, can we say anything about any of the other dancers? Answer the same question if (i) D is a woman; (ii) F, H are husband and wife; (iii) C, F are brothers. What is the maximum number of men at the party?

(10) A group of men are surveying an unexplored region. They

climb a mountain and build a cairn at the top, and from this they take sights with a theodolite on cairns they have already built.

$X \to Y$ if they sight the cairn on Y from mountain X.

Did they go up any mountains twice? Which mountain do you think they climbed first?

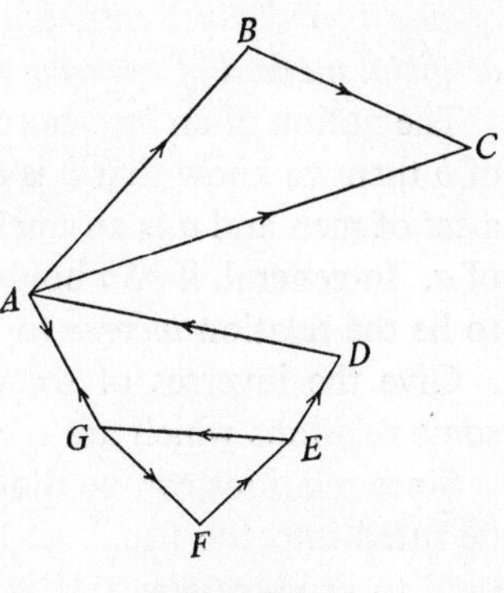

Fig. 6.14

Relations as sets of ordered pairs

In general mathematical terms, if we have two sets X and Y, and if we have some criterion, say R, which enables us, given any element $a \in X$, to find elements in Y associated with a, we say that R is a relation *from X into Y*. If b is such an element of Y, we write aRb. We see that (a, b) is an ordered pair of elements and that R defines a set of these pairs.

If we pair off *each* element of X with *each* element of Y we obtain what is called the *Cartesian product* of the two sets. The relation R, in general, defines a selection of ordered pairs drawn from the Cartesian product—that is, a subset of all possible ordered pairs from the two sets. In the example suggested in Lesson A, the Cartesian product of the set of children and the set of beverages can be written:

(John, tea), (John, coffee), (John, lemonade), ...
(Mary, tea), (Mary, coffee), (Mary, lemonade), ...
(Trevor, tea), ..., ...
etc., etc.

But the relation 'drank..yesterday' will only contain some of these and will exclude such pairs as (Mary, whisky), and so on.

Just as, when we talk about membership of a particular set, we may either define the membership by some *criterion*, or we may *list* the members, we may specify a relation by a criterion or by listing all its ordered pairs. At a more advanced stage, there may be some advantage in defining the word 'relation' to mean a set of ordered pairs, but in the elementary stages it is certainly more natural to think of a rule of formation as being the relation. The distinction between these two views is not in any case very great.

When the relation is *in* a set X, it will obviously lead to a subset of all the ordered pairs formed of a first and second element chosen from X. *A mathematician can say that* all *relations can be thought of as relations* in *a set, because it is always possible to form the union set of any two distinct sets, but the teacher will prefer not to use this artificial method of ensuring uniformity.*

The notion of an *inverse* relation is very important. If a is a factor of b then we know that b is a multiple of a. If we are talking about a set of men and a is an uncle of b then we know that b is a nephew of a. In general, if bSa holds if and only if aRb holds, then S is said to be the relation *inverse* to R.

Give the inverses of some commonly occurring relations. Give some relations which are their own inverses.

Since relations can be thought of as sets of ordered pairs they can be fitted into the usual set terminology, and set operations can be used to produce new relations. For example, it may be useful at a later stage to think of parallelism as the union of the relations 'is identical with' and 'is disjoint from' defined for the lines in a plane; each line being regarded as a set of points. This may conflict a little with the ordinary usage of the word parallel but there are advantages in making parallelism an equivalence relation (see below). As another example, the word 'sibling' is used to denote the union of the relations 'brother' and 'sister'.

In plane geometry similar figures which have the same area are congruent, so 'is congruent to' can be regarded as the intersection of the relations 'is similar to' and 'has the same area as'.

We say that one relation is included in another when all ordered pairs satisfying the first relation also satisfy the second. So it can be said, for example, that 'is congruent to' is included in 'is similar to', or that 'is the father of' is included in 'is a parent of'.

The *negative* of a relation can be defined by writing $aR'b$ whenever it is *not* the case that aRb. Reverting to the star polygons discussed in section 3.10 we may ask, 'For which pairs of numbers of the set $\{2, 3, 4, 5, 6\}$ is there a connected polygon $\begin{Bmatrix} m \\ n \end{Bmatrix}$?' There is such a polygon whenever $m > n$, and when m and n have no factors in common. Colour-fig. 5 shows this as the intersection of one relation with the negative of another.

Composition of relations

Lesson C

If we draw a graph showing the relationships P (is the father of) and M (is the mother of) in a set of people, are there any other relations that can be picked out? We can (in the most general case) find the relations 'is the grandparent of' and 'is the grandchild of' because these two relations are 'compositions' of P and M. We find that the relation 'is the grandparent of' is made up of four included relations, corresponding to the four categories of maternal grandfather, maternal grandmother, paternal grandfather and paternal grandmother. If we draw in the graph an example of the relation 'is the maternal grandfather of' we see that it is only another way of saying the relation 'is the father of the mother of'. A suitable notation for this relationship would be $P*M$.

Using this convention, what would be meant by the relation $M*P$?

If we want to symbolize the relation 'is the maternal grandmother of' (i.e. 'is the mother of the mother of'), we will write $M*M$. And by analogy with a common simplification of notation, this is usually written M^2.

Exercises

(1) Construct a graph showing the relations P and M and at least some of the composite relations: $P*M$, $M*P$, M^2 and P^2.

(2) Construct a graph showing the relations S (is the son of) and D (is the daughter of). Graph all the composite relations formed with S and D.

(3) Show that the relation G (is the grandfather of) is given by the expressions $G = P^2 \cup P*M$, and $G = P*(P \cup M)$.

(4) Discuss the 'squares' of the relations: 'is the brother of' in a set of boys; or reflection in a line for points of a plane; and of a cyclic permutation of a set of letters.

(5) In the Swedish language words denoting relations are used in an especially regular way. Thus 'fader' and 'moder' mean 'father' and 'mother', but in compound words they are shortened to 'far' and 'mor'. In England there are only two words for grandparents, 'grandfather' and 'grandmother', but in Sweden there are four—

'farmor', 'farfar', 'morfar', 'mormor'. The first one means father's mother. How do you explain the other three?

'Bror' means 'brother', and there are two words for uncle. One is 'farbror'; what is the other? The word for 'son' is the same as in English. Can you give *one* of the words for 'nephew'? *The word for sister is omitted because it alters a little in compound words.* If 'dotter' means 'daughter', guess the four words for 'grandchild'.

Fill in the blanks in the following sentences:

'If Ulrika is the farmor of Nils, then Nils is the ..... of Ulrika.'

'If Karl is the brorson of Olaf then Olaf is the ..... of Karl.'

(6) Fill in the blanks in the following sentences:

'If Mr Jenkins is the parent of a pupil of Miss Smith, then Miss Smith is the ..... of a ..... of Mr Jenkins.'

'If Mr Jones is the brother of John's father, then John is the ..... of Mr Jones.'

What is the rule for finding the inverse of a composite relation?

(7)

There were two sisters had two sons,
And these two sons were brothers,
And they were sisters to the sons
And mothers to the brothers.

Draw a graph to explain this. (See Genesis xix 30–8.)

(8) Devise some examples on the relationships 'brother-in-law', 'son-in-law' and 'mother-in-law' showing them as the compositions of two relationships.

In a similar way devise some examples on the relationships step-father, step-daughter, etc.

(9) The Table of Affinity in the Prayer Book lists the people a man may not marry. Using the idea of the composition of relations express the people in this list in terms of as few relations as possible.

(10) Give the relations which are the inverses of 'is a parent of', 'is the husband of', 'is the employer of', 'is the doctor of', 'is a pupil of' and 'is a customer of'.

(11) In Colour-fig. 6 the red lines denote the relation 'a is a brother of b' and the blue lines denote the relation 'c is a parent of d'. Draw in lines indicating 'x is an uncle of y' where it is certain that one may do so. Are there any more cases which are in doubt?

The following examples are more difficult and will tax mature pupils.

It should not be assumed that these questions necessarily have clear-cut answers.

(12) In all of the following questions you are given the relationship of x to y, and of y to z, and you have to decide, if possible, the relationship of x to z. It is not always certain that the letters x and z denote different people; indicate whenever this may be so. x, y and z are all male.

(i) x is the brother of y, and y is the cousin of z.

(ii) x is the cousin of y, and y is the cousin of z.

(iii) x and y have the same grandfather, and y is the cousin of z.

(iv) x is the father of the brother of y, and y is the father of the cousin of z.

(13) Discuss whether or not the logical signs of implication and equivalence are used correctly in the following sentences:

(i) (x is the father of y, and y is the cousin of z) $\Rightarrow$ (x is the brother of the father of z).

(ii) (x is the uncle of y) $\Leftrightarrow$ (y is the child of a sibling of x).

(iii) (x is the grandchild of y) $\Leftrightarrow$ (the brother of y is the uncle of a parent of x).

(iv) (x is the grandfather of a cousin of y) $\Rightarrow$ (x is the brother of a grandfather of y).

(v) (x is a grandson of the same grandfather as the father of y) $\Leftrightarrow$ (x is the father of the great-grandson of the same great-grandfather as y).

(vi) (x is the father of the cousin of y, and y is the father of z) $\Rightarrow$ (x is the brother of the grandfather of z).

(vii) (x is the cousin of the grandfather of y, and y is the uncle of z) $\Rightarrow$ (x is the cousin of the great-grandfather of z).

Order relations

Lesson D

(i) In the set of numbers {1, 2, 3, 5, 10, 15, 120} consider the relation 'is a divisor of'. Illustrate this with a graph (Fig. 6.15).

Can we notice any consistencies in the behaviour of the arrows? Why does every number have a ringlet? Do we ever have arrows between two numbers in both directions? Why not? If we follow a chain of two arrows is there always an arrow from the first to the last number?

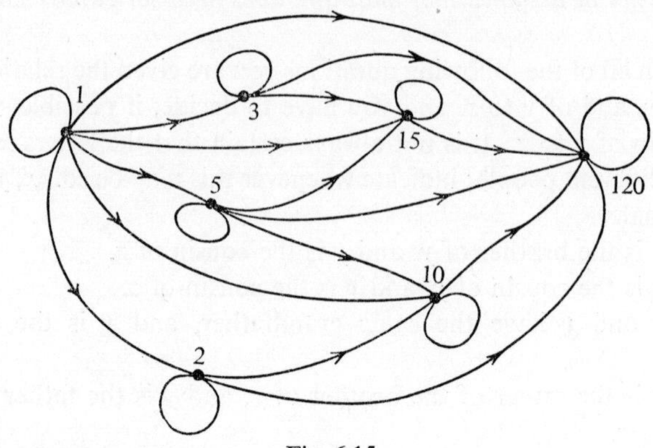

Fig. 6.15

(ii) In the set of numbers {1, 2, 3, 4, 5, 6, 7, 8} consider the relation 'is less than'. Draw a graph. Compare its properties with those of (i).

(iii) For each of the above examples set up a matrix of 0's and 1's to show the ordered pairs. Are there any regularities that become apparent? Are these related in any way to the properties of the graphs?

As a result of this experience, fortified if necessary with other examples, certain key properties begin to emerge. The most immediately apparent is probably the universal occurrence of ringlets in the first example, contrasted with their complete absence in the second. This observation is reinforced by the appearance of a diagonal of 1's in the first matrix and a diagonal of zeros in the second. The ringlets and the 1's occur because each number in the set is a divisor of itself. Any relation in which each element is related to itself is said to be *reflexive.*

The relation 'is less than' is not reflexive. In fact we can say more as it is evident that no element can ever be less than itself. If we have to specify this property, we shall call it irreflexitivity. *It is quite possible for a relation to be neither reflexive nor irreflexive—but this is not very interesting.*

A second property which is present in examples (i) and (ii) is that a chain of two ordered pairs implies another ordered pair—the end-points of the chain. If (a, b) is a pair belonging to the relation, and

(b, c) is another, then (a, c) will also be a member of the relation. Both relations are said to be *transitive* for this reason. Since this property is common to both of the examples it may not seem significant. Perhaps all relations are transitive; in which case it is not a distinguishing property. But it is easy to think of many relations that are not transitive, and this quality assumes an importance for classifying purposes. For instance, if we consider the set of all straight lines in the plane and apply the relation of perpendicularity, it is immediately obvious that this is not transitive. If AB is perpendicular to CD and CD is perpendicular to EF, then AB is not perpendicular to EF (but is parallel to it). This relation is intransitive.

This relation of perpendicularity of lines in a plane suggests another property. If AB is perpendicular to CD then CD is perpendicular to AB—and this is true for any lines in the plane. We say that perpendicularity is a *symmetric* relation. Is this property present in the original examples? What would be the distinctive form of the matrix of a symmetric relation? Symmetric relations are their own inverses.

Exercise

Classify the following relations, saying whether they have reflexive, transitive or symmetric properties:

'is the brother of' in a set of boys;

'is the brother of' in a set of children;

'is parallel to' in the set of straight lines in the plane (see p. 196);

'is the square of' in the set of real numbers;

'has a birthday in the same month as the birthday of' in a set of people;

'is greater than or equal to' in the set of even integers;

'is included in' for the subsets of a set E;

'is similar to' in the set of triangles in the plane.

The relations 'is a divisor of' and 'is less than' are examples of *order* relations. A relation which is reflexive, symmetric and transitive is an *equivalence* relation. These notions are of considerable importance in mathematics but a fuller development does not seem to belong to this book. Almost any modern book of algebra will give a full account of these (and of functional relations).

The following examples will perhaps show how ideas of ordering are present, in a primitive form, in many elementary puzzles.

Examples

(1) There are 10 airports on an air route. They are *G*, *L*, *R*, *C*, *M*, *A*, *K*, *B*, *P*, *S*, but not in that order. A certain airline always uses this route for three of its planes to fly from *G* to *S* and back. The planes either pass over or call at each of the airports, but each plane calls at different places. The first plane goes from *G* to *S* calling at *M*, *P*, *C* in that order and on its return journey it calls at *A*, *P*, *B*, *M* in that order. The second plane flies from *G* calling at *B*, *L*, *P*, *K* in that order and calls at *A*, *C*, *R*, *B* when returning. The third plane calls at *L*, *R*, *K*, *C* in that order on its outward journey and at *K*, *R*, *P*, *M* on the return journey to *G*. Write down the airports in their proper order on the air route.

(2) Peter, Mary, Robert, Susan and Tom are five children who are each given a shilling for each year of their age. Tom, Mary and Peter together get less than Robert, Tom and Peter. Mary, Robert and Tom together get more than Tom, Peter and Robert. Tom and Susan together get more than any other pair. Susan is the youngest. Write them in correct order of age beginning with the eldest.

(3) *A* had invited seven girls to her party. They were *B*, *C*, *D*, *E*, *F*, *G* and *H*. They were to be seated at a round table and *A* was to sit in the place indicated (Fig. 6.16).

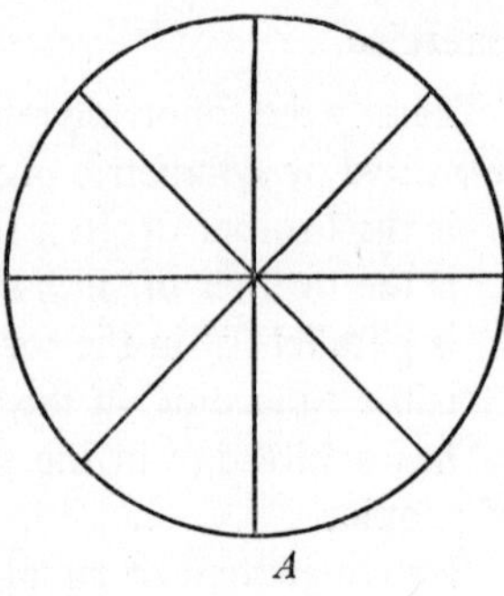

Fig. 6.16

D wanted to sit between *F* and *H*. *C* wanted to sit next to *B* but not next to *G* or *E*. *E* wanted to sit next to *A* and opposite to *F*. *H* wanted to sit next to *A* on her left. Can you put these seven girls in their places so that they will all be satisfied?

(4) If you are at the Regal and you want to go to the King's Head the instructions are: 'Second left, third right, first left and straight on past three cross-roads.' Can you give the instructions for going to the Regal from the King's Head? Will there be any doubt about the instructions?

(5) You wish to go from *A* to *B*, and you do not wish to take an unnecessarily long route. One possible set of instructions is

S L R S S L S. How do you interpret this? Can you give some alternative routes? Is S L S L R R S S a possible route? Can you devise a set of rules which will decide whether or not a string of letters is a possible route, rules which will apply without needing to refer back to the graph?

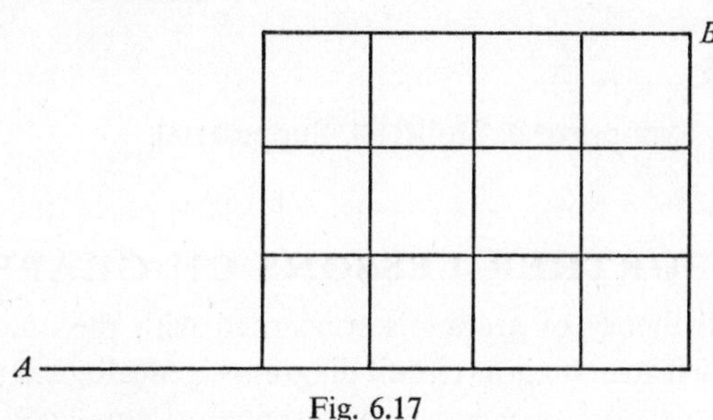

Fig. 6.17

Concluding remark

In textbooks on modern mathematics it is usual to meet a short section dealing with relations between elements of sets. This section is often hurried through so that, with a glance at order and equivalence relations, the particular properties of a functional relation can be developed without delay. It is also customary (as has been done here) to put the work on relations after a fairly full treatment of set language and set operations. This order can easily obscure the fact that the notion of a relation is a very primitive concept well within the range of quite young children. Much of the preparatory work in this section can be discussed with children of eleven (or less) and is quite independent of any kind of formalized set language. It is also, in fact, more directly rooted in the child's everyday experience than operating with sets.

We are suggesting that there is not the slightest need (in the elementary stages) to accept the logical order—sets: relations—as the inevitable pedagogical order and we envisage, as it were, an interleaving of the work on both topics. Within this section itself there is, too, a degree of arbitrariness in the order in which points are taken up, and the teacher should feel free to begin with a lesson or an exercise from any part of the section and see where it leads.

No attempt has been made to formalize the work on relations as extensively as suggested in some elementary undergraduate texts and it is unlikely that such an effort should be made at all at school level.

We have merely tried to show that this work is available in an intuitive form at an early stage; is rich in interest and applicability; and has links with many of the other topics treated in this book.

[Examples B 1–7 and C 1–4 are translations of examples from Papy, published here with the kind permission of the author.]

References

Papy [91], Synopses [90], Stoll [115], Suppes [116].

2. FURTHER LESSONS ON GRAPHS

The general theory of graphs is concerned with the common properties of such structures as circuit diagrams, genealogical trees, communication networks, sociograms, topological simplices, chains of command and the sequence of successive positions in a game. A branch of combinatorial topology, it has an entertaining jargon (e.g. 'every arborescence is a tree') and is an important source of ideas in the behavioural sciences and in operational research.

The following lesson sequences are informal extracts. The theory they imply is summarized at the end; its technical terms could be used in more formal lessons. Of the sequences, the first two are elementary and presuppose no previous knowledge. The second pair are more advanced in that they assume a slight familiarity with the operations of Boolean addition and multiplication. Where appropriate the teacher could derive any identities which are needed by referring back to any previous contexts in which these operations have been studied. The same figure is used throughout.

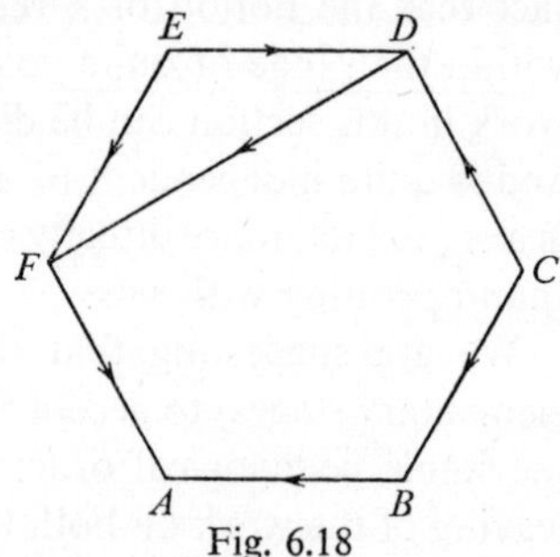

Fig. 6.18

Lessons

(1) You probably all know those diagrams with which people explain the chain of command in the army or in an industrial organization. These are usually like this. We could put arrows in to show the direction in which the commands flow. The arrows show who gives orders to whom. What can you say about the roles played by

the members of this organization? Yes, Charles and Eric both give orders yet do not receive them; perhaps they are on the board of directors. Yes, Alec receives orders only—perhaps he is the office boy. Bernard is a sort of private secretary—he passes on orders from Charles. I think you get the idea. Now come out and draw another figure on the board like mine; let's interpret this one as an organization; what do you notice?

(2) Suppose you wanted to know exactly what goes on in this organization (Fig. 6.18)—perhaps you are a rival firm or an enemy government. What do you do? One way is to infiltrate the system with your agents; where would you place an agent? Will E know what B and C are getting up to? Yes, F hears quite a lot. C and F? Yes, they will not know what B is getting up to with A; you would like your agents at A, B, C and F? Does anybody agree? It does look as if A, C and F would be enough; will someone check? Will it matter what E says to D? We can say that everyone in the organization communicates directly with one of our agents. Are there any other positions we can choose? A, C and E? Yes, it seems as if we must have three key positions at least. But we do not want to use more agents than we need—it looks as if three is enough.

(3) Here is a simple diagram showing trade movements between six small countries (Fig. 6.18). How many would you need to occupy to make sure you received the exports of the rest? Yes, the whole lot would do; yes, we do not actually need to occupy E. A, C and F would be quite enough; yes, there are others. What we want to do now is to find a foolproof and automatic way of deciding the smallest number of countries which need to be captured, for this and other similar figures. Let's start with the earlier suggestion and take the lot—this is a *coup d'état*; we occupy *A and B and C... and F.* Who can suggest a way to symbolize this particular situation? I made it fairly obvious...good...we use the operation of logical multiplication and write $A.B.C.D.E.F.$ Now, think of the last of these, that is F. If it is not captured then A to which it exports must be. We can say that the countries we capture must include *F or A*. Similarly in the case of the factor E. If this is not occupied then *D or F* is; in other words we must take *E or D or F*; and of course you can tell me how to symbolize that; yes, you can use logical addition and write $E+D+F$. This expression will replace E in the original product that described the *coup d'état*; thus in the logical product

$A.B.C.D.E.F$ each factor is replaced by a logical sum; what replaces D? A? Good—the expression now reads

$$A.(B+A).(C+B+D).(D+F).(E+D+F).(F+A).$$

How do I work that one out? Remember that this is not an algebraic expression of the usual sort; well there are some short cuts; let's see what can be done. Consider the first pair in the product $A.(B+A)$, this is a product. Can it be simplified? Where have you met expressions like this before? With punched cards and switching circuits? Then what does $A.(B+A)$ reduce to? It is equivalent to A alone, is it? Since $A.(B+A)$ is A we can leave out the second factor; in fact any factor that includes an A is absorbed by the first A; so the product becomes? $A.(C+B).(E+D+F)$. Now what? Yes, but—thanks for reminding us; we must be sure that the bracket rules hold—well do they? Why? Now evaluate the product; what do you get? What is the smallest combination? Six of them? Well how many in each? We need only capture three. Good.

(4) In any political situation you tend to get alliances. Let us suppose that the markets in this situation are guaranteed and that a country has no need to ally with one it exports to. What alliances are possible here (Fig. 6.18)? Yes, C and E would be an alliance. Well, who would care to define an alliance in abstract terms? A set of vertices, that has no two vertices directly linked. Examples? Yes, we've had C and E; B and F; A, C and E. Is A one? Yes, A alone. Why? Back to the definition. What is the largest alliance? Another? Any more than three? It seems not; but again we shall want a foolproof and automatic way of determining this largest number. That is your problem; see if you can solve it.

Notes

A collection of objects that may be interrelated in some way may be represented by a graph. Formally, a graph is a set X with an associated mapping $f: X \rightarrow X$. An element $x \in X$ is a vertex of the graph. With $y \in fx$ the ordered pair (x, y) is an arc. An arc may be represented by a line with an arrow-head joining two points. A graph is a set of arcs as in the illustrations earlier in this chapter.

An *internally stable* set of a graph is a subset in which no two vertices are directly connected by an arc. In a map where no pair of adjacent countries is coloured the same, any set of countries coloured

red is an internally stable set. Formally, any set $S \subset X$ such that $fS \cap S = \phi$ is internally stable.

An *externally stable* set of a graph is a subset of X which maps every element not in the subset. The radar stations of an early warning system need to be sited at positions which form an externally stable set. Formally, any set $T \subset X$ such that $fT' \cap T = \phi$, where $T' = X - T$ is externally stable.

The lesson sequences consider the problem of determining externally stable sets which contain the minimum possible number of elements. An algorithm was obtained using Boolean algebra, in particular the absorption law. The final problem is concerned with maximum internally stable sets. If a logical product is formed by taking at each vertex either the vertex itself or all the vertices adjacent to it then the complement of this product set includes the required sets.

A classical example is Gauss's problem of determining the number of ways in which eight queens may be placed on a chess-board so that no queen can take any other. This requires the number of maximum internally stable sets of a (symmetric) graph with 64 vertices, each pair of vertices being joined by an arc if the squares they represent lie in the same row, column or diagonal. A similar problem is to find minimum externally stable sets of this same graph. These determine the least number of queens which can be put on a board to control all of the squares.

References

The material for the lesson sequences is suggested by chapter 4 of Berge [13] where there are other examples. The well-known books on mathematical recreations may also be consulted. Further references to linear programming, game theory and network flow are given in another section.

7

LINEAR PROGRAMMING

Will someone describe the game of Twenty Questions to me? Yes, you narrow it down. What is it that you are narrowing down? Possible answers—until? Good; now I am going to think of a number and you have twenty questions to find out what it is. Yes, any number, and that's one question. No, it's not a million. You are wasting your questions. Yes, it's less than a hundred; yes, it's less than ten—that's four questions. No, it is not 4. No, it is not 7. No, it's not 1. May I advise you not to make wild guesses? Talk over what you want to ask me for a few moments. Remember that I am probably trying to make it hard for you. No, it is not a whole number. No, it is not positive—that was a good question. Yes, it is bigger than -10; yes, it's bigger than -100; you were trapped there. Yes, it is bigger than -5; that is twelve questions, by the way; yes, it's bigger than -4. No, it's not bigger than -3; no, it's not $-3\frac{1}{2}$. Yes, it is minus a number between 3 and 4. No, it's not $-3{\cdot}1$; no, it's not $-3{\cdot}2$. There are millions of numbers between 3 and 4 aren't there? You have only two questions left, so choose the right one, incidentally it's quite a famous number. No, the number I was thinking of is not plus π. Yes! but you only just made it. Let's play again, only this time don't start guessing too early.

At the beginning of this game we had many choices but after a question and answer the number of choices was reduced. We hoped to reduce to one possibility before twenty attempts.

In the lesson that follows, which has been given to 12-year-olds, we see that once again we work from many possibilities to a smaller number, and sometimes to one, by imposing restrictions. This is the essence of the topic—*linear programming*.

Lesson 1

Here are the contents of two boxes of chalk. What can you tell me about them? One is green and the other red; there is more red than green; there is more green than red. Can the last two facts *both* be true? Yes, but we must say what particular property of the green and red chalk we are thinking about. The weight of chalk, number of pieces or the total length of chalk in each box.

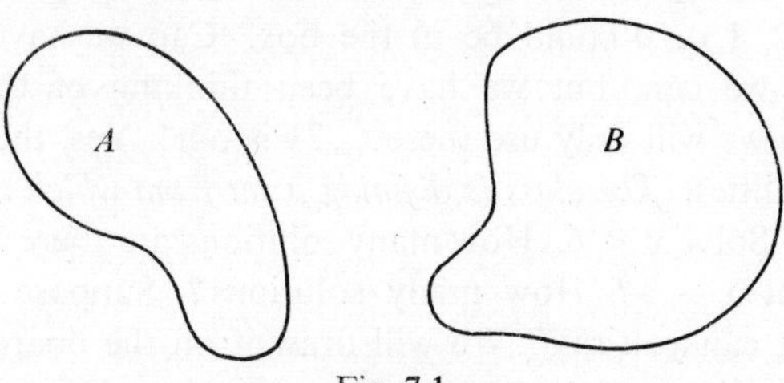

Fig. 7.1

Now consider these two pieces of card (Fig. 7.1). What can you tell me? *B* is bigger than *A*. *A* is smaller than *B*. Turn over the shapes. Now what? There are more squares than circles.

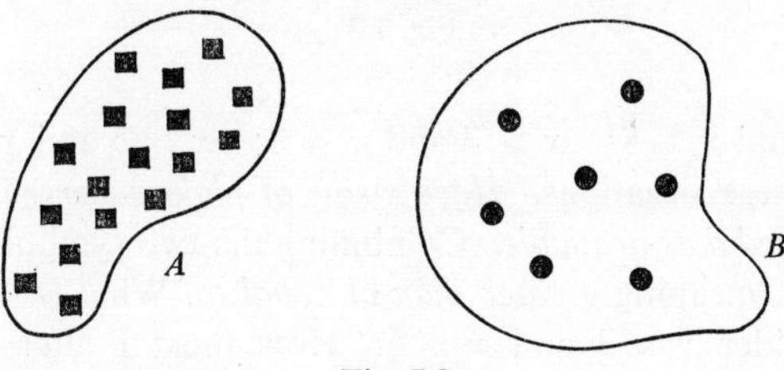

Fig. 7.2

With similar examples establish the notion of *more* and *less* (greater, smaller, bigger) stressing the *objects* and the *property* being considered.

Compare the money two pupils have with them—the number of coins—the total value. The question of equality will, no doubt, be raised. It can be considered neither *more* nor *less*.

We can introduce the symbol $<$. For $6 < 10$ we say 'six is less than ten'—an *inequality*. It also tells us that 10 is more than 6. We can read inequalities *both* ways.

Write on the blackboard statements such as:

$$10 < 8; \quad 3 < 4; \quad 4 > 2; \quad 5 < 5; \quad 5 > 5; \quad \text{and} \quad 5 = 5.$$

Discuss these by asking questions such as—'What does this say? Is it true? Is there another way of saying this?'

On the blackboard draw a box, followed by '< 4'. If the box contains a number, what number might be in the box? 2, 3, 1, ... yes, 4? Think about this. Is $4 < 4$? Any others? 0? Good. We will replace what is in the box by x. Now we have $x < 4$, and this tells us that 3, 2, 1 or 0 could be in the box. Can we have $2\frac{1}{2}$ for x? Sometimes we can, but we have been thinking of the counting numbers so we will only use these...$2\frac{1}{2}$ is out! Yes, there are only four possibilities. *The class is defining a set from which the solutions must come.* Solve $x < 6$. How many solutions are there for $x < 57$? What about $x > 3$? How many solutions? Suppose $x < 6$ *and* $x > 3$ what can you say? We will draw it on the board. *Build up Fig. 7.3.*

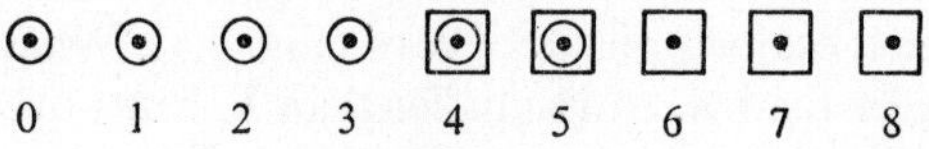

0 1 2 3 4 5 6 7 8

$x < 6$ and $x > 3$

Fig. 7.3

Solve

$$x < 10 \text{ and } x > 8; \quad x > 7 \text{ and } x < 8; \quad x > 5 \text{ and } x < 3.$$

Illustrate these situations. *Have strips of plain paper with a line of equally spaced dots available.* Combining the two symbols $>$ and $=$ we have $\geqslant$, meaning *greater than or equal to*. What is $\leqslant$ a symbol for? Consider $x \geqslant 3$ and $x \leqslant 6$. How must I alter the earlier figure so that it illustrates these new relations? (Fig. 7.4).

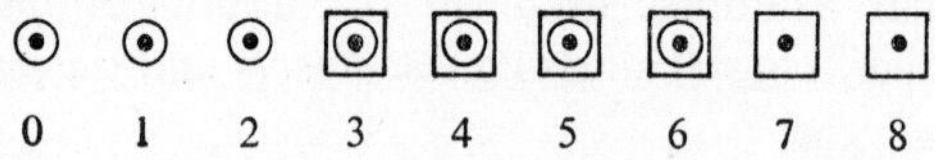

0 1 2 3 4 5 6 7 8

$x \leqslant 6$ and $x \geqslant 3$

Fig. 7.4

Solve and illustrate:

$$x \leqslant 9 \text{ and } x \geqslant 5; \quad x \geqslant 6 \text{ and } x \leqslant 7;$$

$$x \leqslant 4 \text{ and } x \geqslant 7; \quad x \geqslant 8 \text{ and } x \leqslant 8.$$

What are the greatest and least values of the solutions for each of the above inequalities?

Examples

(1) Joyce is a pupil in this school (secondary) and she can travel on the railway at half fare. How old is she? Is there something more you want to know? Yes, you can travel at half fare until you are 14 years old. 12 or 13? Yes, there are two possible answers.

(2) A farmer had a flock of 28 sheep. In the spring, 25 of the sheep each had either one or two lambs. What size is the farmer's flock now?

This first lesson involves the use of one variable. The next uses two variables but is still restricted to the natural numbers.

Lesson 2

A square point-lattice is drawn on the blackboard and the pupils have a similar lattice on paper or nail-boards.

I have 5 red counters and 5 green counters. I will hide them and mix them up, then put 5 in this box. Now, what have I got in the box? 5 counters. Yes, but what colours? Perhaps 3 red and 2 green, or 5 red. How many different possibilities are there? Consider this point lattice—a square point-lattice. *Colour-fig. 8 is built up.*

We think of a point as representing a set of red and green counters.

This point represents 2 red and 2 green. We can write (2, 2).

This point represents 3 red and 0 green. We can write (3, 0).

What point represents 1 red and 3 green?

What does this point represent?

The class mark in the possible contents of the box on the lattice.

Suppose I say I am putting not more than 5 counters in the box. How many points can we mark on the diagram this time?
The class draw their own.
In the first case $r \leqslant 5$ and $g \leqslant 5$ with $r+g = 5$; and secondly, $r \leqslant 5$ and $g \leqslant 5$ with $r+g \leqslant 5$. What can we say about the points? They lie on a line or they form a triangular shape.

Take similar examples varying the numbers of red and green counters and also the number of counters selected. Establish the fact that an equality gives points on a line and an inequality points in an area.

Exercises

(1) If $x > 3$, $y > 1$ and $x+y \leqslant 7$, find all the possible values of x and y.

(2) If $x \geqslant 0$, $y \geqslant 2$ and $x+y \leqslant 6$, how many solutions are there?

(3) (See Fig. 7.5.) Area C is given by $x \geqslant 0$, $x \leqslant 2$, $y \geqslant 1$, and $x+y \leqslant 6$. Give the set of inequalities that define the other areas. These areas are called *domains*.

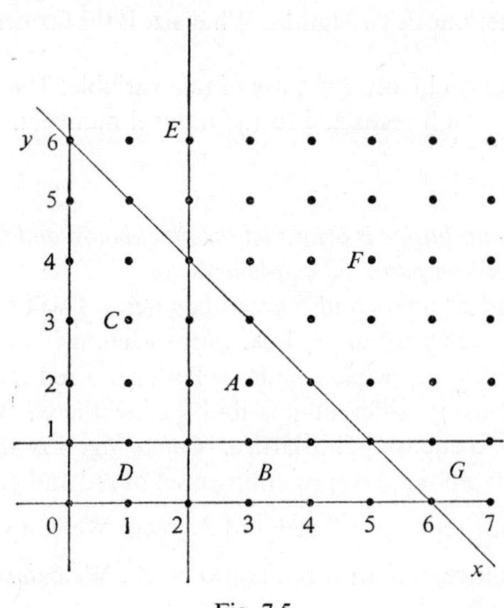

Fig. 7.5

Example

Each year a shop sells up to 60 boxes of large bars of chocolate. A box contains 2 dozen bars. At least twice as many bars of milk chocolate (profit 5*d.* a bar) as bars of plain chocolate (profit 6*d.* a bar) are sold. Find the maximum possible profit and the number of boxes of each kind of chocolate that must be sold for this profit.

The possible quantities of chocolates are restricted. Let one axis, the x-axis, show the number of boxes of milk chocolate sold and the other, the y-axis, show the boxes of plain chocolate sold.

What are the restrictions? We must sell not more than 60 boxes altogether. x and y must be positive numbers. We must sell at least twice as many bars of milk as plain. Now let us find the domain in which the solution lies. It is defined by the inequalities $x+y \leqslant 60$, $x \geqslant 0$, $y \geqslant 0$ and $x \geqslant 2y$.

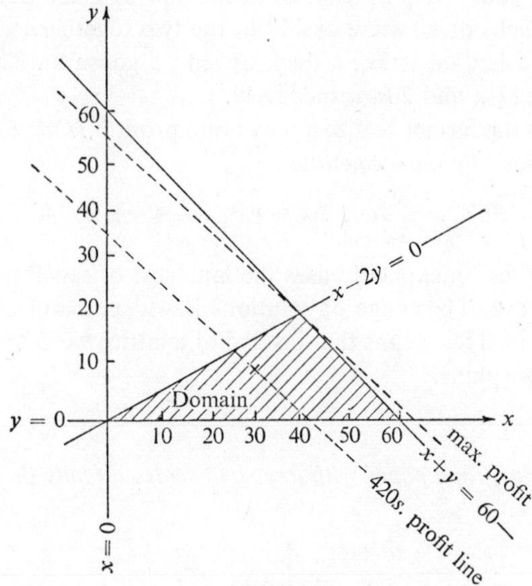

Fig. 7.6

State a pair of values for x and y which satisfy all the restrictions. $x = 30$ and $y = 10$. Yes, that point is in the domain. What is the profit in this case? Find other values for x and y that give the same profit. Good, they all lie on a straight line. We shall call this a profit line; the 420*s*. profit line.

Put other profit lines on the diagram. All the lines are parallel and have equations of the form $10x+12y = P$, where P is the profit in shillings.

The maximum profit line will be the one furthest from the origin and having at least one point in common with the domain. It is the 640*s*. profit line. The equation is $10x+12y = 640$, with the point $x = 40$ and $y = 20$ in common with the domain.

Interpreting this, the maximum profit possible is 640*s*. or £32 and is made by selling 40 boxes of milk chocolate and 20 boxes of plain.

The best or *optimum* quantities of each variety have here been obtained by a process called *linear programming*.

Exercise

A poultry farm carries a stock of 600 birds—geese, ducks and hens. The farmer must keep at least 20 ducks and 20 geese but not more than 100 ducks or 80 geese or 140 of the two together.

Rearing a hen costs 3*s*., a duck 6*s*. and a goose 8*s*. They can be sold at 8*s*., 13*s*., and 20*s*. respectively.

How can the farmer realize a maximum profit? *Hint*: *Consider the domain defined by the inequalities*

$$20 \leqslant x \leqslant 100, \quad 20 \leqslant y \leqslant 80, \quad x+y \leqslant 140.$$

The next lesson explicitly uses the language of sets implied in the previous work. The range of solutions is widened and includes the real numbers. This means that instead of a lattice we consider *all* the points of the plane.

Lesson 3

The class has graph paper with axes and scales already drawn.

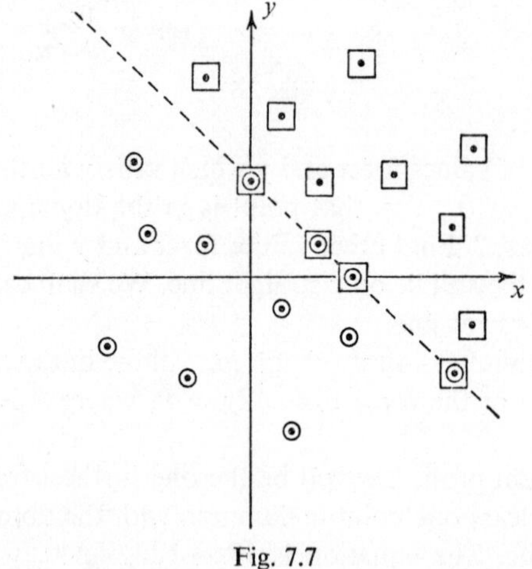

Fig. 7.7

Mark at least ten points anywhere on your graph paper. Note their coordinates. Put a small circle round a point if the sum of its coordinates is less than or equal to 1, and a small square if it is greater than or equal to 1. What do you notice? Mark ten more points. Now what do you notice? Yes, the circles are together and the squares are together. What about points that have a circle and a square? Yes, they lie on a line. Can you tell me anything about it? Good, the coordinates of all points on it add up to 1. Draw in the line. The circles and squares are on different sides of the line. How can we describe the points with circles? They are points (x, y) such that...? They are points that satisfy the inequality...? Good. Can you solve the inequality $x+y \geqslant 1$? Solve means 'find a solution'; solve $x-1 = 0$; solve $x+y = 1$. Find a solution, I said. Solve $x+y = 1$. Is that the only solution? What is the set of *all possible* solutions of $x+y \geqslant 1$? Precisely; in fact we call this part of the coordinate plane the solution set of the inequality. What is the solution of the inequality $x+y \leqslant 1$? What is the intersection of the two sets we have defined? The union? Now come out to the board and shade with red chalk the solution set for $x \geqslant 0$ and with blue chalk the solution set for $y \geqslant 0$. What colour is the intersection of these sets? The complement of their union? Now shade with yellow chalk the solution set for $x+y \leqslant 1$.

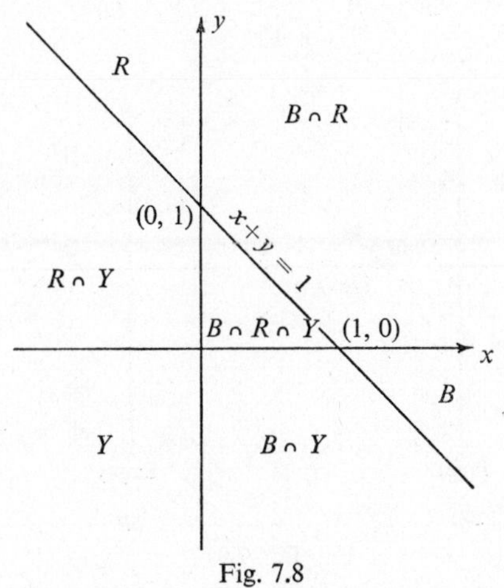

Fig. 7.8

What inequalities define the solution set that is coloured green? Coloured orange? Purple? What colour is the intersection of all three original sets? How many inequalities are satisfied within this intersection? What shape is it? (See also Colour-fig. 7.)

Indicate on your graph paper the solution sets for the following linear equations and inequalities:

(*a*) $x \leqslant 1$, $y \leqslant 1$, $x+y \geqslant 0$, and $y-x \leqslant 1$.
(*b*) $x \geqslant 0$, $y \geqslant 0$, and $x+y = 1$ (note the equality).
(*c*) $x+y = 3$ and $x-y = 2$.
(*d*) $x+y \leqslant 1$, $x \geqslant 0$, $y \geqslant 0$, and $x+2y = 3$.

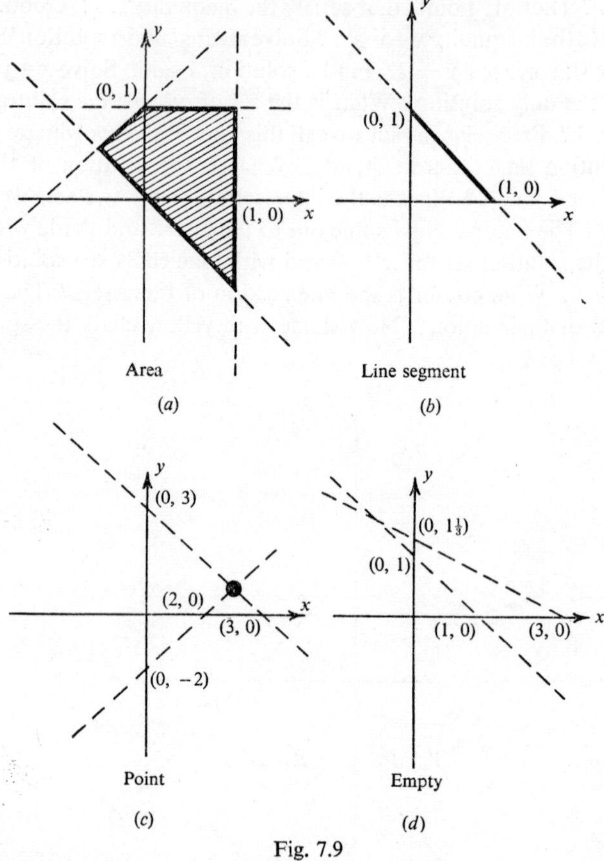

Fig. 7.9

How are the solution sets of (*a*), (*b*) and (*d*) altered by replacing $\leqslant$ by $<$ and $\geqslant$ by $>$?

The inequalities considered in the last lesson are all linear and so the solution sets are half-planes. Solution sets for non-linear inequalities (Fig. 7.10) should also be studied.

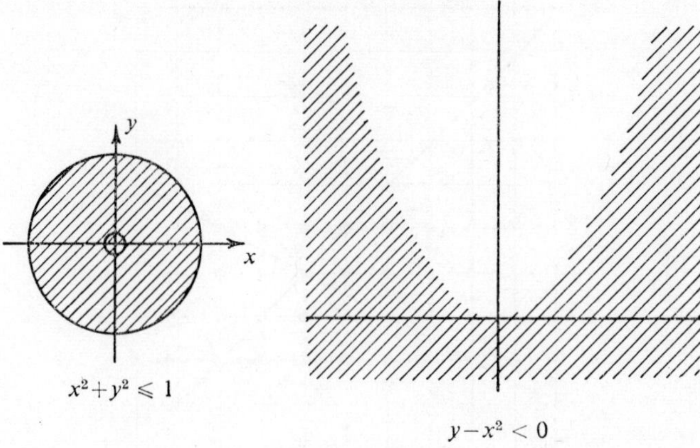

Fig. 7.10

Meanwhile we shall be particularly concerned with linear equalities, for the solution sets of such systems are always *convex*.

An intuitive understanding of convexity is easily formulated and extended to other than polygonal figures; a plane figure is convex if it wholly includes the line segment that joins any two points of the figure. This definition does not exclude unbounded figures; a half-plane, for example, is convex. This idea is explored more fully in chapter 9.

This section has been concerned with the values of a linear function defined over a convex polygon. The inequalities in the problem on chocolate bars, for example, defined the convex polygon *ABC*.

The values of $P = 10x+12y$ can be calculated for points in the polygon:

	x	y	$P = 10x+12y$
A	0	0	0
B	40	20	640
C	60	0	600
D	20	0	200
E	20	10	320
F	30	10	420
G	50	10	620

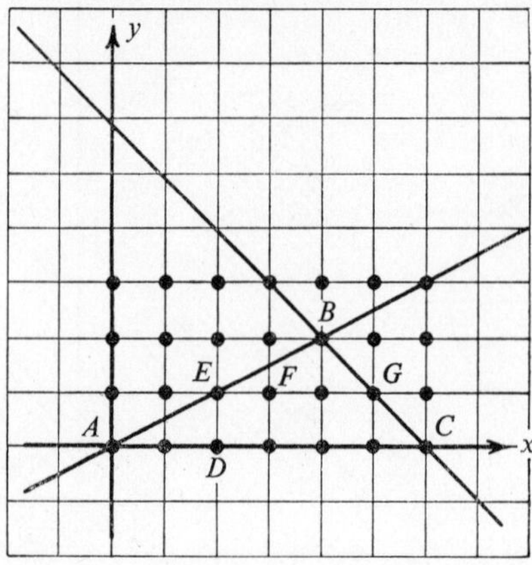

Fig. 7.11

A model can now be made by erecting perpendiculars, for example, with wire or thin dowelling at these points, with heights proportional to the values of P. (See Colour-fig. 9.) It will be seen that the upper ends lie in a plane. Thus the optimum values of P are at points on the edges of the polygon. In this case one vertex of the defining plane gives a single optimum value.

What is the position of the defining plane if there is more than one optimum value?

Genesis of a linear programming problem—a lesson sequence

Suppose you are the manager of a refinery which can normally supply any blend of three grades of fuel. The sulphur content and cost of producing each grade are given in the table on the blackboard.

Grade	*A*	*B*	*C*
Percentage weight sulphur	2·3	1·8	1·3
Price per ton	29	31	35

What do you notice about these figures? What mixture of the three grades could you make for a customer who asked for fuel with not more than 1% sulphur? Not more than 3%? Not more than 2%? Who else would just supply grade *B*? But in that case how much grade *A*? Would you expect to use any grade *C*? You would use as much A and as little B as possible in a mixture that is 2% sulphur? Good, that sounds reasonable; but can you explain that more precisely? One of them is ·3 more than 2. The other? ·2 less; what follows? We need two of A to three of B. How can we summarize this?

5 parts of mixture must contain:

2 parts of grade A fuel,

and 3 parts of grade B fuel.

Can we turn this into percentages of A and B? We need 40% of A and 60% of B. That is easy enough. Now what happens if there is a temporary shortage of grade B fuel, perhaps due to some political crisis? Yes the price would go up and it might cost more than grade C fuel; the prices might now be 29, 40 and 35, respectively. What is the best plan now? A mixture of A and C? In what proportion? (70% of A and 30% of C).

Now what happens if there is another change in prices? Now they read 32, 30 and 35. Can everyone find the best thing to do? The requirement was not *more* than 2% sulphur, so grade B alone is the cheapest way out. When would we have to use all three fuels in one blend? Can you give an example? It does look as if two will suffice. Let's look at this question more closely. Until we know the actual prices we certainly do not know which grades to choose, and we may still need all three. If in the final mixture there is x% of grade A and y% of grade B then there must be $(100-x-y)$% of grade C.

How can we express the fact that the mixture has not more than 2% of sulphur?

$$\frac{2{\cdot}3x+1{\cdot}8y+1{\cdot}3(100-x-y)}{100} \leqslant 2,$$

which gives

$$x+\tfrac{1}{2}y \leqslant 70.$$

So if y is 20, x must be less than or equal to ...? If x is 80, y must be ...? Why can't x be 80? Why can't y be negative? Does that hold for grade A? Yes; and for grades B and C. Therefore x and y are restricted by a number of inequalities. Let us list them: $x \geqslant 0$, $y \geqslant 0$, $x+y \leqslant 100$ and $x+\frac{1}{2}y \leqslant 70$. What do you expect the solution set for these four inequalities to be? Let us plot the graph and confirm it.

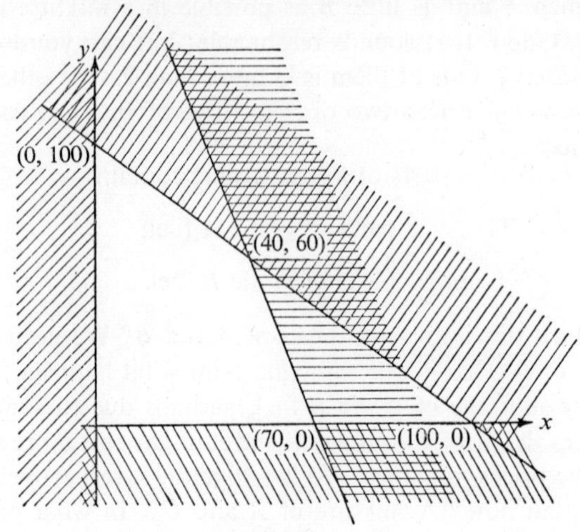

Fig. 7.12

When constructing these diagrams it is more convenient to shade the excluded domains, leaving the required convex polygon which represents the solution set.

There is the convex quadrilateral that defines the solution set in this case. Each point inside or on it represents a possible blend with not more than 2 % of sulphur. What are the costs? In the original case 100 tons of mixture will cost $29x+31y+35(100-x-y)$, which is $3500-6x-4y$. And so the cost per ton cannot be more than ...? To make it as low as possible we have to make $6x+4y$ as large as possible. Let's tabulate some values from the solution set.

x	0	10	20	0	40	70
y	0	10	30	40	60	0
$6x+4y$	0	100	240	160	480	420

Will everyone now choose his own set of prices and work out a table of costs for various points of the solution set? What do you notice?

Exercises

(1) A boy finds, when preparing for a weekly science test, that 6 minutes spent on Chemistry earn him an extra mark and that

7 minutes spent on Physics earn him two extra marks. To pass the test he needs 12 extra marks and he only has 60 minutes available for preparation. He wants to do more Chemistry than Physics although it is not so rewarding so far as marks are concerned. How should he allot his time? Can he spare any of the 60 minutes for other interests?

(2) A market gardener intends to split a 20-acre field between lettuces (L) and potatoes (P). The relevant details for the two crops are:

	L	P	Maximum available
Cost per acre, including labour, in £	5	3	75
Labour per acre, in man-days	$2\frac{1}{2}$	1	35
Estimated profit per acre, in £	3	2	—

How should he allocate the land for maximum profit? *Hint: He need not use the whole 20 acres. Do not use the profit figures until the three half-planes have been determined. Then consider lines parallel to* $3L+2P = 30$, *that is, lines parallel to* $3L+2P = C$ *with* C *as large as possible.*

(3) A tobacconist blends two types of tobacco, one costing 3*s*. per oz. and the other 5*s*. per oz. In each pound of mixture there is a whole number of ounces of each. The mixture is too harsh if it contains more than 11 oz. of the cheaper, and too dear if it costs more than 74*s*. per pound.

State the two expressions which determine the ordered pair (c, d), where c is the number of ounces of the cheaper tobacco per pound, and d is the number of ounces of the dearer. Sketch or illustrate on a nail-board. How many different blends can he sell?

If he sells at a whole number s shillings per ounce such that, if the cost per ounce is p shillings, $s-p \leqslant 2$, which blend produces the most profit per ounce? Does this same blend produce the highest percentage profit?

(4) A gardener wants to plant from 18 to 20 sq. yd. with fruit—partly blackcurrants B costing 5*s*. each and partly cordon apples C at 10*s*. each. He intends to allow 2 sq. yd. for each B and 1 sq. yd. for each C. He wants at least one third to be C. He does not intend to spend more than 100*s*. How many of each may he buy?

If he has to allow 10*s*. of the 100*s*. for packing and carriage how many of each must he order?

These four exercises are taken, with permission, from draft textbooks of the School Mathematics Project.

Supplementary information for the teacher may be found in [82, 125, 126].

Theory of games

The theory of games, a recently developed branch of applied mathematics with important economic applications, is a related topic offering scope for experiment in the school classroom. It would seem to have a special educational value because problems with a few very simple numbers often demand extremely careful thought and interpretation. Material suitable for introductory lessons and references for further reading may be found in [46].

8

PATTERNS AND CONNECTIONS

The lessons suggested and described in this section are aimed at helping to create a mathematical attitude. Simple relationships are used and time is taken to consider the various assumptions that are made as some of the properties of the material are developed. The section may be considered as a stage A topology but this depends on how it is developed.

The work has been used mainly in the age range of 11–13 as an introduction to geometry but with extensions to a much later age. It was developed in the first instance from combinatorial work on line patterns in an effort to build up confidence in analytical thinking before dealing with the geometry of the Euclidean plane. As work progressed it became clear that the material had its own mathematics and several aspects are discussed in more or less detail. It should be emphasized that though much of the material has been used in schools over a number of years there is much to be developed. It remains experimental.

1. DRAWING PATTERNS

In this sequence we are concerned to look at the number of strokes of the chalk, pencil, etc., needed to draw any given pattern (referred to later as network).

Lesson

Draw a series of lines on the blackboard, for example,

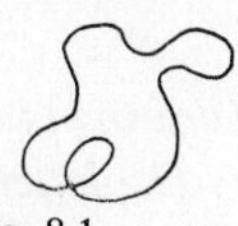

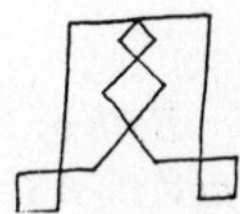

Fig. 8.1

(*use one stroke only*)

and then,

Fig. 8.2

(*draw naturally with as many strokes as are appropriate*).

Is there any difference between the two sets of lines? *Draw a similar series, distorting the shapes but leaving the crossings unchanged. Thus*

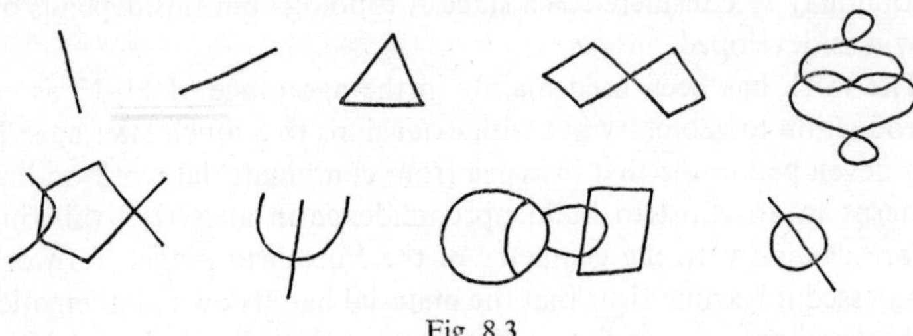

Fig. 8.3

Can you say what things have changed and what things have not changed?

Watch these two patterns carefully as I make them.

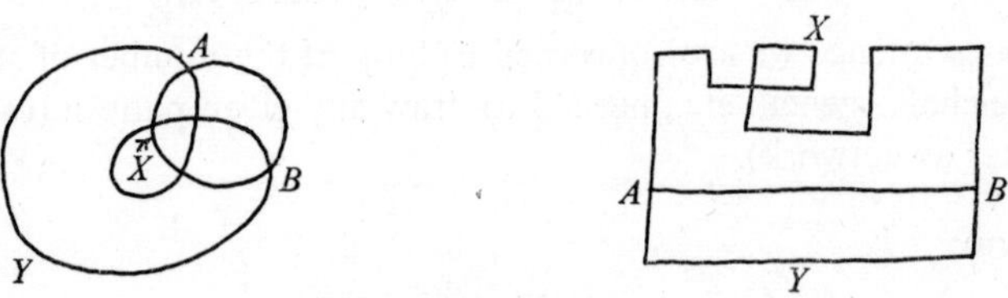

Fig. 8.4

Both of these should be drawn in two strokes making the circuit ABX and then the circuit ABY. The letters do not appear on the board.

The first set is drawn with one stroke of the chalk and the second set with more than one stroke. Class activity can proceed whereby patterns are drawn and the number of strokes used is noted. The problem implicit in this activity is that if a pattern or network is drawn with n strokes, is it possible to copy the pattern in less than n strokes? A discovery may be made that it is not necessary that a curve that was drawn in n strokes must *be drawn in n. Increased interest is developed if it is suggested that a pattern may need only one stroke.*

*But the first stage is to classify the patterns produced by the number of strokes. Secondly, the least number of strokes necessary to draw a particular pattern should be discovered, and finally an intuitive understanding of some of the theorems involved can be reached. Gradually, by investigating patterns from different points of view, attention can be concentrated on the junctions of the network. By experiment with their own patterns, reduction in the number of strokes needed will quickly take place. One example of a helpful activity is to consider whether if strokes are added to a pattern of, say, power 3 (a word has to be invented for ease of communication—*power *was one word suggested by a child referring to the number of strokes needed for a pattern) it follows that the power increases. It cannot increase beyond the number of additions; but in any case a class discussion soon reveals that some kind of limitations have to be imposed to decide what is acceptable as an alteration of pattern.*

Such discussion points that arise are:

(i) *Are we talking about open or closed patterns?*

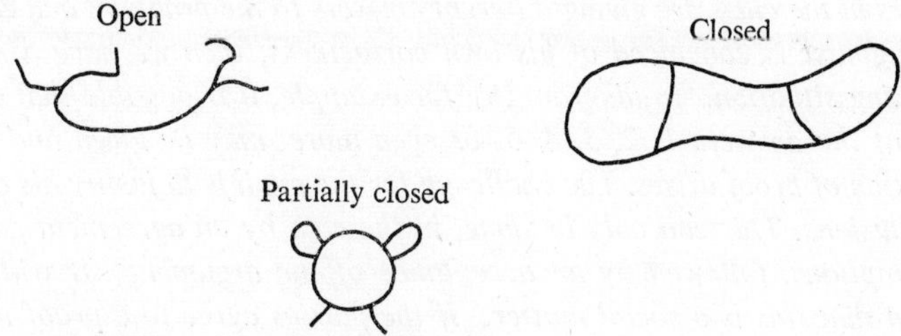

Fig. 8.5

(ii) *Is connectedness the rule, that is, is one point of the network always accessible from any other?*

(iii) *What rules for the addition of strokes do suggestions from* (*i*) *and* (*ii*) *imply?*

Having obtained a network which apparently has a power greater than 1, can one line be added to reduce it to a network which needs only one stroke? If not why not? If it can, then why?

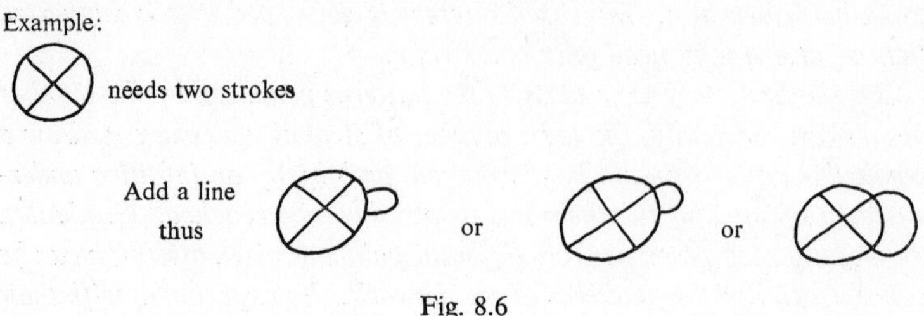

Fig. 8.6

Are there any more ways of adding a line to this pattern that are essentially different?

Let us look at the problem in another way. *Present a collection of networks. The collection given in Fig. 8.7 is drawn with a deliberate variation in the number and type of junction.* What is the least number of strokes needed to draw each of the patterns in Fig. 8.7?

If it is known that a network can be drawn in one stroke, the pupil tends to be satisfied when this task is accomplished. He looks for the next task. A more fruitful situation arises in those cases where a network has a power greater than one. When each pupil discovers that others in the class are giving different answers to the problem, and each protagonist is convinced of his own correctness, then we have a rich learning situation. In diagram (h), for example, it is possible that any one of the answers 1, 2, 3, 4, 5, or even more, may be given and the question of proof arises. The challenge for the pupil is to justify his own conclusions. This can only be done, in the end, by an agreement as to assumptions, followed by an acceptance of the argument. It will be noted that this is a social matter. If the parties agree to a proof then this is a proof—until a third party working on the same assumptions demonstrates to the communal satisfaction (and possible irritation) that it was inadequate. This gives an introduction to the notion of proof which appears more acceptable to pupils than some of the theorems in Euclid, which prove what to them is 'obvious'.

By work of this sort we should have created a situation where there is a motivation for investigation.

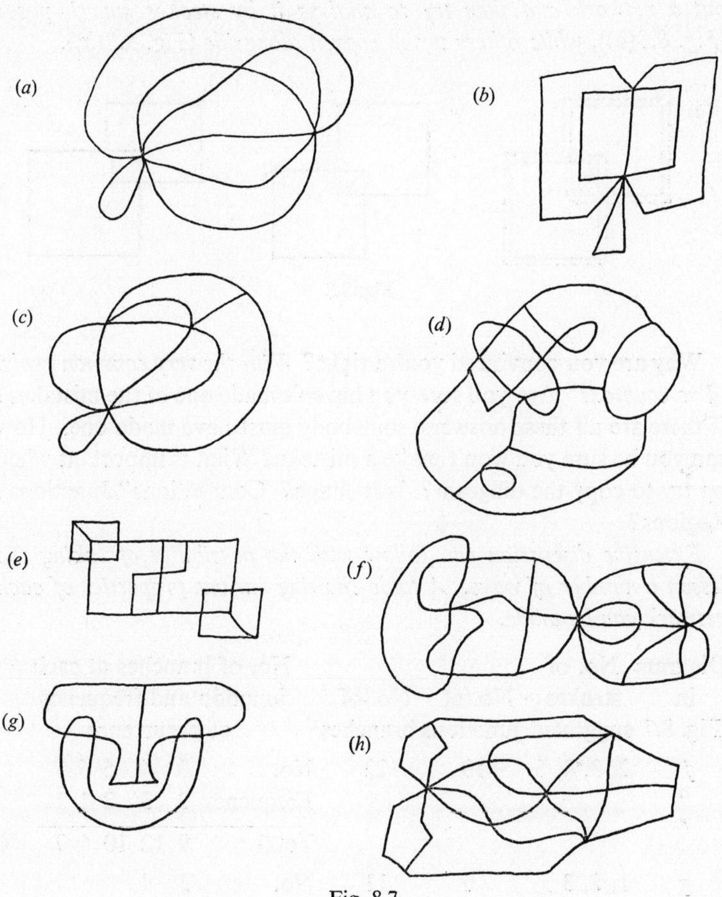

Fig. 8.7

How can the number of strokes be organized? What accounts for the differences in the answers given? To this may come the replies:

'going over sections twice';

'not completing certain sections';

'completing too many independent paths without checking whether they can be connected';

and so on.

There is a question of 'diagram dazzle' here; and it may also be observed that children work in a number of different ways. Some copy out a network and then try to analyse it by strokes superimposed (Fig. 8.8(a)), while others try to copy it alongside (Fig. 8.8(b)).

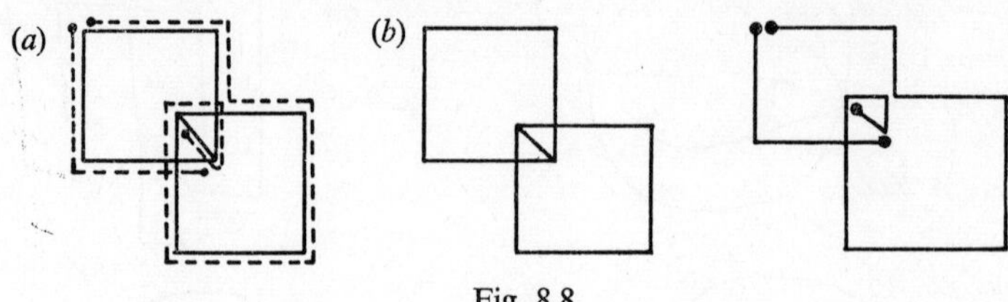

Fig. 8.8

Why are you convinced you're right? *With the very common reply,* '*I've counted!*' Are you sure you haven't made one of the mistakes? If there are all these answers, somebody must have made one. How can you be sure you won't make a mistake? What is important when we try to copy the diagram? Is it shape? Connections? Junctions? Regions?

Extensive discussion can follow with the possibility of taking the lesson a number of ways. A table showing various properties of each network can be made.

Diagram in Fig. 8.7	No. of strokes suggested	No. of junctions	No. of branches	No. of branches at each junction and frequency of occurrence						
j	2, 3, 4, 5	10	22	No.	3	4	5	6	7	
				Frequency	3	3	2	1	1	
				Total	9	12	10	6	7	44
g	1, 2, 3	6	11	No.	3	4				
				Frequency	2	4				
				Total	6	16				22

It should be possible to see the outlines of Euler's laws for networks emerging:

1. *A network can be drawn in one stroke if there are only two junctions or none, with an odd number of branches. Such a network is said to be unicursal.*

2. *If a network contains n 'odd junctions' then ½n paths will be needed.* References: Goodstein [56]; Berge [13]; Arnold [9]; Rouse Ball [107].

There are a number of minor results, such as—'no closed network has an odd number of odd junctions', 'all closed networks with all the junctions even can be drawn with one stroke', etc.

After making the table look again at the patterns. Look at line endings and junctions. Look again at the example on Fig. 8.6 and see whether with more knowledge we can analyse the different cases. As a further example consider the network shown in Fig. 8.9; and the following, each of which is the above network with one closed stroke added.

Fig. 8.9

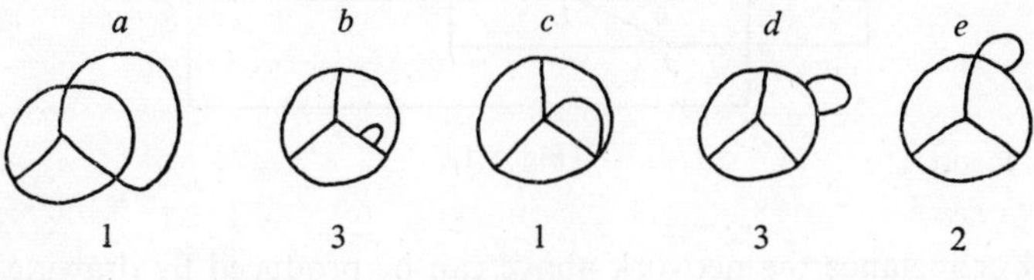

Fig. 8.10

Put them in order of power. Is there more than one order? What is the difference between odd and even junctions as far as their effect on the number of strokes is concerned?

While no proof has been established, some hypotheses will have been formed, and Euler's laws will have been confirmed experimentally. The proof implies that any network must satisfy the laws, and doubts as to the necessity of being able to draw the given network in the appropriate number of strokes will remain, until it can be shown that a network made by an arbitrary number of strokes can be reduced to the minimum by a set of rules.

Draw a network. Make each stroke or, its more usual name, *path*, in a different colour. Can the paths be joined, that is, can two or more be joined in an obvious way to make a unicursal path? Let us call them *connectible* if they can. Compare

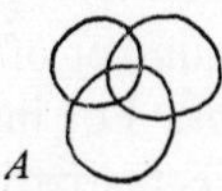

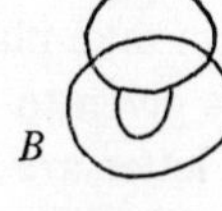

Fig. 8.11

It should be possible to make (*A*) with paths that are closed, or that can be closed by reduction. In (*B*) and (*C*) any set of paths will be a mixture of open and closed paths. Why? Let us now be systematic and see what can be gained by naming the junctions, and then looking at the paths identified by the junctions through which they pass.

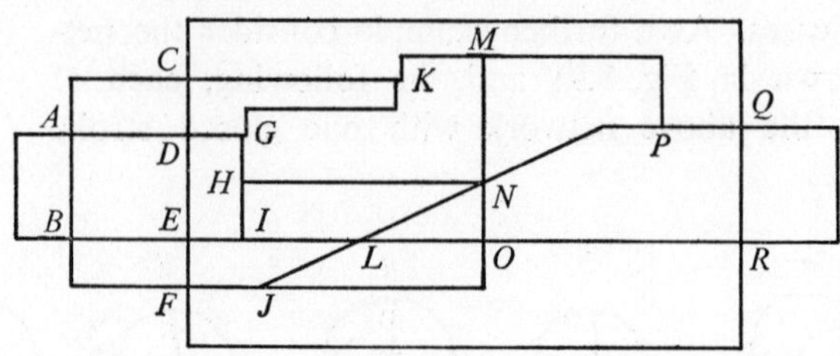

Fig. 8.12

For instance the network above can be produced by drawing the seven paths with the following labels:

(1) *C D E F R Q C*
(2) *A B F J L O N M K C A*
(3) *A B E I H G D A*
(4) *G K*
(5) *H N P Q R O*
(6) *I L N*
(7) *M P* — 7 paths

Paths 1, 2, 3, 4 can be joined by starting with (4) at *G*, going to *K*, moving round part of (2) to *C*, looping round (1) back to *C*, continuing (2) to *A*, looping round (3) back to *A* and continuing (2) to *K*. This reduces the power to 4.

On the other hand consider:

(1) *C Q R F E D C*
(2) *M N H G K M*
(3) *B E I L O J F B*
(4) *B A D G*
(5) *B A C K*
(6) *H I*
(7) *J L N O R Q P M*
(8) *N P* — 8 paths

In this case reduction can again take place. A number of things can be discovered about the labels given to the paths. For instance:

(*a*) When the first and last letter are the same, the path is closed.

(*b*) If not (*a*), then the path is open. It may be partially closed.

(*c*) If two open paths have the same end letter, then they are consecutive. What happens if the pair of end letters on each open path are the same?

(*d*) What can be said about two closed paths with points in common?

(*e*) Closed paths which are not connected can be connected if there exists an open path such that the two intersections of the set of letters describing the open path, with each of the sets describing the closed paths, are both non-empty.

These are not exhaustive but they can be made so. Note that set language will be the most efficient way to describe the complex (e). A difficulty here is that the set is ordered and the same points occur twice or more when the path is closed or partially closed. Accepting this difficulty the statements below can be written, using a notation similar to that described in chapter 5. The reduction can then be effected. Numbers in circles refer to the list of the paths in the second example above.

What can be said about each of the eight paths in the second example? ①, ②, ③ are closed and all the others are open. Are there any intersections? Can we connect paths?

$\cap$ indicates the letters in common and produces a set; $\oplus$ indicates the ordered join, when this is possible in the way defined above, of two paths.

Then we can say

$$①\cap② = \phi,$$

$$①\cap③ = \{E, F\},$$

so $$①\oplus③ = C\ Q\ R\ F\ B\ E\ I\ L\ O\ J\ F\ E\ D\ C.$$

Now $$④\cap⑤ = \{B, A\},$$

where B is an end point in both.

Thus $$④\oplus⑤ = G\ D\ A\ B\ A\ C\ K.$$

Then as $$(④\oplus⑤)\cap② = \{G, K\}$$

and $$(④\oplus⑤)\cap(①\oplus③) = \{B, C, D\},$$

finally $[(④\oplus⑤)\oplus②]\oplus(①\oplus③) = G\ K\ M\ N\ H\ G\ D\ A\ B\ E\ I\ L\ O\ J\ F\ E\ D\ C\ Q\ R\ F\ B\ A\ C\ K$. As this is not a closed path none of the other open paths ⑥, ⑦, ⑧ can be adjoined, and hence four paths remain. *At this stage, which may be considerably later than the work at the beginning, the proofs of the theorems can be established.*

Some simple examples:

(i) Try the traditional trick of drawing the envelope in one stroke.

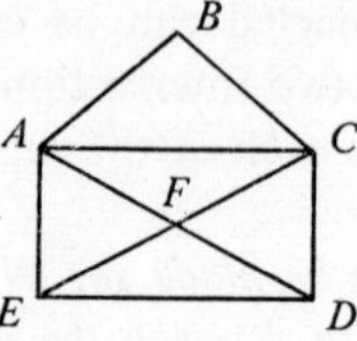

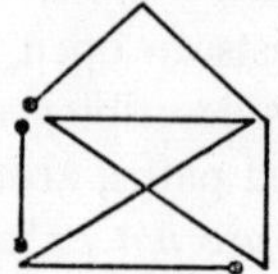

Fig. 8.13

Arbitrary paths:

1. *A B C D A C F E D.* 2. *A E.*

This can be solved automatically; ① ⊕ ② = *E A B C D A C F E D.*

(ii) Take the even junctioned figure made by a series of paths in Fig. 8.11. See also Colour-fig. 10.

1. *B A C B D C E.* 3. *A F D.*

2. *B F E A.* 4. *E D.*

⊕ all paths = *D F A E F B A C B D C E D.*

(iii) Euler's bridge network

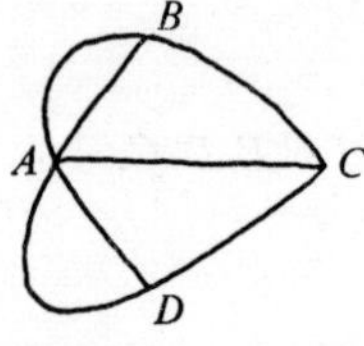

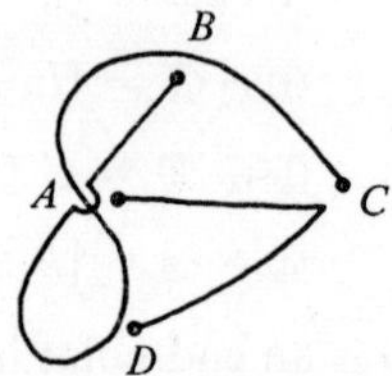

Fig. 8.14

1. *C B A D A B.* 2. *D C A.*

Irreducible.

(iv) Investigate the networks that can be drawn from a given set of paths. Consider, for example, the two paths ① *A B C D A D*, ② *B C.*

Further examples and work

(1) The conditions for a unicursal network can be used in different situations. Reference is made to p. 24 on the chain code. See also Petersen [95].

(2) Another neat example is the Domino problem. This is—given a chain of ordinary dominoes matched as in the ordinary game, what can we say about the possibility of closure of the chain in an arbitrary sense? That is, if a matching chain is built up in any order does it follow that the last domino completes the chain? This can be extended to sets of dominoes of order other than seven. One solution involves the unicursal theorem on p. 228; or rather the less general statement which maintains that any network in which all the junctions are even is unicursal.

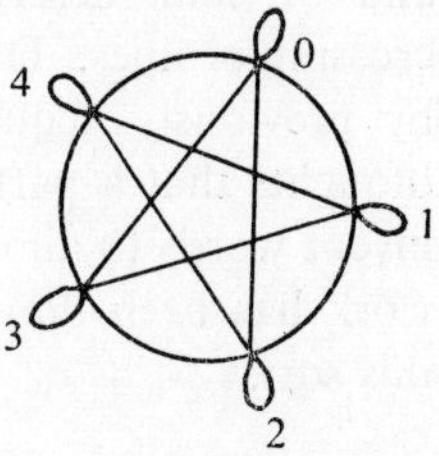

Arrange the numbers round a circle using, for example, a set of order five and join each number to every other number, including itself.

Fig. 8.15

Ignoring the junctions inside the outer boundary, we can make the immediate generalizations that for odd order domino sets the closure is possible, while for even order greater than two it is not. In the first case all junctions are even and a set of paths can be reduced to one; while in the second case any set of paths will reduce to $\frac{1}{2}n$ paths. What do the loops in the diagram represent?

(3) The interlacings on Celtic ornaments are intriguing examples of networks with even junctions, but they introduce the 'under-over' notion [122].

Fig. 8.16

(4) The *incidence matrix* discussed in chapter 12 is a more efficient means of handling a network when the information contained therein has to be used. By appropriate operations on the matrix the order of each junction can be established.

2. INVESTIGATION OF LINE PATTERNS

The second approach with this material is concerned with combinatorial properties of networks of lines and some of the situations that can be derived from them.

These situations can arise from the replies of the pupils to questions, and we shall describe here one that started from discussing the crossing of lines. Discussions of this sort are going to be hampered by previously acquired notions but encouraged by any original thoughts that a particular child may have. It may be necessary to invent words to allow ease of handling. The age range in which this work has been done is 13–16 but there are no real limitations of this sort.

Lesson

Q. How many times do two lines cross each other?
A. 1. Only once.
A. 2. It depends whether they are straight.
Q. How?
A. If they are straight they go on after the crossing.
Q. If they are not straight? *Drawing a* C *shape and a line across.*
A. Twice! No! As many times as you like.
Q. What do you mean 'as you like'?
A. Well you can go on drawing.
Q. What about going on drawing here? *Pointing to the two straight lines crossing.*
A. 1. Only once.
A. 2. No! They could bend too—afterwards.
Q. After what?
A. 1. If you went on drawing like this. *Comes up to board and continues the lines in such a way that they meet again.*
A. 2. But those aren't straight any more.
A. 3. They only looked straight.
A. 4. We can decide what to draw.

Q. What about my original question? How many times do two lines cross?

A. As many times as we like.

Q. Suppose we say they don't cross, what will that look like?

A. Parallel.

Q. Like this? *Draw two parallel segments.*

A. Yes.

Q. Or like this? *Draw two non-intersecting segments.*

A. 1. No.

A. 2. Yes! Because we could turn away.

A. 3. They haven't got to be parallel. It is what we do.

Q. So if we want to talk only about lines that either cross once or not at all can we call them straight?

A. 1. No.

A. 2. They could be.

A. 3. But they needn't be.

So the lesson proceeds. In one such lesson the term 'strone' line was invented for lines which under such operational definitions crossed each other once or not at all. We shall not continue the dialogue but the procedure will be in a similar spirit. Once having chosen a definition we can look, for instance, at those patterns of lines where all lines cross each other once and once only. Given more than two such lines then what modifications have to be made to the definitions? Crossing separately or not? *Each of these issues leaves another avenue to explore at a later date. In this case it is the matter of crossings of more than two lines at one point.* If the crossings are all separated then what is the total number of crossings?

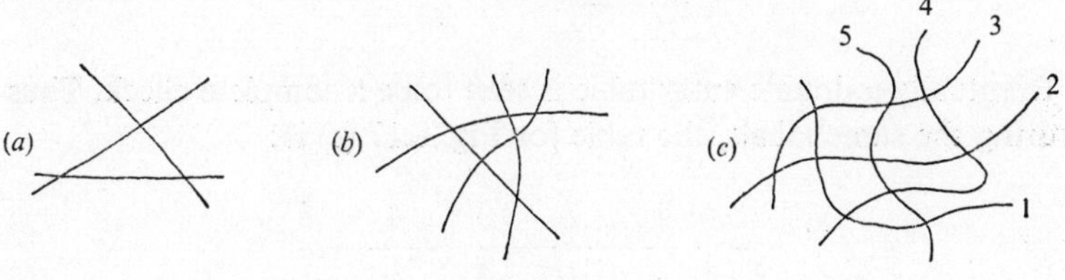

Fig. 8.17

All these patterns are drawn according to our definition and we know that if we continue to draw them beyond their present limits they must not cross again.

Is the last one drawn in accordance with our definition? How do we check? We need a system. If there are five lines, each will have four crossings. So let us label the lines and check:

Lines	1	2	3	4	5
Number of crossings	4	4	4	4	4.

Q. Does this confirm our definition?

A. 1. Yes.

A. 2. No! Because one line could cross more than once...it's difficult to see.

Exploration for patterns which obey the table but not the definition follows. An example is given in Fig. 8.18.

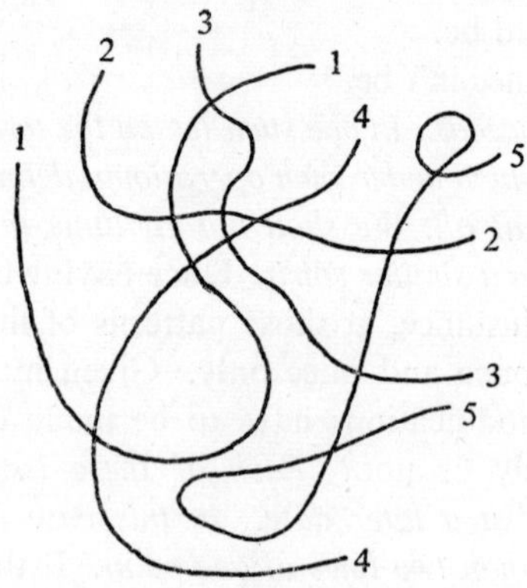

Fig. 8.18

Eventually a double entry table is seen to be a complete check. Thus using the same labels, the table for Fig. 8.17 (*c*) is:

	1	2	3	4	5
1	–	1	1	1	1
2	1	–	1	1	1
3	1	1	–	1	1
4	1	1	1	–	1
5	1	1	1	1	–

When this is complete with a blank diagonal and every other entry of value 1, the definition is satisfied. *This is another use of the 'incidence matrix' referred to in §8.1 and discussed in chapter 12.* Compare the double entry table for Fig. 8.18:

	1	2	3	4	5
1	0	1	1	2	0
2	1	0	1	1	1
3	1	1	0	1	1
4	2	1	1	0	0
5	0	1	1	0	2

The exploring and the abstraction go together and it can be observed how the abstraction, itself done for convenience, leads to the fact that the nth triangular number gives the number of crossings for $n+1$ lines. This is followed up and can be taken as far as the class allows. The abstraction can be used first to discuss 100 lines and then to consider a proof by induction.

Questions for further work

(1) Given that any line must cross any other once and once only, how many more crossings are implied by the following patterns, where some lines and some crossings are given?

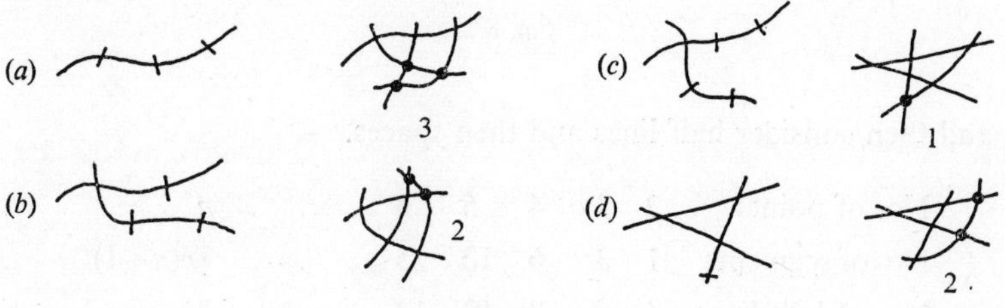

Fig. 8.19

(2) Given that lines can either cross once or not cross, state the possible implications for extra crossings, assuming that no more lines are drawn than are implied by the crossings given.

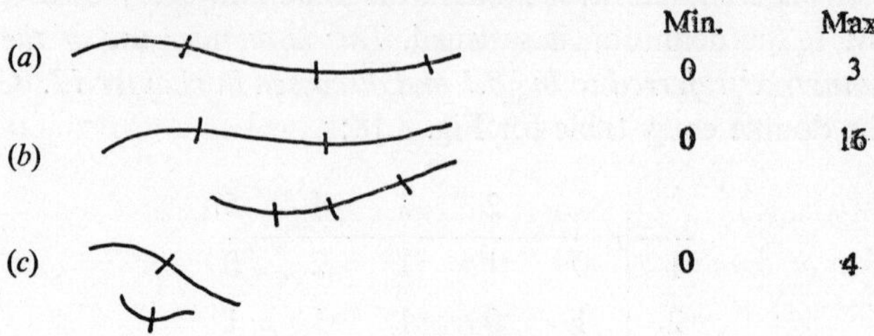

Fig. 8.20

The minimum is always zero as lines can be drawn which do not meet. The maximum is $\frac{1}{2}(m+k)(m+k-1)-m$, where m is the number of junctions and k is the number of lines.

	m	k	No. of possible extra junctions
(i)	3	1	$\frac{1}{2}(4.3)-3 = 3$
(ii)	5	2	$\frac{1}{2}(7.6)-5 = 16$
(iii)	2	2	$\frac{1}{2}(4.3)-2 = 4$

(3) Number of segments on a line made by points:

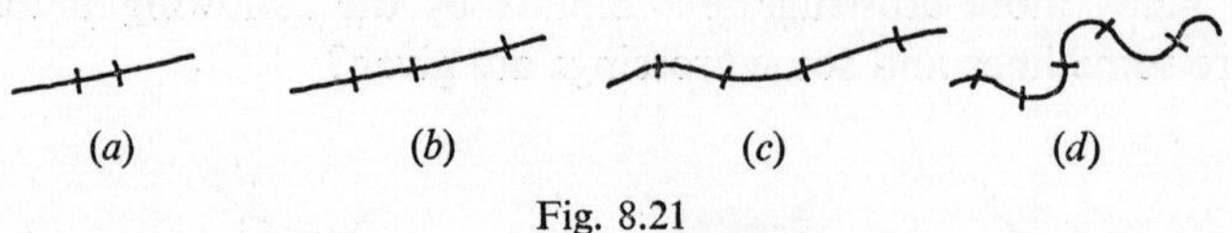

Fig. 8.21

and then consider half lines and then spaces.

No. of points	2	3	4	5	6	. . .	n
No. of segments	1	3	6	10	15	. . .	$\frac{1}{2}n(n-1)$
No. of half-lines	4	6	8	10	12	. . .	$2n$
No. of spaces	5	9	14	20	27	. . .	$\frac{1}{2}n(n+3)$

(4) Given that there are m segments with n points, and assuming that two adjacent segments produce another, how many segments must be identified in order to produce them all? See Fig. 8.21.

Questions that can be investigated are: under what conditions can this be done? Which segments must be chosen? In order to identify all ten in the last example, Fig. 8.21 (*d*), what are the conditions that the four chosen must obey? This introduces a metric property of the line.

The examples 5–8 are particular cases of inductive work which can be treated at a number of levels. The emphasis is on exploration and recognition of relationships, exemplified here by number patterns in relation to line and region patterns.

(5) (*a*) A given set of lines is such that each line crosses every other line separately. How many regions are formed?

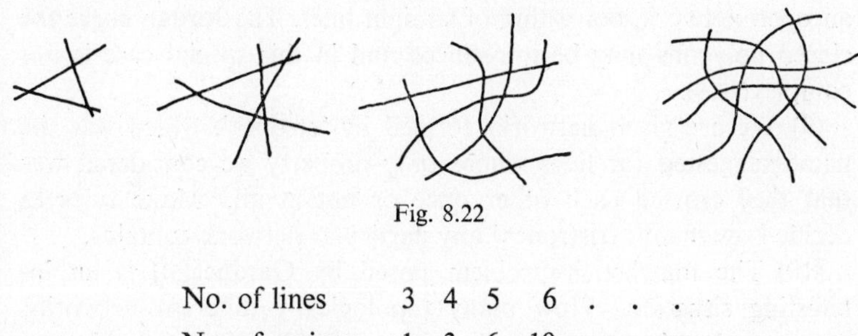

Fig. 8.22

No. of lines	3	4	5	6	. . .
No. of regions	1	3	6	10	. . .

(*b*) How many triangles? How many four-sided figures? What happens after this? What for instance do we allow as a four-, five-, six-, etc., sided figure? *This is a problem associated with 'Convexity', see chapter 9.* A problem raised here is what happens to one set of dissections when a line is drawn across.

(6) Suppose every line can cut every other twice.

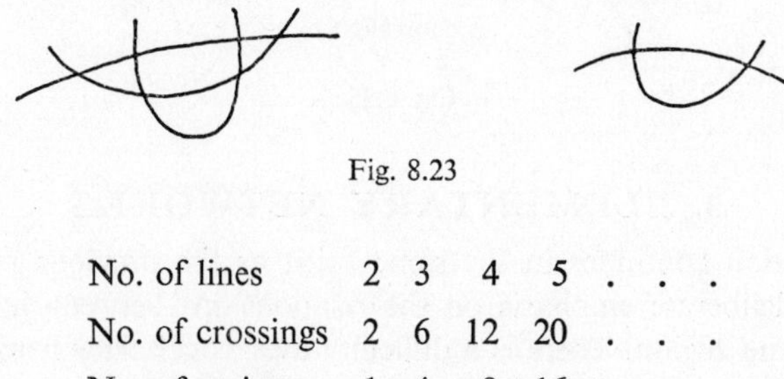

Fig. 8.23

No. of lines	2	3	4	5	. . .
No. of crossings	2	6	12	20	. . .
No. of regions	1	4	9	16	. . .

(7) Suppose every line can cut every other three times.

No. of lines	2	3	4	5	. . .
No. of crossings	3	9	18	30	. . .
No. of regions	2	7	15	26	. . .

(8) How many regions, including overlapping and infinite, are made by any network? A distinction should be made between open and closed networks. In the former any open path would be assumed infinite and then we could say that the number of regions is $2(n+n_0)$, where n is the number of open, non-self-crossing paths, and n_0 is the number of closed paths. In the case of a closed network $n = 0$. A special case is the subset which defines all the angles of an open network, consisting of straight lines. The Jordan curve for closed polygons may be introduced and in this special case is not complex [29, 9].

(9) We are given networks formed by 'strones', which was the name suggested for lines whose only property we considered was that they crossed each other once or not at all. Make rules to decide how many 'tristrones' any particular network contains.

(10) The matchstick problem posed by Gardner [51] is an interesting situation. How many topologically different networks, open or closed, can be made with n matches (lying flat)? For example, $n = 3$.

Fig. 8.24

Note that for $n = 6$

is equivalent to

Fig. 8.25

3. ELEMENTARY NETWORKS

This section continues in the same spirit as the previous two but with a deliberate emphasis on the relationships between junction, branch and region. There is a difficulty over vocabulary here which has to be accepted. We shall attempt to be consistent and in this section we shall use *branch, node* and *region*; though elsewhere branch may go under the name *edge, side, segment, simple link*, etc. A node may variously be called *junction, intersection, crossing, terminal, vertex*, etc. Region refers to a discrete non-overlapping space which is not necessarily enclosed. It may be the infinite region. In some studies it may be referred to as *face.*

In particular, this work extends into network theory wherever this occurs. References are given at the end to other parts of the book and to other literature which take up the notions for more advanced use, both in application and in theory. The work is with older pupils and the lesson form adopted earlier is not followed here. As an example of an investigation of the network the notion of *duality* is used. This notion is important for the light that is thrown on a structure *A* by its transformation into a structure *B*, the properties of the latter being perhaps more susceptible to investigation than the former.

As with many such concepts which are introduced at an advanced stage, the student often finds difficulty because given a hard problem (structure *A*), he is then expected to use an unfamiliar notion which has only been introduced to solve the problem. In this case, the unfamiliar notion is 'duality' used to convert to structure *B*. This situation justifies the case for introducing an elementary aspect of the unfamiliar notion at an earlier stage.

The duality of networks is discussed here for experiment and use as a creative material.

Lesson

There are a number of initial definitions which can be decided:

(i) A *branch* has a *node* at each end.

(ii) A *node* is a point at which any number of *branches* end.

(iii) A *region* is the space enclosed by a sequence of branches and nodes.

The *order* of a branch, node or region is

(i) *branch*—the number of nodes on it (always 2);

(ii) *node*—the number of branches on it;

(iii) *region*—the number of nodes on the boundary, which is also equal to the number of branches.

It should be decided whether to allow:

(*a*) Nodes of order 1, as these imply a branch which is not a segment of a curve enclosing a region.

(*b*) Nodes of order 2, as this merely increases the number of branches without increasing the number of regions.

(*c*) Branches of order 2 which have only 1 region.

(*d*) Branches with only one node, which are necessarily loops.

Examples:

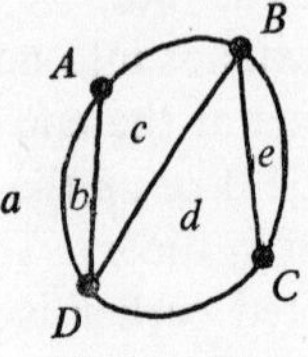

Fig. 8.26 (*a*)

Fig. 8.26 (*a*)	Nodes:	*A*	*B*	*C*	*D*	
	Order:	3	4	3	4	
	Regions:	*a*	*b*	*c*	*d*	*e*
	Order:	4	2	3	3	2
	No. of branches:	7				

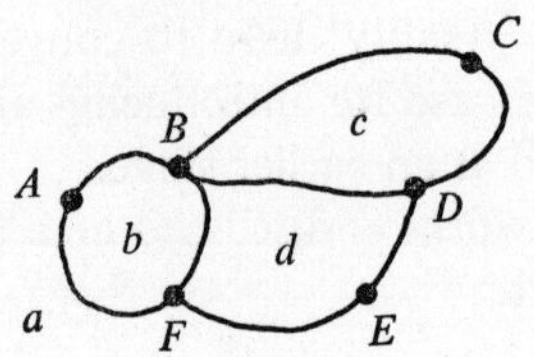

Fig. 8.26 (*b*)

Fig. 8.26 (*b*)	Nodes:	*A*	*B*	*C*	*D*	*E*	*F*
	Order:	2	4	2	3	2	3
	Region:	*a*	*b*	*c*	*d*		
	Order:	4	3	3	4		
	No. of branches:	8					

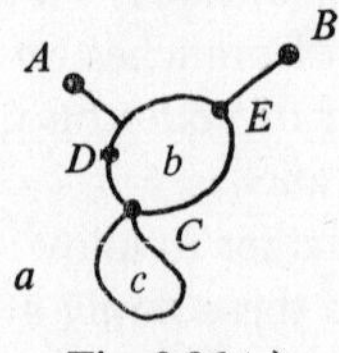

Fig. 8.26 (*c*)

Fig. 8.26 (*c*)	Nodes:	*A*	*B*	*C*	*D*	*E*
	Order:	1	1	2	3	3
	Regions:	*a*	*b*	*c*		
	Order:	5	3	1		
	No. of branches:	6				

The problem is now given of constructing topologically different networks, given *N* nodes of given order. We shall decide not to allow the possibilities *a*, *b*, *c* and *d*.

Example 1:

$N = 2$, both of order 3.

Stage 1.

Fig. 8.27

Fig. 8.28

Stage 2. Choose junctions for connection. Here we depend on our assumptions.

We have Fig. 8.28 as the only case.

Stage 3. As a matter of interest, look at the figure obtained if we draw so that in this second figure the nodes correspond to the regions in the first, and the regions correspond to the nodes. This figure is called the *dual* of the first, and vice versa.

The branches between the new nodes in Fig. 8.29 (ii) are inserted to satisfy the conditions observed in the first figure; that is, a region *a* in Fig. 8.29 (i) has nodes *A*, *B* in its boundary and so in Fig. 8.29 (iii) node *a* will have regions *A*, *B* surrounding it. This is the criterion by which branches are connected. Occasionally a free choice is available and this may lead to a discussion on the imposition of an order. In more complex figures problems arise when two regions have two separate boundaries. The dual can be obtained, as in Colour-fig. 11, by having the new nodes within each region and drawing the dual superimposed in a different colour. In this way all boundaries must be crossed once.

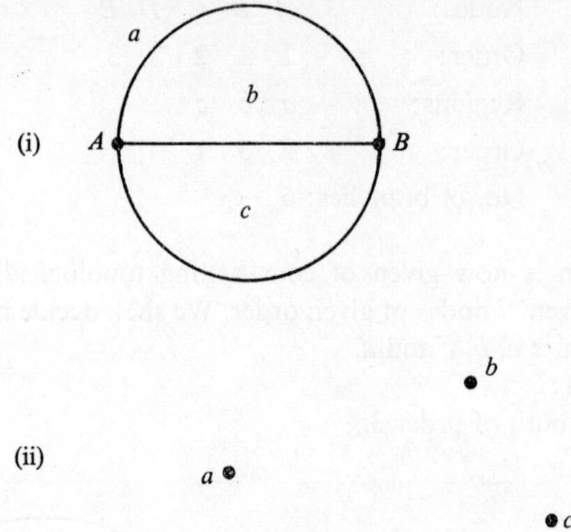

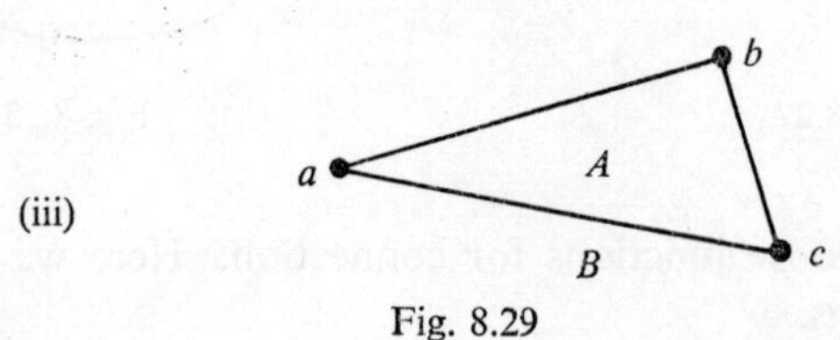

Fig. 8.29

Example 2:

$N = 4$, all of order 3.

Again, excluding *a*, *b*, *c* and *d* we have only

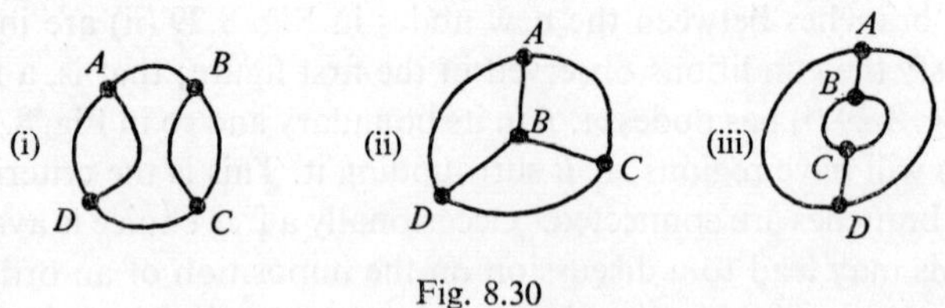

Fig. 8.30

$N = 4$, $R = 4$, $B = 6$ in each case.

All other possibilities seem to be equivalent by inspection. But consider the duals

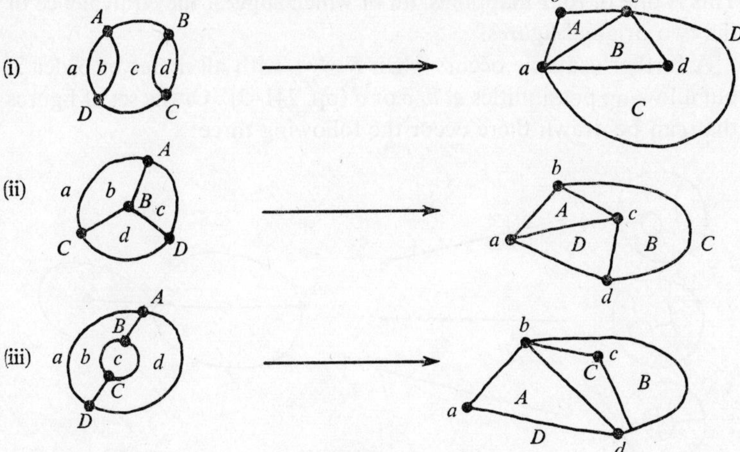

Fig. 8.31

and we see that (i) and (iii) have the same duals. By putting into 1–1 correspondence the nodes of the dual figures, a mapping of the two original figures can be effected.

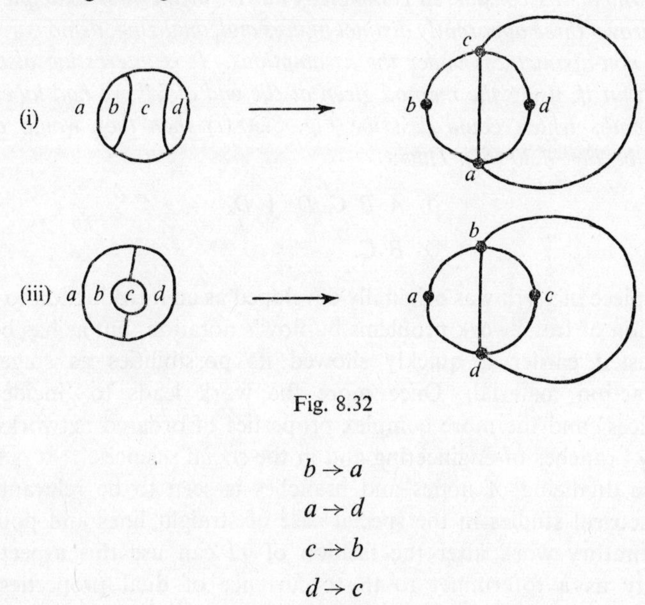

Fig. 8.32

$b \to a$

$a \to d$

$c \to b$

$d \to c$

This is one of four mappings, all of which suggest the equivalence of the two original figures.

A further example occurs when $N = 4$, with all nodes of order 3, but allowing possibilities *a, b, c* or *d* (pp. 241–2). Of the set of figures that can be drawn there occur the following three:

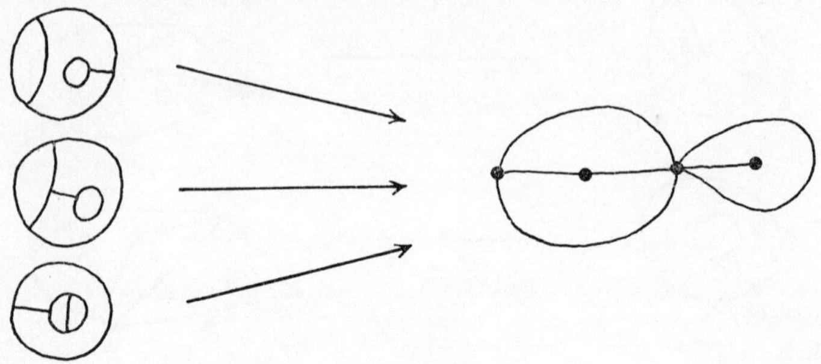

Fig. 8.33

all with the same dual.

This discussion is not intended to do more than show some possible implications of the duality property. Here we have opened up a situation which was considered complete. That is, in the first example we had drawn three apparently distinct figures and, dualizing, demonstrated their non-distinctness under the assumptions. It is interesting also to note that if, using the method given at the end of §*1, we had labelled two paths which could describe Fig. 8.30* (*i*) *then they would also describe Fig. 8.30* (*iii*). *Thus*:

① *A B C D A D*,

② *B C.*

This piece of work was originally developed as an introduction to the solution of framework problems by Bow's notation, but as has been suggested earlier it quickly showed its possibilities as stage *A* abstraction material. Once more the work leads to 'incidence matrices' and the more complex properties of ordered networks in many branches of engineering and in the social sciences.

The dualizing of nodes and branches is seen to be relevant to geometrical studies in the special case of straight lines and points. Exploratory work after the fashion of §2 can use this aspect of duality as a forerunner to the occurrence of dual properties in projective geometry. The number of triangles in one figure can be checked against the number of trilaterals in the dual, etc.

A further reference can be made to the dual properties of the five Platonic solids, where an operation in three space, similar to that which we have discussed, transforms a cube into an octahedron, an icosahedron into a dodecahedron, and a tetrahedron into itself.

4. OTHER INTRODUCTORY WORK IN TOPOLOGY

(1) Extend Euler's formula to the problem of the classification of surfaces.

(2) Study triangular maps and the theorems concerning their numbering. A proof of the basic tiling or plaster theorem in two dimensions depends on Sperner's Lemma [7]. This shows that given (i) a triangular mapping labelled so that there are three main vertices 1, 2 and 3; (ii) the segments joining the vertices i, j can be subdivided by points labelled i or j; and (iii) that triangles can be formed inside the main triangle with any number of vertices labelled 1, 2 or 3 at random, then there exists at least one triangle with vertices 1, 2, 3. For example,

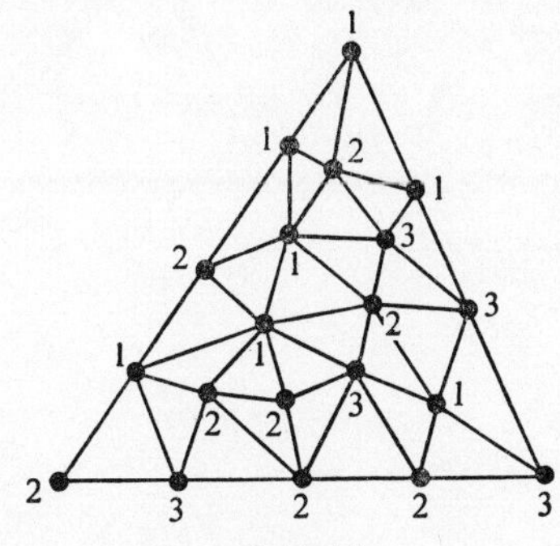

Fig. 8.34

A proof follows.

Consider all the segments joining points labelled 2 to points labelled 3. Let there be B segments inside the main triangle and B_0 on the perimeter. Consider all the triangles containing these (2, 3) segments. They are labelled (1, 2, 3), (2, 2, 3) or (2, 3, 3). Let their number be A_1, A_2, A_3 respectively. Then the total number of (2, 3) segments is $A_1+2A_2+2A_3$. But this is the same as B_0+2B (since all the interior segments have been used twice). Hence

$$B_0+2B = A_1+2A_2+2A_3.$$

But it can easily be shown that the number of (2, 3) segments on the main side (2, 3) must be odd. That is, B_0 is odd. Hence A_1 is odd and cannot be zero.

The kind of deduction based on the labelling procedures illustrated here has been used at all levels of the secondary course as a neat example of logical structure, with very little sophistication needed for understanding.

(3) Map colouring problems. Goodstein [56] discusses two-, three- and four-colour problems as well as contiguous region theorems.

This work on the elementary topological properties of Euclidean space provides an intuitive background to more advanced work on topological transformations of a much more general character. See Arnold [9]; Alexandroff [7]; Patterson [93], National Council of Teachers of Mathematics [87]; Coxeter [30]; Johnson and Glenn [67]; Kemeny, Snell and Thompson [70]; Rademacher and Toeplitz [102].

9

CONVEXITY

1. INTRODUCTORY LESSONS

Whilst it is truly an elementary topic, convexity has attracted the attention of mathematicians only in comparatively recent times, and many very simple theorems have a short history. Its effectiveness as a subject for elementary teaching was shown in a demonstration lesson by Krygowska in Cracow in 1960. We may justify the introduction of the subject because:

(i) it is part of the physical science of the space in which we live;

(ii) as it is formalized, it offers scope for the use of arguments based on sets and vectors; simple but powerful arguments which can be extended to n-dimensions later in a mathematical career;

(iii) it is a growing part of mathematics, and there is a stimulus in working in parts of the subject which are of interest to creative mathematicians.

Our work will be entirely in two dimensions, but the methods extend immediately to more. We give below the bare outline of a way in which an intuitive approach to this subject might be made. This same material could be used with appropriate modifications for a wide range of ages.

Lesson

As many children as possible are given blackboard space, and they draw diagrams to illustrate the discussion as it proceeds. The other children work on paper of their own. The children are invited to draw outline shapes, geometrical shapes, pictures of objects or just 'doodles', and to shade the interior. The teacher inspects these and singles out some but not others—he may say that he 'likes' some or 'chooses' some. What is the basis for his choice? It transpires that he 'chooses' figures like

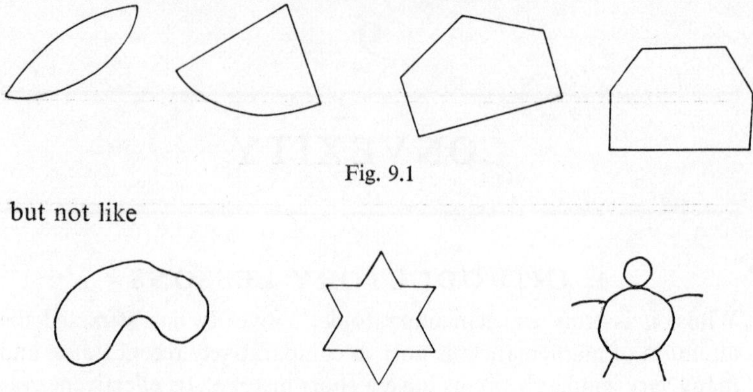

Fig. 9.1

but not like

Fig. 9.2

The distinction should first be established at a non-verbal level, so that everyone eventually knows to what class fresh shapes belong, but cannot necessarily put the distinction into words. We can then introduce the terminology *convex* and *non-convex* to describe the two types of figure.

Exercise

Draw some figures extending off the paper 'to infinity' which are (*a*) convex, and (*b*) non-convex. Can you divide the whole plane into two regions which are (*a*) both non-convex, (*b*) one convex and the other non-convex, and (*c*) both convex? *The last question is slightly subtle, and prepares for something which comes later.*

The teacher should observe that convex sets can be *open* or *closed*, using the terms in the technical sense which they usually carry in analysis. This distinction might only confuse the issue at an early stage and we suggest that it should be avoided unless pupils raise the matter. One should, however, introduce the notion of the *boundary* of the set of points and the *interior*. These technical terms should cause no difficulty.

It is now time to seek formal definitions; and the rest of the lesson can be spent in comparing different possible definitions of *convex*, and seeing how the different definitions imply one another. There are perhaps four possible definitions which might be considered. The lesson should go as the class are inclined—but as there are now so many possibilities we will merely indicate the information which might be expected to come out at the end. The purpose of the discussion is to take the initial suggestions in their crude form and to develop precise language in which to express them accurately.

Possible definition 1

A figure is convex if, and only if, given any two points P and Q of the figure, it contains all points on the segment PQ.

Possible definition 2

A figure is convex if, and only if, every straight line through any arbitrary interior point cuts the boundary in just two places.

A straight line which meets the figure, and is such that all the points of the figure are either on the line or on one side of it is called a *support line*. Note that tangents to the figure are support lines, if they do not cut the figure somewhere else, but a support line through a 'sharp corner' is not correctly described as a tangent. Draw different cases.

Possible definition 3

A figure is convex if, and only if, through each point on the boundary there is at least one support line.

Possible definition 4

A figure is convex if and only if for every exterior point P there is just one point of the figure which is *nearest* to P. *The teacher should note that this definition implies a measurement of distance and is therefore metrical, whereas the previous definitions are affine. More strictly, they involve only notions of 'straightness' and 'betweenness', they do not involve parallelism or distance. To reconcile this definition with the others it may be useful to observe that a circle is a convex figure. The more brilliant among the pupils might appreciate the question 'Why?'. Also, for them alone, there are such questions as* why *is the interior of a parabola, or an ellipse, convex? The definition may be discussed in practical terms by considering a ship's captain trying to get information about a rocky coastline by listening for echoes of blasts on the ship's siren.*

Exercises

(1) If two convex regions do not overlap then there is at least one line which divides the plane into two parts, with one region in each.

(2) Given two convex islands in the sea, if they are seen from a ship then always the view of one island, at least, is not masked at all by the other.

A further lesson

Another subject for inquiry is the derivation of further convex figures from given figures.

The intersection of any two convex sets is convex. (*Why?*) In particular a half-plane is convex. This allows another possible definition, related to Definition 3 above. A region is convex if it is the intersection of half-planes. The regions of interest in linear programming problems (chapter 7) are convex.

Exercises

(1) Draw diagrams to show various convex figures, circles, etc., as the intersection of half-planes. This is related to the idea of drawing a curve as an envelope.

(2) An island forms a plane convex region of a flat earth. In times of emergency civilians are not allowed within d miles of the coast, where the value of d depends on the emergency. Prove that the region to which at any time civilians are restricted is a plane convex region. The capital is to be so sited as to lie in all possible civilian regions. Prove that, if the boundary of the island is a quadrilateral Q, this rule determines the site completely, provided Q is not a trapezium. (Cambridge Scholarship.)

Another more sophisticated result is that the vector sum of two convex regions is convex, that is, if $\mathscr{X}$ and $\mathscr{Y}$ are convex, and $\mathscr{Z}$ is the set of all points of the form

$$\mathbf{z} = \mathbf{x}+\mathbf{y} \quad \text{with} \quad \mathbf{x} \in \mathscr{X} \quad \text{and} \quad \mathbf{y} \in \mathscr{Y},$$

then $\mathscr{Z}$ is convex. The teacher may appreciate a proof of this.

Let $\mathbf{z}_1 = \mathbf{x}_1+\mathbf{y}_1$ and $\mathbf{z}_2 = \mathbf{x}_2+\mathbf{y}_2$; with $\mathbf{x}_1, \mathbf{x}_2 \in \mathscr{X}$ and $\mathbf{y}_1, \mathbf{y}_2 \in \mathscr{Y}$. Then any point $\mathbf{z}$ on the segment joining $\mathbf{z}_1$ and $\mathbf{z}_2$ is of the form

$$\mathbf{z} = \lambda\mathbf{z}_1+\mu\mathbf{z}_2 \quad \text{with} \quad \lambda+\mu = 1,\ \lambda \geqslant 0,\ \mu \geqslant 0, \text{ (see p. 305)},$$
$$= (\lambda\mathbf{x}_1+\mu\mathbf{x}_2)+(\lambda\mathbf{y}_1+\mu\mathbf{y}_2).$$

That is, it is the vector sum of a point on the segment $(\mathbf{x}_1, \mathbf{x}_2)$ and a point on the segment $(\mathbf{y}_1, \mathbf{y}_2)$. This means that $\mathbf{z} \in \mathscr{Z}$, and so $\mathscr{Z}$ is convex, using Definition 1.

Exercises

(1) If $\mathscr{X}$ is a circle, with its interior, and $\mathscr{Y}$ is a square, with its interior, draw the region $\mathscr{X}+\mathscr{Y}$. (See Papy [92] for many other examples.)

(2) Show that the vector sum of two non-convex regions may, in special cases, be convex.

(3) The figure which arises as the vector sum of two regions depends on the point chosen as origin. But if different points are selected as origin then the resulting figures differ from one another only by translations. Illustrate by diagrams.

(4) If an island is convex then the region made up of the island plus its territorial waters is also convex.

(5) Investigate situations involving both the intersection and vector sum of convex regions.

A further topic, and one which can be done in early lessons, is the notion of the *convex closure* (or convex cover, or convex hull) of a set. Given any set of points $\mathscr{X}$ (even a set of isolated points) we may imagine an elastic band surrounding them which is allowed to contract as much as possible. The elastic band surrounds a region which is convex, which contains the original set of points, and which is the smallest convex region which does so. This is called the convex closure of $\mathscr{X}$. Give a definition of this region which does not involve reference to 'elastic bands'.

Exercise

Investigate the notion of convex closure in connection with the intersection and vector sum of sets.

For further possible classroom topics, such as the in-circle and circumcircle of convex sets, the diameter of a set and sets of constant width see Rademacher and Toeplitz [102], Yaglom and Boltyanskii [135] and Eggleston [38].

2. HARDER TOPICS

In this section we touch briefly on some harder questions in convexity which indicate some of the directions in which the simpler topics which we have just discussed might lead. For fuller details of this very exciting field and a formal statement of the theorems involved, with proofs, we must refer to [135].

By the *diameter* of any set of points we mean the maximum distance between two points of the set. With this understanding we may state the theorem:

Any spot of diameter d on a table cloth can be covered with a circular napkin of radius $d/\sqrt{3}$. This assertion is typical of certain theorems about convex figures and such theorems can usefully illustrate some important mathematics in an elementary way. The methods are essentially geometric rather than analytic; the results have wide applications, for example, in crystallography and the approximation of functions by polynomials. The following discussion of the opening assertion is directed to the sixth-former. Proof by induction, the notion of convexity and some elementary triangle geometry is assumed. The informal and colloquial form of a classroom dialogue is not reproduced, and many of the assertions would normally be arrived at with the full participation of the class.

The class tries to solve a particular problem. In groping towards a solution the need for more general theorems arises. These are formulated and attacked in the same way. A summary reverses the order of discovery.

Lesson

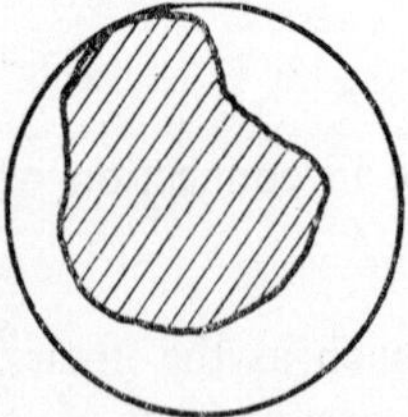

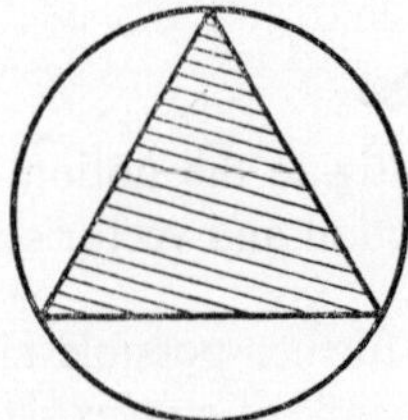

Fig. 9.3

We can think of the spot as a set of points, in this case an infinite set. It will make things a little simpler to consider in the first instance a finite set of, say, n points such that any point is within unit distance of the others. What circle will enclose all these points? Any one point can be enclosed by a circle of any radius. Two points could be enclosed by a circle of radius at least $\frac{1}{2}$. How can we enclose three points, say A, B and C?

The triangle ABC has no side greater than 1; suppose the greatest side is BC. If angle $A \geqslant 90°$, then the triangle is completely enclosed by the circle with diameter BC. (*Why?*) This circle has radius $\frac{1}{2}$ (see Fig. 9.4 (i). Meanwhile $A \geqslant 60°$. (*Why?*) So that $BC \geqslant 2R \sin 60°$, where R is the circumradius of triangle ABC. Since $BC \leqslant 1$ it follows that $R \leqslant 1/\sqrt{3}$ (see Fig. 9.4 (ii)). Thus in any case we can always enclose three points by a circle of radius $1/\sqrt{3}$. Shall we need circles of larger radius to enclose more than three points?

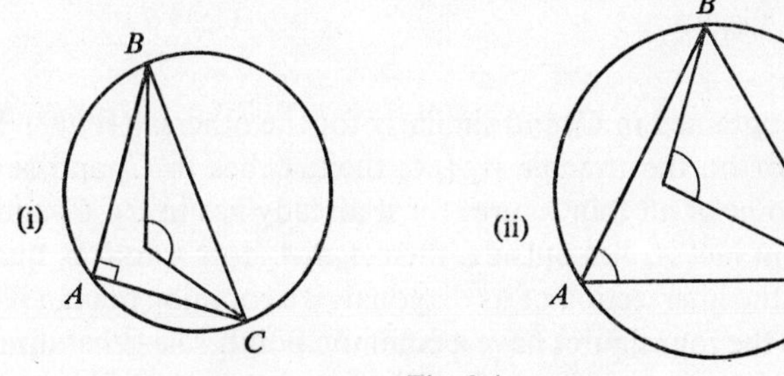

Fig. 9.4

Any three points A, B and C may be enclosed by a circle of radius $1/\sqrt{3}$. Let the centre of this circle be X. Now draw circles of radius $1/\sqrt{3}$ with centres at A, B and C. These all contain X. (*Why?*) In other words, X lies in the common part of the circles with centres at A, B and C (see Fig. 9.5). If all n points are to be contained in a circle of radius $1/\sqrt{3}$ then its centre will lie inside each circle of radius $1/\sqrt{3}$ centred on the n points. Thus we have to find out whether these n circles have in fact a common point.

Formally we have to find out whether n circles have a common point when any three of them have one. (*Is this plausible?*) A generalization to convex figures suggests itself, but in any case it is obviously profitable to study the special case in which $n = 4$. *Try some cases for circles...convex figures...any figure...see Fig. 9.6.*

Suppose then that there are four convex figures C_0, C_1, C_2 and C_3 such that any three (not C_i) have a common point (A_i). The triangle

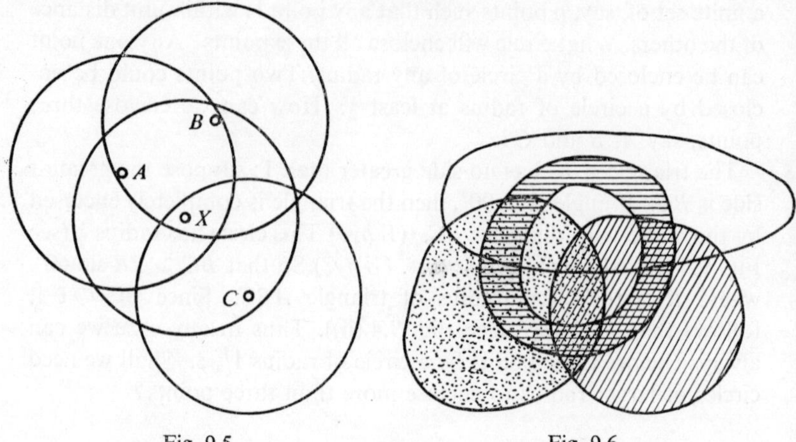

Fig. 9.5

Fig. 9.6

$A_1A_2A_3$ is contained in C_0 and similarly for the others. (*Why?*) If A_0 lies inside or on the triangle $A_1A_2A_3$ then A_0 lies in C_0 and so is a common point of all four figures for it already lies in C_1, C_2 and C_3. (*Why?*) If in fact A_0 lies outside, then $A_0A_1A_2A_3$ is a convex quadrilateral and the intersection of its diagonals is a common point. (*Why?*) In any case the four figures have a common point. The generalization to n convex figures each three of which have a common point is easily made. (*How?*) If the arguments used remain valid as $n \to \infty$ (*do they?*) we have shown that any figure of unit diameter can be covered by a circle of radius $1/\sqrt{3}$. *State formally the theorems that have been proved.*

The theorems implied by the discussion are:

(i) Helly's theorem: n plane convex figures, any three of which have a common point, all have a common point;

(ii) n points in a plane, any three of which can be enclosed in a circle of given radius, can all be enclosed in such a circle;

(iii) Jung's theorem. n points in a plane, any pair of which are within unit distance of each other, can all be enclosed in a circle of radius $1/\sqrt{3}$.

There is a sting in the tail when $n \to \infty$ for Helly's theorem may then be untrue if the figures are unbounded. An example of this failure is the set of half-planes $y \geqslant n$ for $n = 0, 1, 2, \ldots, \ldots, \infty$. The theorem is true, however, for an infinite number of bounded, plane, convex figures. Jung's theorem is true as $n \to \infty$ and can then be formulated as the assertion that every plane figure (*note*: *not necessarily convex*) of unit diameter can be enclosed in a circle of radius $1/\sqrt{3}$. An unsolved problem is to determine the figure of least area which covers every plane figure of diameter 1.

A diameter is the maximum distance between parallel supporting lines of a bounded, convex figure. A width is the minimum distance between such lines. Another theorem that may be deduced from Helly's asserts that every bounded, convex figure of width 1 contains a circle of radius 1/3 (Blaschke). Another theorem is due to Krasnoselski: if the rooms of a polygonal picture gallery are such that for any three paintings there is a point from which all three can be seen, then there is a point from which all the paintings of the gallery can be seen.

In addition to the previous references see also Steinhaus [114].

10

GEOMETRY

In geometry a pupil is learning about space, and at the same time learning about mathematics and about logical argument. In what remains of Euclid in the schools, these aims are often confounded and obscured. He learns about plane figures without achieving any mastery of the plane; mathematicians say they are no longer concerned with this kind of mathematics. And as to logical argument, while matters of question-begging, and of confusion between theorem and converse, still arise, there is usually no attempt at an axiomatic treatment, while much is heard of the word 'proof'.

The problem for the schools is so to conduct the discussion of fundamental geometrical configurations that (i) the pupil's spatial imagination is stimulated and developed, and (ii) he learns to think in terms and in modes that will support, and not conflict with, his later mathematical activity.

In various parts of the world several different attempts are being made to reconcile these aims. See, for example, the geometry section of the 'Program for College Preparatory Mathematics' of the College Entrance Examination Board in the U.S.A. for a taste of rigorous argument combined with an early use of analytical and vector methods. Or see the proposal, that vectors and vector-spaces be given a central place, made by Dieudonné at the O.E.E.C. Royaumont Seminar [89] and supported, with perhaps fuller awareness of its implications for the schools, by Servais [110]. And consider the German and Swiss programmes of 'movement' or 'transformation' geometry with their emphasis on reflections, rotations, dilatations, etc., as expounded by such writers as Jeger [66] and Botsch [16].

The first thing to realize is that school geometry could be different from what it has been, and what it has become, in this country. In the following pages an attempt is made to indicate possible alternative methods and lines of development, at what we have become used to calling stage A and stage B.

1. AN APPROACH TO GEOMETRY

It may be argued that, while geometry for a mathematician like Euclid or Hilbert begins with undefined, elementary, abstract entities such as points and lines and planes, geometry for a child today begins with the physical manipulation of real things, apprehended globally, used before analysed, and often 'special' (e.g. rectangle) rather than 'general' (quadrilateral).

The first geometry lessons can be centred on the handling, development and construction of interesting objects (Toblerone packets, Oxo cube wrappings); the making of these and other shapes—using square-ruled paper until skills with instruments are acquired; and making skeleton models of the same shapes using Meccano, or balsa wood, or milk straws joined by short lengths of pipe-cleaner.

Then attention may be focused on the *faces* of the solids—the squares, rectangles, isosceles and equilateral triangles that have become familiar shapes.

Suppose we have made, from paper, an open box in the shape of a cuboid, and we decide to fit a lid, which we cut out of paper.

In how many different ways can the lid be placed over the opening? One? Two? Four? Eight? It depends. Certainly four; but for some boxes it can be eight. Once we have got our pupils to appreciate this, they have grasped the essential difference between the rectangle and the square. This goes to the heart of the matter: in the language of chapter 3, the group of the symmetries of the square is of order eight, that of the rectangle of order four; and of the rhombus, four; parallelogram, kite and isosceles trapezium, two; the general quadrilateral, one. The master wall-diagram for the quadrilateral family is, in the first instance, that shown in Fig. 10.1, where the numerals are understood by the pupils as indicating 'the number of ways I could put it back in the hole if I cut it out'.

Later, Fig. 10.2, with the symmetries suggested, replaces Fig. 10.1, and the numerals have rather greater significance.

The idea of symmetry as derived from experience of 'filling the holes', is not quite a sufficient foundation for our geometry; we require also the notion of space-filling, of covering the plane with

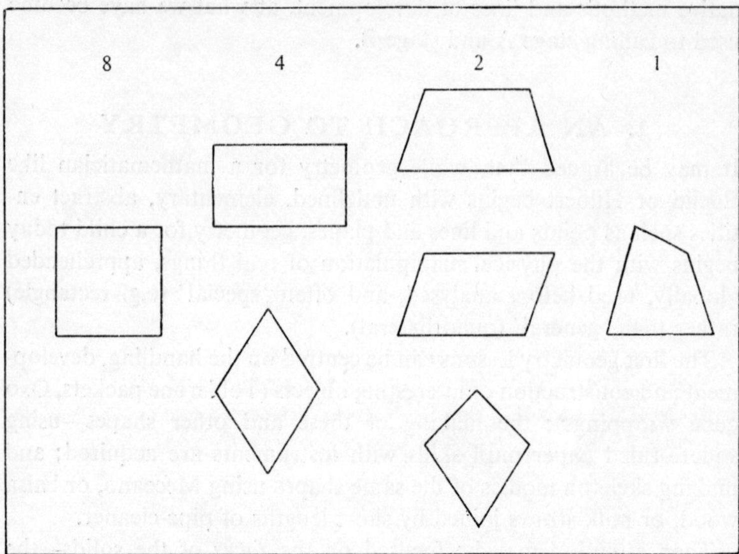

Fig. 10.1

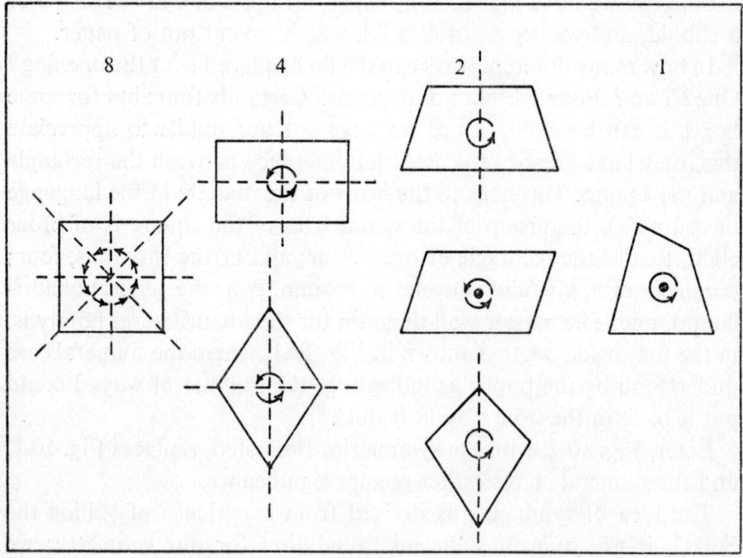

Fig. 10.2

congruent tiles. We can build an endless wall with bricks, can cover a boundless floor with square lino-tiles or a boundless hearth with rectangular tiles. The two ideas are connected by the fact that the hole the tile fills can be a hole in a tiling (see Fig. 10.3).

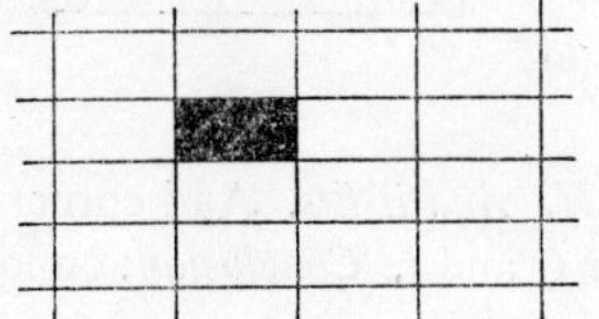

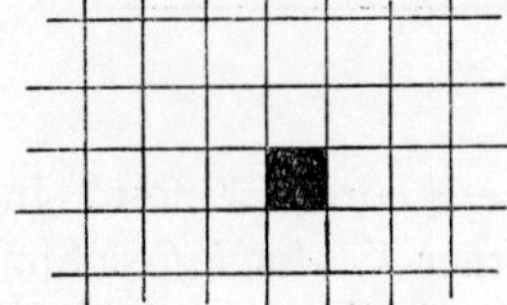

Fig. 10.3

A lesson on the properties of the rectangle

Each pupil is provided with two identical thin, stiff, rectangular cards, about $3\frac{1}{2}$ inches by $2\frac{1}{2}$ inches. We remind ourselves how one card 'fits exactly' over the other, and of the things we can do with the top card to repeat the 'fit'—(1) put it down in the original position, (2) turn it *over* (*top to bottom*), (3) turn it *over* (*side to side*), (4) turn it *round* (*through 180° in the plane, we shall say later*). Also we remind ourselves about 'tiling'—covering the desk with our own and neighbours' tile cards.

We draw the outline of a card, twice over, on a sheet of paper. *If it will amuse the class, we are glaziers, fitting a pane of glass into a frame.*

Fig. 10.4 (*a*)

We label the corners of one outline, as shown; each corner of each card is similarly labelled, front and back.

We begin with a card within the second outline, in the original *HOXI* position. Turn it *round* (*no. 4 in the list above*) into the *XIHO* position. What we have before us now is Fig. 10.4 (*b*).

Fig. 10.4 (*b*)

Where is corner *X* now? In corner *H*. *And it fits.* And corner *H*? In corner *X*. *And it fits.* Similarly for *O* and *I*. *Conclusion*: opposite corners fit (*later*, *opposite angles are equal*).

Now think about edges. Where has edge *HO* gone? It lies on top of edge *XI*. *And it fits.* And so on. *Conclusion*: opposite edges fit (*opposite sides are equal*).

Now back to the original position. This time, turn it *over* (*top to bottom, no. 2 above*; or *side to side, no. 3 above*).

Summary of conclusions: opposite edges fit (*once more*); consecutive corners fit. And this means that corner *H* fits *X* and fits *O* and fits *I*—all the 'corners' are equal. But are we entitled to say each is a square corner (right angle)? Yes! because tiling is possible, and corner *H* and corner *I* (or *O*), juxtaposed in adjacent tiles, give a straight join.

For diagonal properties, it is best to begin with one diagonal. In the left-hand outline, draw the diagonal from *H* to *X*; and draw this diagonal on *each* face of one of the cards.

Begin in the original position (*HOXI*). Turn round in the plane—no discovery; the line that stretches from *H* to *X* also stretches from *X* to *H*. *But we note the interchange, implying congruence, of the two triangular portions of the figure.* Back to the original: now turn over (*side to side*). That's funny! The line that stretches from top left to bottom right also stretches from top right to bottom left. *Conclusion:* one diagonal is as long as the other—the diagonals are equal.

Now draw both diagonals, in the left-hand outline and on one face of a card. Turn the card round from the original position *HOXI* to position *XIHO*; its diagonals change places, the point of intersection is in its original position, each segment of a diagonal is interchanged with the other segment—the diagonals bisect one another.

Notes

(1) The procedure has been 'experimental' because the pupils are immature. Strictly, physical manipulation of the cards is not necessary: if one grants (as 'axioms'!) the symmetry and space-filling properties, the side-angle-diagonal properties may be deduced. The purpose of the experimentation is to make the deductions explicit, and also to provide a store of mental images which will constitute the pupil's knowledge of 'rectangle'. This is far removed from 'measure the sides of the rectangle and tell me what you notice'.

(2) Properties of the rectangle have not been separated out from those that belong to the parallelogram. This separation will take place later, when the parallelogram is examined. On the other hand the square will be explored right away—the essential feature being that it is 'just like a rectangle only you only need to give it a quarter-turn'.

(3) There has been no use of congruence, no dependence on 'angle as an amount of turning', no reference even to parallel lines.

Something of the concept of parallelism emerges from observation of 'tilings'—the straight 'joins' march across the plane, and however far we go there is always room for the width of one of the tiles between adjacent joins; and each member of each of the two sets of joins ('up-and-down', and 'across') makes square-corners with each member of the other set.

(4) We have used both axial and rotational symmetry, but without reference to either axes or centres of symmetry. These will impose themselves in due course, possibly when experience of other tilings has been gained. In the long run the idea of a symmetry as a one–one mapping of the set of points of the plane on to itself is important, but we have been discussing the very early stages. An introduction to 'reflection geometry' is suggested in §2 of this chapter.

(5) The notion of 'composition of operations' is at hand during the manipulations, and a suggestion from the class may prompt some discussion of this.

(6) The reader will have realized why the letters *H*, *O*, *X* and *I* were used to designate the 'corners', and may amuse himself by choosing labels for the corners of a square for a similar exercise.

Lessons based on the exploration of a tiling

After experience of 'covering the plane' with congruent tiles of various shapes—rectangles and squares (and their triangular halves), isosceles triangles (made from these halves), kites and rhombuses (made from isosceles triangles) and parallelograms and scalene triangles (made from half kites), much can be discovered from an examination of a drawing of a tiling:

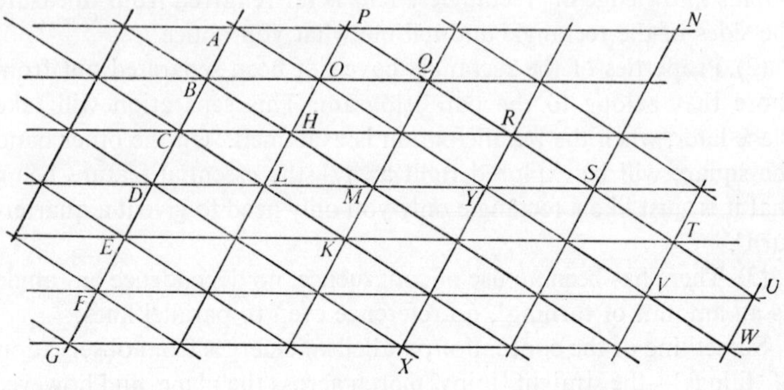

Fig. 10.5

For example, the suggestion that the angle-sum of a triangle is a straight angle can hardly be ignored. Three lessons from rather later stages are outlined.

The ratio of segments on two transversals

What fraction is *AC* of *AG*? Of *AF*? What fraction is *PR* of *PU?* What is the ratio of *AD* to *AF*? And the ratio of *PS* to *PU*? Show me $\frac{3}{5}$ of *VF*, starting from *V*; and $\frac{3}{5}$ of *VA*, starting from *V*. Show me $\frac{3}{5}$ of *FV*, starting from *F*; and of $\frac{3}{5}$ *AV*, starting from *A*.

Towards the left side of a sheet of ruled paper rule a line crossing consecutive printed rulings at *A*, *B*, *C*, *D*, etc.; and towards the right side rule another line, and on it mark points 'like' the points *P*, *Q*, *R*, *S*, ..., of Fig. 10.5. *In what sense 'like'? We are making explicit the correspondence* $A \rightarrow P$, $B \rightarrow Q$, *etc., which the set of equally spaced parallels establishes. Ask questions, as above, on this figure. Then have the pupils draw a similar figure on plain paper; and discuss it.*

Some properties of similar triangles

Look at the triangles *AFV* and *PLS* in Fig. 10.5. They differ—in position and in size; they are alike—in shape; the angles are equal, the sides are in ratio 5:3. What about triangles *AFV* and *QMS*? *AFV* and *PSN*? Show me another triangle equiangular to *AFV*, and comment on its side-lengths; another; another; .... Show me a triangle whose sides are $\frac{4}{5}$ of the sides of *AFV*: what are its angles? One with sides $\frac{2}{5}$ ...? Shall we risk a generalization?

Now what about the areas of these triangles? Counting the 'elementary triangles' like *CDL*, we collect results like

'Ratio of sides = 5:3 ⇒ ratio of areas 25:9.'

Collinearities and a concurrency

How many sets of parallel lines does Fig. 10.5 contain? Look now at the parallelograms they form—the smallest ones you can see. Name typical ones—three different shapes, *CDLH*, *CLMH*, *BCLH*. In each type can you see the role played by the third set of parallel lines? It provides one of the two diagonals. What about the other diagonals? Consider for example the diagonals *AH*, *HK*, *KX*; is there one line or three? *AHK looks* straight: the proof is by angle-summation at *H*; or by considering the fate of the 'tile' *AOHB* as we slide it 'south-westwards' into *BHLC*, then 'south-eastwards' into *HMKL*, along the guide lines *AG* and *CK*; or by sliding *AOHB* along *AH* produced, which must take *AB* into *HL* and *AO* into *HM*.

Thus we discover, or rediscover, whole sets of collinearities like *AHKX*, like *CMU*, like *DHQ*—three more sets of parallel lines.

Now if we draw in *AX*, *WD* and *GY*, these are the medians of the triangle *AGW*, we see them to be concurrent and to trisect one another.

And everything we have done with this tiling we could do with any other such tiling.

Lessons on the circle

We turn now to the most symmetrical of all figures, the circle.

The basic fact, that a circle is conserved by any rotation in the plane about its centre, may be made manifest by rotating a tracing of a circle over the original circle. It helps if one radius of each circle is drawn.

Similarly, we exhibit the connection between equal chords, equal arcs and equal angles at the centre, and the property that equal chords are equidistant from the centre.

The diametral symmetries may be examined in terms of the symmetry of the isosceles triangle, or may be explored experimentally.

Typical exercises involving tracing paper and folding are: 1. Given a circle on tracing paper, find its centre by folding. 2. Given a part of a circle drawn on tracing paper, complete it by folding and tracing, without finding its centre.

Tangents appear, in the limit, as a secant recedes from the centre *either* remaining parallel to itself *or* rotating about some external point. *The disappearing isosceles triangle, in this latter case, guarantees the perpendicularity of tangent and radius.*

The treatment of the 'angles at centre and circumference' property should include discussion, in symmetry terms, of the related arcs, perhaps on the lines of the following lesson sequence.

Angle properties of the circle

Pupils draw on their papers a pair of intersecting lines, and move over them a circle drawn on tracing paper. We first explore the situation by moving the circle freely over the intersecting lines, noting, and classifying the main possible configurations, Fig. 10.6 (*a*)–(*h*).

We concentrate on positions like (*d*) to (*h*) in which the circle meets each of the two lines, and attend to the lengths of the arcs intercepted within the acute angle. As we move from (*d*) through (*e*) and (*f*) to (*g*) we observe the arc *AB* decreasing steadily as *DC* first decreases to zero, and then increases within the vertically opposite angle.

Go to position (*e*) and mark on the circle points *B* and *C*. Slide these along the secant until *C* reaches *O*, as in (*f*). In this position trace the other secant on to the tracing paper, and then slide *B* and *C* back to their original position in (*e*). This gives (*e*′).

Note that the movement which we have just performed is, throughout, simply a sliding of the tracing paper to and fro along the line *BC*. The difference between arc *AB* and arc *DC* in (*e*′) is measured by arc *A*′*B*, which is the arc *AB* of (*f*). In fact, in the transition (by

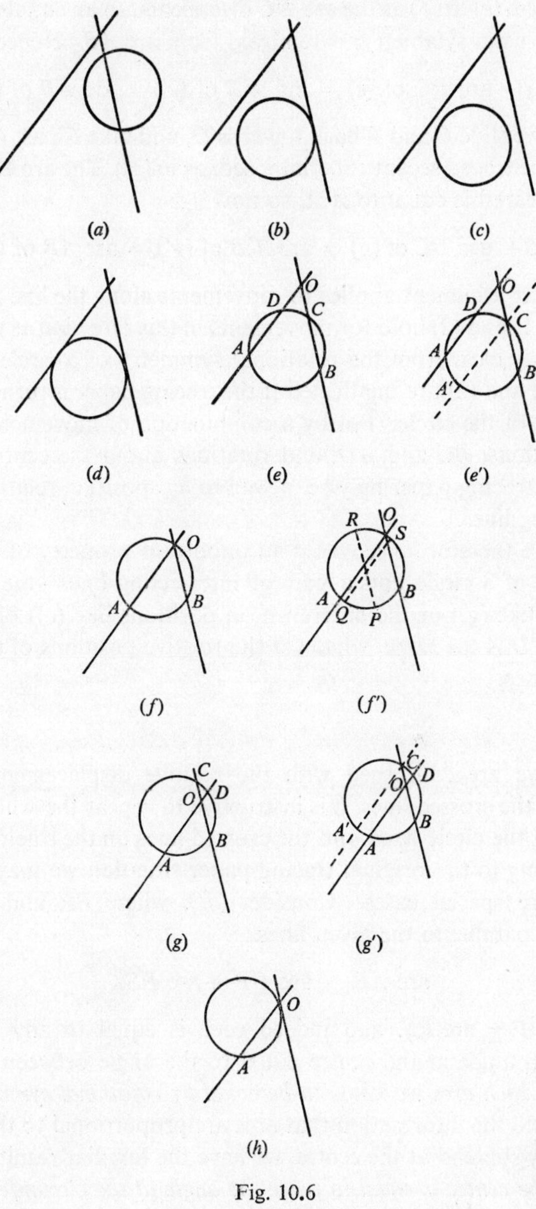

Fig. 10.6

sliding) from (e) to (f) as the arc DC disappears, so an equal part AA' (equal, by axial symmetry) is removed from arc AB. Hence

$$\text{arc } AB - \text{arc } DC \text{ of } (e) = \text{arc } A'B \text{ of } (e') = \text{arc } AB \text{ of } (f). \quad (1)$$

If now we slide C and B back towards O, and take C past O, we get (g) with the traced secant superimposed, as in (g'). The arc CD which has reappeared is equal to $A'A$, so now

$$\text{arc } AB + \text{ arc } DC \text{ of } (g) = \text{arc } A'B \text{ of } (g') = \text{arc } AB \text{ of } (f). \quad (2)$$

A similar argument applies to movements along the line AD. Relations (1) and (2) hold for movements in this direction as well. It is immediately clear from the rotational symmetry of a circle that relations (1) and (2) are unaffected if the tracing paper is turned about the centre of the circle. But by a combination of movements in the two directions, BC and AD, and rotations about the centre of the circle, the tracing paper may be moved to *any* position relative to the intersecting lines.

We have therefore discovered an important property of the configuration of a circle and a pair of intersecting lines—the sum (in positions like (g)) or the difference (in positions like (e)) of the arcs AB and CD is the same, whatever the relative positions of the circle and the lines.

Exercise

Since we are concerned with the relative displacement of the circle and the crossed lines, it is instructive to repeat the whole argument with the circle fixed and the crossed lines on the tracing paper.

Returning to the original tracing-paper situation we may look at some more special cases. Consider (f'), where PR and QS are diameters parallel to the given lines.

$$\text{arc } AB = \text{arc } QP + \text{arc } RS.$$

But arc QP = arc RS, and indeed each is equal to any arc subtending an angle at the centre equal to the angle between the two lines. *All such arcs we know to be equal by rotational symmetry.* If we now add the information that arcs are proportional to the angles which they subtend at the centre, we have the familiar result that *the angle at the centre is equal to twice the angle at the circumference.*

Exercise

Taking another special position, move into position (*h*) by sliding the circle, keeping the diameters parallel to the secants, until the secant *CB* becomes the tangent at *C*, as in (*h*). Deduce that the angle between a tangent and a chord is equal to the angle in the alternate segment.

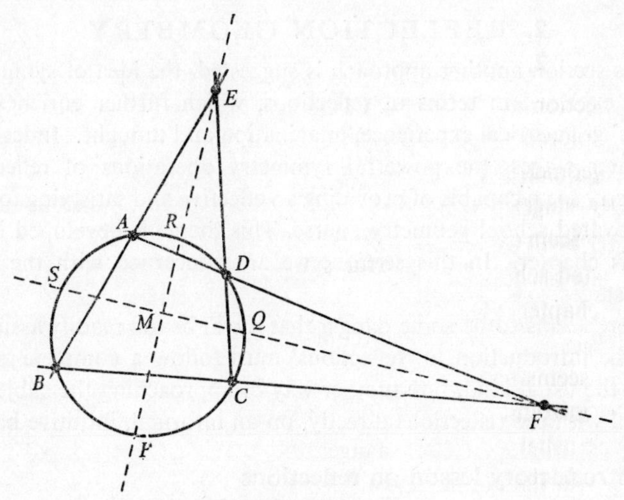

Fig. 10.7

Exercise

In Fig. 10.7 *ABCD* is a cyclic quadrilateral; the bisectors of the angles *E* and *F* meet at *M*. Prove that these bisectors are mutually perpendicular.

Because $\quad$ angle BEP = angle PEC,

$$\text{arc } BP - \text{arc } AR = \text{arc } PC - \text{arc } RD,$$

and so $\quad$ $\text{arc } BP + \text{arc } RD = \text{arc } PC + \text{arc } AR.$

Similarly, $\quad$ $\text{arc } BS + \text{arc } DQ = \text{arc } CQ + \text{arc } SA.$

Adding, $\quad$ $\text{arc } PS + \text{arc } QR = \text{arc } PQ + \text{arc } RS,$

and so $\quad$ angle SMP = angle PMQ,

and each is a right angle.

Note

In lessons such as this the teacher's counterpart to the pupil's tracing paper is a sheet of cellulose acetate, 0·005 inches thick, moving over a sheet of paper fixed to the blackboard or wall; the acetate adheres to the paper by static electricity.

2. REFLECTION GEOMETRY

In this section another approach is suggested, the idea of symmetry being explored in terms of reflections, which further enriches the pupils' geometrical experience, imagination and thought. Indeed, in the later stages, the powerful symmetry operations of reflection geometry seem capable of providing an effective and satisfying tool in a renovated school geometry course. This theme is developed in §3 of this chapter. In this section, we are concerned with the very early stages.

There seems to be some danger that it will be too readily assumed that the introduction to 'reflections' must follow a complete 'stage A' of the usual kind, but another way of approaching the subject is to deal with the reflections directly, on an informal, intuitive basis.

An introductory lesson on reflections

Naturally, in this context, one thinks of mirrors. They present the child with the immediate experience and allow many of the main points to be examined. Questions can be asked about reversal, the distances of the object and the image from the mirror, what happens when the object moves and when the mirror moves, or when it rotates. On the other hand, the mirror reflects the whole background, and particular images are not easily isolated. Moreover, with the mirror situation, we are confronted with the plane face of the mirror, whereas in reflection geometry we wish to deal with the reflection of a plane figure in a line; that is, *looking at the mirror edge-on.*

Now, in general, by restricting ourselves to the plane, we cannot slide a 'copy' of the object to cover the image. Before this can be done, the 'copy' may have to be turned over. If, however, movement out of the plane is allowed, then a rotation of 180° about the line of the mirror as axis places the object over the image. How can we make a simple representation of this process? A possible aid is

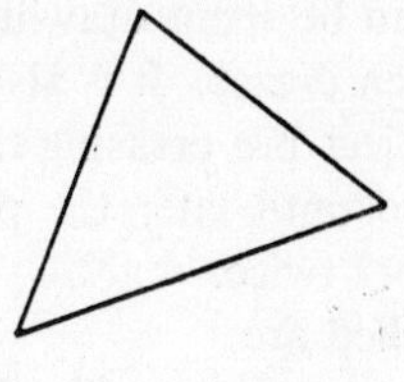

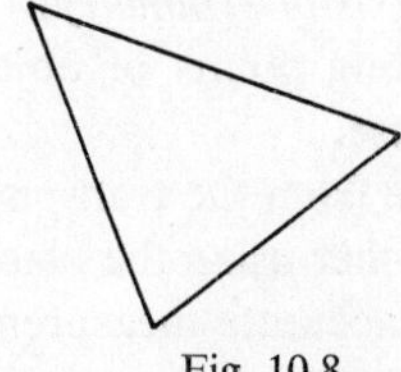

Fig. 10.8

transparent paper, usually available in rectangular sheets. Cardboard cut-outs might also be useful.

Suppose any figure, say a letter *F*, is heavily drawn in pencil near one corner (Fig. 10.9). Then if the corner containing the figure is neatly folded over and a firm crease made, this can be our 'mirror line'.

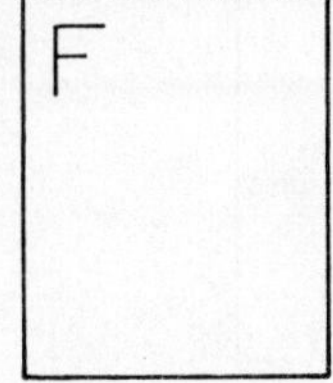

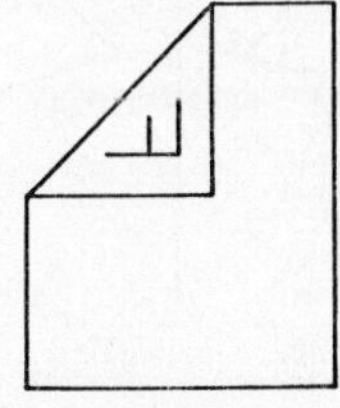

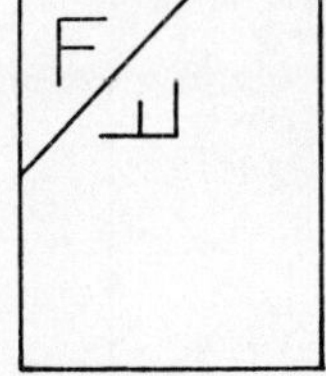

Fig. 10.9

Since the paper is transparent, the 'object' can still be seen through the paper, but of course in the image position. A copy of the object is now made by drawing over the outline of the object. When the fold is reopened an indented image will be seen in the usual reflection position. This itself can now be pencilled in. In this way the children can produce the images of quite complicated objects in a short time and without instruments. This is the starting-point of the lesson.

A fair amount of time can be spent drawing different designs and finding an image for a given crease. It is also worth examining the effect of using the same object but creasing the paper in a different place. It is better to leave until later the production of multiple images by creasing the paper twice.

The points to be established are:

(i) That the figures are congruent; this simply means that one figure covers the other.

(ii) That the image is 'turned over' (*so it is important not to select only objects which have an axis of symmetry*).

(iii) That the corresponding points of object and image are the same distance from the crease.

The next phase is to move from the transparent paper to ordinary paper and pencil via the teacher using the blackboard. The intention is not so much to produce accurate measurements but correct concepts. A line, vertical at first, is drawn on the board and a simple figure drawn on the left. A pupil is then invited to draw the image 'as it would look if you had used the paper-folding method.'

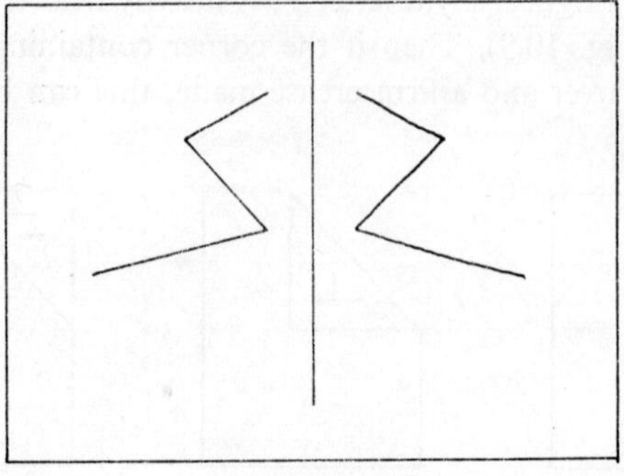

Fig. 10.10

Next the class may be asked to copy various 'objects' and then sketch the images. Examples should bring out the special feature involved when the object crosses the crease. The position of the mirror can also be varied.

At this stage, for a pupil in difficulties, the check is to return to the transparent paper.

Next, the children may be asked to work in pairs, on the same piece

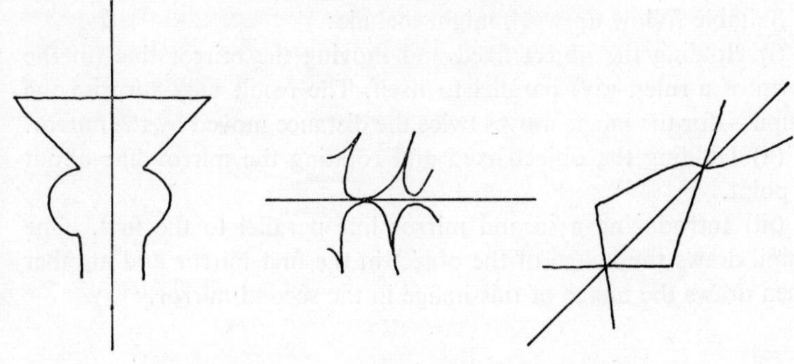

Fig. 10.11

of paper. A line is drawn on the paper and one pupil slowly draws an object, while the other draws the image. Results of the following type are to be expected.

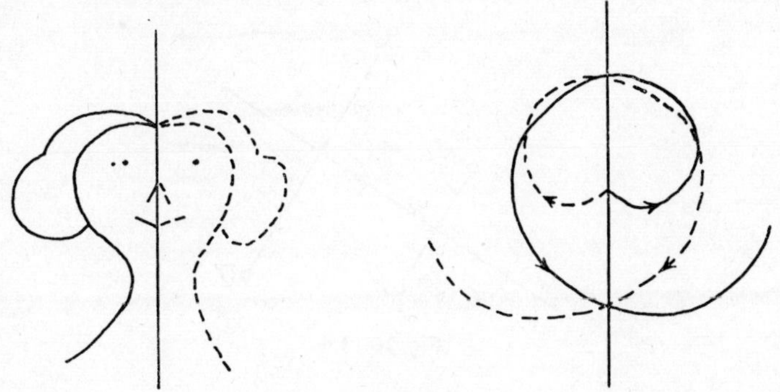

Fig. 10.12

An important element missing so far has been the effect of moving the object. For this an object and image may be cut out of cardboard. The mirror line is held fixed and a pupil attempts to move the image in response to movements of the object. The points to be established are:

(i) When the object moves a certain distance away or towards the mirror, so does the image.

(ii) When the object is rotated, *the image turns in the opposite direction*—this may cause some difficulty.

Suitable follow up work might include:

(i) Holding the object fixed and moving the mirror line (in the form of a ruler, say) parallel to itself. The result may surprise the pupils—for the image moves twice the distance moved by the mirror.

(ii) Holding the object fixed and rotating the mirror line about a point.

(iii) Introducing a second mirror line parallel to the first. One pupil draws the image of the object in the first mirror and another then draws the image of this image in the second mirror.

Further lessons on reflections

Construction of a 'mirror'

Let us look again at the paper-folding method of producing reflections in a line. P, Q are two points whose images in the mirror line m are P', Q'.

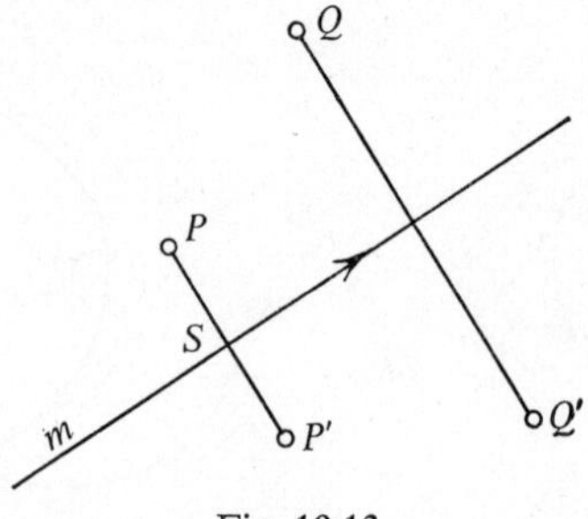

Fig. 10.13

Notice that since we may fold the paper from left to right or from right to left, the image of P' is P. Thus if we reflect a point and then its image in the *same* mirror, we get back to where we started. It is also clear that $PQ = P'Q'$.

Suppose that PP' meets m at S. Then *from the way we have folded the paper*, $PS = SP'$ and PP' is perpendicular to the mirror.

We have seen that PQ folds on to $P'Q'$. Can we see what happens to the whole line PQ? It folds on to the line through $P'Q'$ (see Fig. 10.14).

In general, the line PQ will meet the mirror at O, but since O is *on* the mirror, when reflected it stays where it is. In that case the line

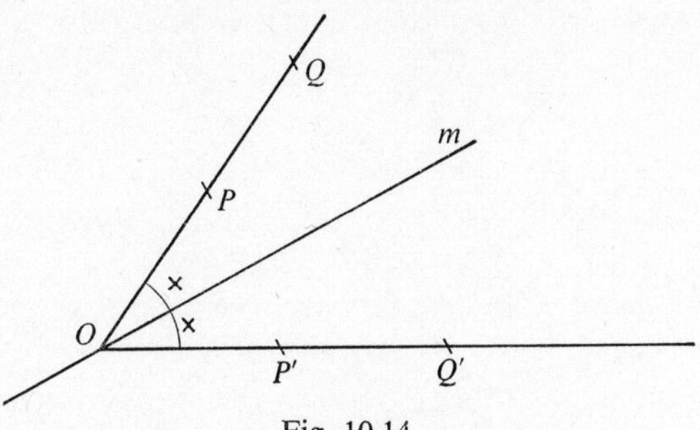

Fig. 10.14

$P'Q'$ must also pass through O. Furthermore, as the region between line PQ and m folds over on to the region between line $P'Q'$ and m, the mirror bisects the angle between a line and its image. *What happens if the line is parallel to the mirror?*

Next consider the line PQ'; what happens when it is reflected in m? (see Fig. 10.15).

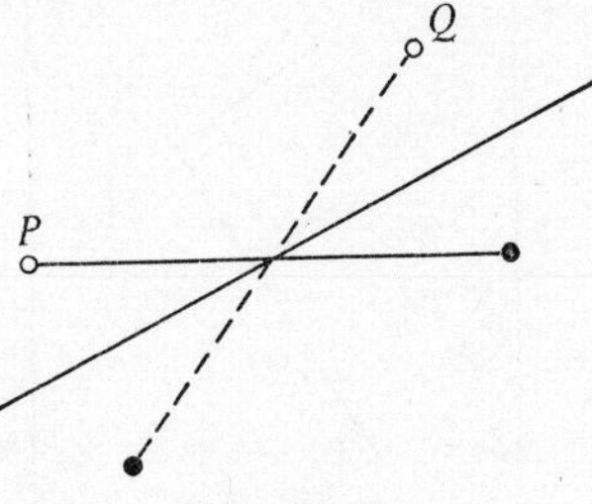

Fig. 10.15

Since the image of P is P' and the image of Q' is Q, line PQ' has as its image, $P'Q$. Thus PQ', $P'Q$ intersect on the mirror, since this point of intersection stays put on reflection. This suggests that if we are given two points and their images, we have an easy way of finding the position of the mirror using a ruler alone.

Can you do it? Look at Fig. 10.16. Will the construction always work?

Reflection in two mirrors

We next wish to find out what happens when we reflect the image of a figure in a mirror, m_1, in a second mirror, m_2. There are two cases to consider, when the mirrors are parallel and when they are not parallel.

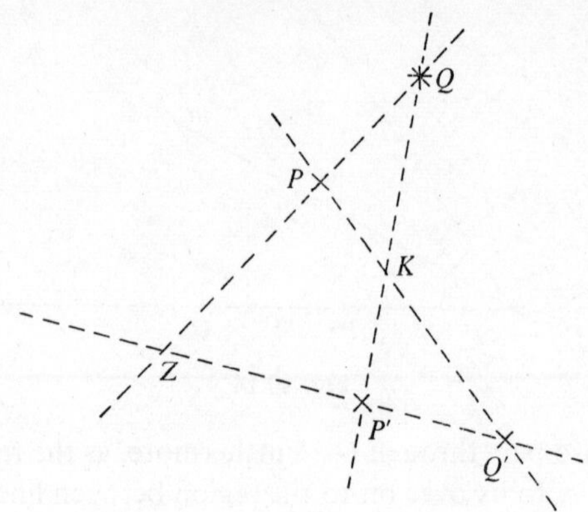

Fig. 10.16

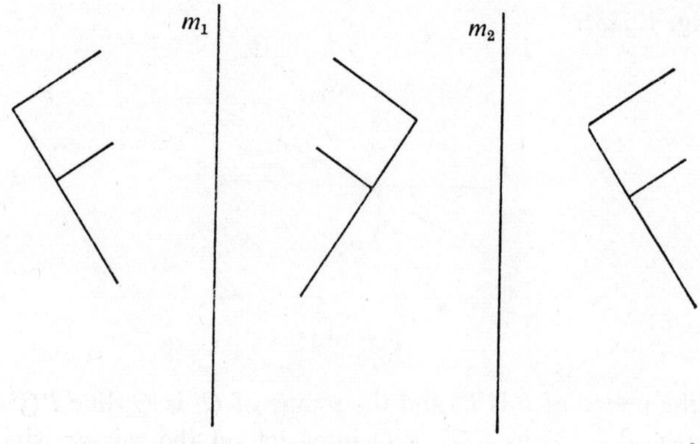

Fig. 10.17

Mirrors parallel (Fig. 10.17)

With the transparent paper, we should first fold about mirror m_1, to produce an image of the figure. Then we unwrap the paper and fold again along the mirror m_2. There are two things to notice, first, the final image is the same way up as the figure; secondly, since the creases are parallel, we folded the paper twice in the *same direction*. Thus the effect is that of shifting the figure at right angles to the mirrors; that is a *translation*. By just picking out a point of the figure, we notice something else (Fig. 10.18).

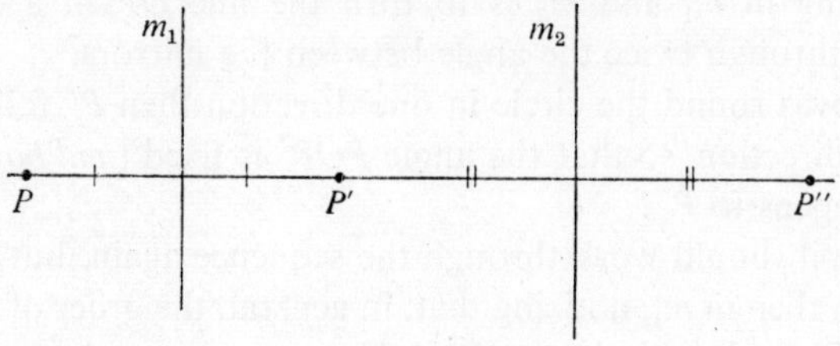

Fig. 10.18

P folds on to P' and P' on to P'', and as an object is always as far from the mirror as its image, the total translation is twice the distance between the mirrors. As an exercise, the pupils may verify this result when the point P is between the mirrors.

Mirrors not parallel

Next suppose the mirrors meet at a point O (Fig. 10.19).

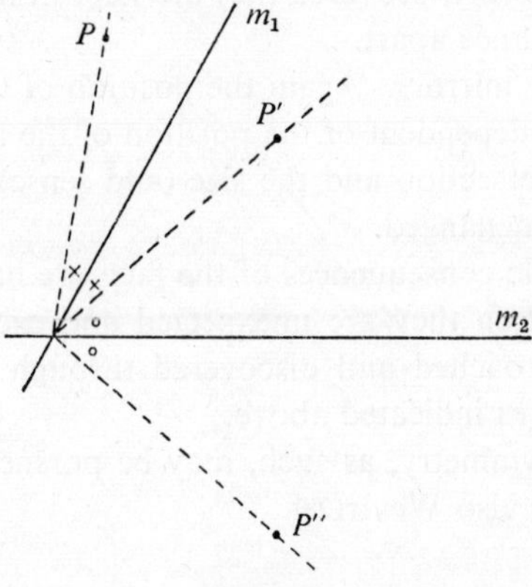

Fig. 10.19

P is first reflected to P' in m_1 and then P' is reflected to P'' in m_2. It is worth examining the effect as P is moved about. In order to fix ideas, let P move on a fixed circle with centre at O, and observe what happens to P''.

Now we recall that the mirror bisects the angle between a line and its image. By examining the diagrams above, it is seen that the effect of reflecting in m_1 and m_2 is to turn the line OP in a clockwise direction through twice the angle between the mirrors.

If P moves round the circle in one direction then P'' follows it in the same direction, so that the angle POP'' is fixed (*and how large?*). What happens to P'?

The pupil should work through the sequence again, but reflecting first in m_2, then in m_1, noticing that, in general, the order of reflection matters. They should also explore the special case when m_1 and m_2 are at right angles to each other. They may well examine the case of reflection in four mirrors which all pass through a single point, first by painstakingly considering the four separate reflections, then by taking the mirrors two at a time.

Two important results

In §3 we shall make much use of the following facts:

(i) Parallel mirrors. If a fixed point P has image P' in m_1, and P' has image P'' in m_2, then the position of P'' is independent of the position of the mirrors, provided they are kept in the same direction and the same distance apart.

(ii) Intersecting mirrors. Again the position of the image P'' of a fixed point P is independent of the position of the mirrors, provided their point of intersection and the size (and sense) of the angle between them are unchanged.

These are simple consequences of the facts we have just been discussing; for children they are unexpected and exciting new knowledge, to be approached and discovered through further practical activity on the lines indicated above.

The study of symmetry, as such, may be pursued along the lines of chapter 3. See also Weyl [128].

3. COMPOUNDING REFLECTIONS

The purpose of this section is to show one way of developing previous experimental investigations of reflections, rotations and translations, and to link them together in a closed system. Technically this system is the group of isometries in the Euclidean plane, which is a closed set of transformations of the plane which preserve size and shape. After the examples of the preliminary lessons we shall also be developing an algebra of these operations, or transformations, which gives an example of the universality of algebraic structure. We are in fact attempting in a few pages to show the link between the elementary work of §2, and an overall view of geometry itself. Work at this level would prepare the way for an axiomatic approach later.

We are concerned with reflections in a line, reflections in a point (also referred to as *half-turns*), rotations and translations (also treated under vectors in chapter 11). Reference can be made to chapter 3 for work on symmetry groups and to chapter 12 on matrices.

A reminder is necessary here that when we talk of these three types of transformation we are discussing their effect on *the plane as a whole*. That is, when we speak of reflection in a line, we have in mind reflecting the whole plane in the line. Every point of the plane is mapped on to its image in the line. To think concretely we may consider the whole plane to be turned through 180° about the line as axis; but if we think this way, care is needed when this work is extended to three dimensions. Likewise, when we speak of a rotation, the whole plane is turned through a given angle about a fixed point. This means that we interpret the last result stated in §2 as asserting that successive reflection of the plane in two concurrent lines at angle θ is equivalent to a rotation of angle 2θ, and is independent of the orientation of the axes of reflection. The intuitive picture of translation is that the whole plane moves a given distance in a definite direction.

We shall study the composition of these transformations, paying attention to special cases and solving some instructive problems. Drawings should be made at every stage, the corresponding algebraic and geometric treatments should be compared, and where appropriate the work should be illustrated by moving cardboard shapes over a table and other concrete activities. The main aim is not the learning of theorems but the expression of our experience in an algebra which we build up as we go along. In this outline we proceed fairly directly to the goal of establishing the closed system of isometries of the plane, that is of being able to reduce any number of transformations, which are performed successively, to one single transformation. In the classroom many of the supplementary questions at the end of the section would have to be studied as examples on the way.

We will start by remembering what is involved in the notion of *reflection in a line.* It will be convenient to denote reflection in some particular line by a letter such as **l**. Note that **l** is not the name of the line, it is used to denote the operation of reflection in the line denoted by *l*.

Consider the case of reflection in two lines, referred to at the end of §2. First the special case. If the lines are parallel we have a translation, its magnitude being twice the perpendicular distance between the lines. *It may help to regard this as a special case of a rotation, the centre being at infinity; but this point of view should not be imposed on a class. If they suggest it themselves it should be continually re-appraised throughout the lesson as further points are raised.*

Reflection **l** followed by reflection **m** will be denoted by **lm**.

Q. 1. Is **lm** = **ml**?

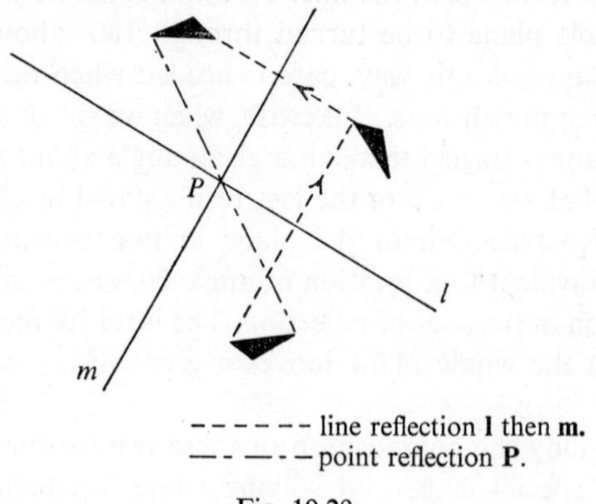

Fig. 10.20

For parallel axes of reflection this is not so (unless of course the lines coincide); the translations are in opposite directions. For intersecting axes, **lm** represents a rotation in one sense, **ml** a rotation of the same magnitude in the opposite sense. In fact, **lm** = **ml** if, and only if, the lines are perpendicular, or if they coincide. Neglecting the trivial case of coincidence, if **lm** = **ml** the corresponding rotation is a *half-turn*, that is, a rotation of the plane of two right angles about an axis, perpendicular to the plane, through the point of intersection of the axes. We denote by **P** the operation of a half-turn about the point *P*. In another sense this particular transformation can be seen as a reflection in a point.

The importance of the half-turn concept is that a point reflection **P** can be represented by successive reflections in two perpendicular lines. Thus **P** = **lm**.

Again, is **Pl** = **lP**?

If, and only if, *P* is on *l*. In this case, a sketch will show that the resultant operation is a reflection. (*In what axis?*)

Q. 2. What about **ll** or **l**^{2}, **PP** or **P**^{2}?

Obviously these restore the original situation and we have identity operations. We write **l**^{2} = **I**, **P**^{2} = **I**.

Q. 3. What is the meaning of an equation like **ab** = **cd**? Each member of the equation is a composition of two line-reflections, which is equivalent to a rotation about a point through twice the angle between the lines. The equation asserts that **ab** and **cd** produce the same rotation, which is only possible if the four lines are concurrent, and the angles between the lines *a*, *b*, and between the lines *c*, *d*, are equal and in the same sense (with an obvious modification if we are dealing with parallel lines). (See Colour-fig. 12.)

Observe from the diagram that if **ab** = **cd** then **abd** = **c** (i.e. reflecting in turn in lines *a*, *b*, *d* is equivalent to reflecting in line *c*), and **abdc** = **I**. Also, **cab** = **d** and **dcab** = **I**. Algebraically, we may proceed thus: if **ab** = **cd**, then **(ab)d** = **(cd)d**,

hence

$$\mathbf{abd} = \mathbf{cd}^2 = \mathbf{cI} = \mathbf{c}\text{; etc.}$$

It is often convenient to remember that if *a*, *b*, *c* are any three concurrent lines there is a fourth line *d* through the same point, such that **abc** = **d**.

The composition of two rotations

Q. 4. What is the resultant of two rotations?

In general it is a third rotation, and it is easy to see that the magnitude is equal to the algebraic sum of the magnitudes of the two component rotations.

But where is the centre of the composition of the rotations (A, θ) and (B, ϕ)? We use the notation (A, θ) to denote the rotation about the point *A* through an angle θ.

We can use the fact that a rotation (A, θ) is equivalent to reflection in any two lines through *A* inclined at $\frac{1}{2}\theta$ to one another, and similarly for *B*. The sophisticated procedure is to use the line *AB* as the *second* axis of reflection through *A* and also as the *first* axis of reflection through B. Thus: draw line m through the points A and B, draw line l through point A such that the angle from l to m is $\frac{1}{2}\theta$, and line n through point B such that the angle from m to n is $\frac{1}{2}\phi$.

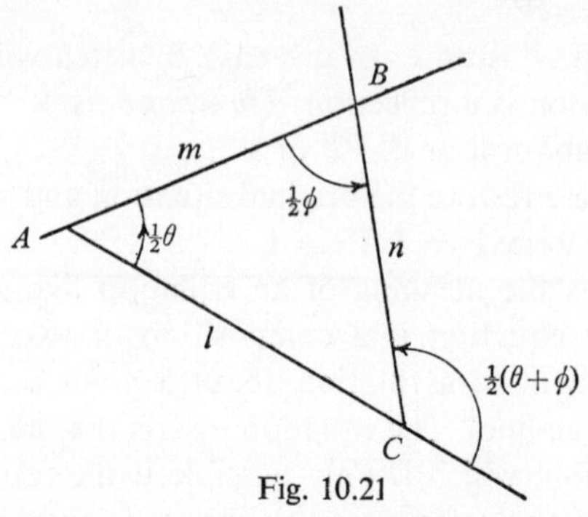

Fig. 10.21

Then the resultant of the rotation (A, θ) followed by (B, ϕ) is given by $\mathbf{lm}.\mathbf{mn} = \mathbf{lm}^2\mathbf{n} = \mathbf{ln}$, which is (in our figure) a counterclockwise rotation round C through an angle of twice $\frac{1}{2}(\theta+\phi)$.

In Fig. 10.22 the sophisticated procedure is led up to in stages. Fig. 10.22 (*a*) shows the primitive rotations; 10.22 (*b*) shows each rotation replaced by two reflections in axes at the correct angle but with no particular orientation; 10.22 (*c*) shows two of the axes in the line of AB, and two of the images superposed; 10.22 (*d*) shows the resultant rotation.

It is instructive to draw Fig. 10.22 (*a*) to (*d*) on clear tracing paper and superpose the tracings two, three, four at a time.

Q. 5. Are there any exceptional cases?

When $\theta+\phi = 0$ or 2π, the lines l and n are parallel, and we have a translation.

When $\theta+\phi = \pi$ or 3π, l and n are at right angles, and we have a half-turn or point-reflection.

The composition of two half-turns

Q. 6. What is the resultant of two half-turns?

If we think of a half-turn as a rotation through π, then the resultant of two, by Q. 4, should be a rotation through 2π. If we think of it, as in Fig. 10.22, as produced by reflection in l, m and n, where l is perpendicular to m and m is perpendicular to n, then the resultant $\mathbf{ln}$ (as may be seen in Colour-fig. 13) is a translation. If we

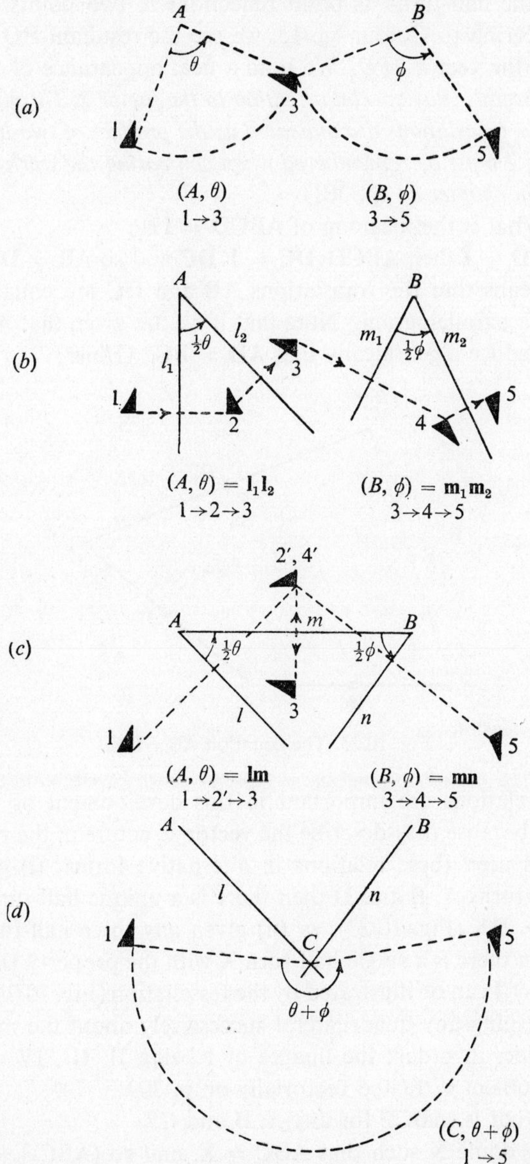

Fig. 10.22 (*a–d*). Composition of rotations (*A*, θ) and (*B*, φ).

think of the half-turns as point reflections in two points P and Q, again referring to Colour-fig. 13, we see the resultant **PQ** is represented by the vector $2PQ$. We note a neat appearance of the 'mid-point theorem'. *Pay special attention to the factor 2. Throughout this chapter our translations are indicated as the product of two half-turns; this factor 2 must be remembered when comparing the work here with the work in chapter 11* (p. 301).

Q. 7. What is the meaning of **ABCD** = **I**?

If **ABCD** = **I** then **ABCD**.**DC** = **I**.**DC** and so **AB** = **DC**.

This means that the translations **AB** and **DC** are equal, or that $ABCD$ is a parallelogram. Note that if we are given that **AB** = **DC** we may deduce algebraically that **AD** = **BC**. (*How?*)

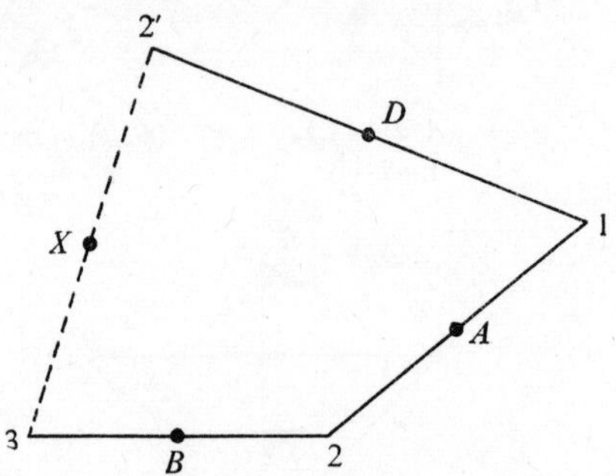

Fig. 10.23. The equation **AB** = **DX**.

These relations are important in this development of reflection geometry because they describe the vector structure of the plane. We sometimes need these relations in alternative forms: (i) given any three half-turns **A**, **B** and **D** then there is a unique half-turn **X** such that **AB** = **DX** (Fig. 10.23); *or* (ii) given any three half-turns **D**, **A** and **B** then there is a single half-turn **X** with the property **DAB** = **X**.

ABCD = **I** can be illustrated by the tessellation (Fig. 10.24) formed by half-turning any quadrilateral successively about the mid-points of each side, in order; the images of I being II, III, IV and I. A related problem is treated vectorially on p. 309.

Q. 8. What is $(\mathbf{ABC})^2$ for any **A**, **B** and **C**?

There is some **X** such that **ABC** = **X**, and so $(\mathbf{ABC})^2 = \mathbf{X}^2 = \mathbf{I}$. This is illustrated in Fig. 10.26 (*a*). Generalize to any odd number of half-turns.

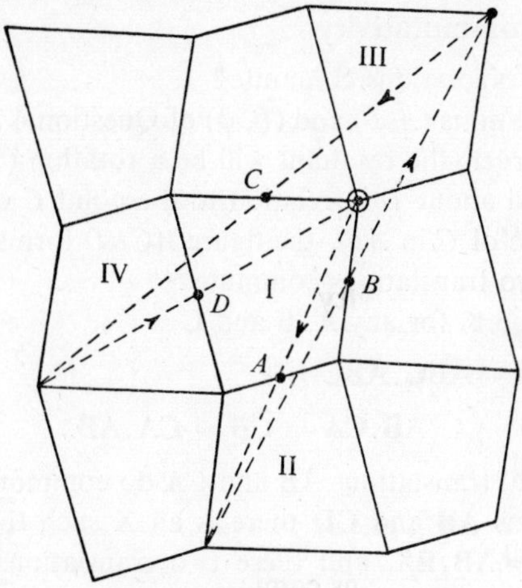

Fig. 10.24. **ABCD** = **I.**

The composition of two translations

Q. 9. What is the resultant of two translations?

First consider two special translations **PQ** and **QR**.

$$\mathbf{PQ}.\mathbf{QR} = \mathbf{PIR} = \mathbf{PR}$$

(which is another translation). This corresponds to the triangle law for the addition of vectors. More generally consider the two translations **PQ** and **RS**. By Question 7 there is an **X** such that **RS** = **QX**. Then **PQ**.**RS** = **PQ**.**QX** = **PX**, and again we see that the resultant is another translation.

The composition of a rotation and a translation

Q. 10. What is the resultant of a rotation and a translation?

The translation can be regarded as the resultant of reflections in two parallel axes *m* and *n*, the first of which passes through the centre of the rotation; and the rotation as the resultant of reflections in two axes *l*, *m*, through the centre, the second of which is also the line *m*. Then we have $\mathbf{lm}.\mathbf{mn} = \mathbf{lm}^2\mathbf{n} = \mathbf{lIn} = \mathbf{ln}$, which is a rotation of the same angle as the original, but with a different centre. (See Colour-fig. 14.) The centre of the resultant rotation is the point of intersection of the first and third axes.

Questions of commutativity

Q. 11. Do two rotations commute?

If the two rotations (A, θ) and (B, ϕ) of Question 4 are performed in the reverse order, the resultant will be a rotation $(D, \theta+\phi)$; and what can be said about D in relation to the point C of Fig. 10.21?

It is the image of C in AB—the figure $ACBD$ forms a kite.

Q. 12. Do two translations commute?

Using Question 8, for any **A**, **B** and **C**

$$\mathbf{ABC}.\mathbf{ABC} = \mathbf{I}$$

and so
$$\mathbf{AB}.\mathbf{CA} = \mathbf{CB} = \mathbf{CA}.\mathbf{AB}.$$

Hence the special translations **AB** and **CA** do commute. In general, given translations **AB** and **CD** there is an **X** such that **CD** = **BX**. Then **AB**.**CD** = **AB**.**BX**, and these two translations commute as we have just proved. So **AB**.**CD** = **BX**.**AB** = **CD**.**AB**. Hence any two translations commute.

Q. 13. Does a rotation commute with a translation?

The reader may easily verify that it does not. In terms of Colour-fig. 14 the *second* reflection axis for the translation, which will serve also as the *first* reflection axis for the rotation, will be the line m; the other two reflection axes will be the images in line m of the lines l and n. The new image, 5′ say, of 1 will be seen to be obtainable from 5 by a translation; what translation?

The composition of a number of reflections

We have combined rotation with rotation to produce a rotation, translation with translation to produce a translation, and rotation with translation to produce a rotation. We proceed to consideration of the resultant of any number of reflections.

Q. 14. What can we say about any four reflections?

There are a number of cases to consider.

(i) If all the axes are parallel, or if the first two are parallel and the last two parallel, the resultant is the resultant of two translations, which is a translation, which is equivalent to two reflections in parallel axes.

(ii) If the first two are parallel, and the last two non-parallel, we have a translation followed by a rotation, which gives a rotation which is equivalent to two reflections.

(iii) If neither the first pair nor the second pair are parallel, we have a rotation followed by a rotation, and again the resultant is a rotation, i.e. two reflections.

Thus any four reflections can be reduced to two.

Q. 15. Generalizing, what about any even number of reflections? They can be reduced to two.

Q. 16. What can be said about the resultant of any *odd* number of reflections?

From Question 15 they may always be reduced to three reflections, but only in special cases can three line-reflections be reduced to a smaller number. We treat these special cases first. We have seen already that for any three *concurrent* lines *a*, *b*, *c*, the operation **abc** is equal to a fourth reflection **d**. Likewise, reflections in three parallel lines are equivalent to a reflection in a fourth, suitably chosen line. But in general three reflections cannot be reduced to a smaller number.

Glide reflections

We now prove that reflection in three given lines is equivalent to one reflection combined with a translation in a direction parallel to the axis of this reflection. Such an operation is called a *glide-reflection.*

Consider the reflections **a, b, c,** and the composite operation **abc.** **bc** is equal to **b′c′**, where *b′* and *c′* are any two lines through the intersection of *b* and *c*, inclined at the same angle. Choose *b′* as the line perpendicular to *a*, and meeting *a* at *L*.

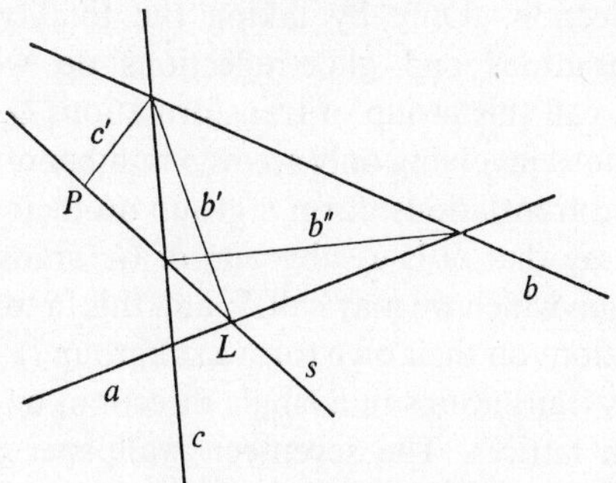

Fig. 10.25. **abc = LPs.**

Let *s* be the line through *L* perpendicular to *c′*. Then from the perpendiculars we have **ab′ = L**, and **c′s = P** or **c′ = Ps.**

Then

$$\mathbf{abc} = \mathbf{ab'c'} = \mathbf{LPs} \text{ (with } L \text{ and } P \text{ on } s).$$

This is a translation of magnitude $2LP$ along s, followed by a reflection in s; i.e. a glide-reflection in s. s goes through L, and a similar argument would show that it also goes through the foot of b'', the perpendicular from the meet of a and b to c. That is to say the axis of the glide is a side of the pedal triangle. This is illustrated very well by the diagram of Schwarz's proof that the pedal triangle has the smallest perimeter of any triangle inscribed in a given triangle [29]. This figure also makes it immediately apparent that the magnitude of the glide is equal to the perimeter of the pedal triangle. Note that the magnitude is therefore independent of the order in which **a**, **b** and **c** are carried out.

Since **abc** is equivalent to a single reflection if, and only if, a, b and c are concurrent (or parallel), one may use the picturesque phraseology '**abc** is a line' to mean 'a, b, c are concurrent (or parallel)'.

Some general remarks

We have now developed a closed set of transformations, called the group of isometries of the plane. The system is said to be *closed* because the resultant of any two transformations of the set is another member of the set. Note that it was necessary to proceed as far as this before the system became closed. The resultant of two reflections is not a reflection, neither, in general, is the resultant of three reflections a reflection. Only by taking the totality of reflections, rotations, translations and glide-reflections do we get a closed system. Let us call this group of transformations G.

Transformations involving only an even number of reflections, that is, rotations and translations, form a group on their own. If we call this H we can say that H is a subgroup of G. Translations on their own form a group which we may call K, and this in turn is a subgroup of H. (Do rotations on their own form a subgroup?) K itself has subgroups, such as translations in a single direction, or the translations associated with lattices. The seventeen wallpaper groups and the seven strip-pattern groups are discrete subgroups of our group G. There is a note on the seven-strip ornaments in §5 of this chapter; the seventeen wallpaper patterns are illustrated in [127].

So far all the groups which we have discussed in this section are infinite groups. Has G any finite subgroups? Yes, it has. These alone form an interesting topic which is fully treated in [30]. The group of the symmetries of the plane which leave one point fixed is discussed in our own §3.5.

This hierarchy of groups and subgroups extends in the other direction. The classic account is by Klein [72], who showed how geometry is the study of invariants under certain groups of transformations. The group of similarities contains the group of isometries which is our group G.

This is as far as we wish to take the present study at the moment. But the system which we have constructed exemplifies many ideas which are of mounting importance as the pupil proceeds with the study of group theory and other branches of abstract algebra. For example, operations of the form $\alpha\beta\alpha^{-1}$, where α and β denote transformations of some kind, reflections, half-turns, etc., are used in chapter 12 to express the matrices of operations in terms of other simpler, previously known operations. We have here an example of similar, or conjugate, elements in a group, and of the related notion of conjugate subgroups within a group [75, 92]. In this context these ideas arise naturally, because of their geometrical significance and not out of any desire to elaborate theory for its own sake.

The systematic approach to geometry which we have been following seems to be due in the first place to Hjelmslev, and a very readable discussion, which makes extensive use of these ideas, can be found in Coxeter [30] and Yaglom [134]. A full axiomatic treatment is in Bachmann [10], but this is very much at university level and is concerned with the development of projective geometries as well as Euclidean geometry. The interesting expository article by Thomsen [120] contains many good examples, but this describes the theory at a comparatively early stage of evolution and many of his axioms are unsatisfactory, pedagogically speaking, as they are a very long way from being obvious. One remark of his deserves quotation. 'The aim is to make the group calculus into an instrument so perfect as to take a rival place by the side of cartesian analytical geometry.' The work of the last thirty years has made significant steps towards the fulfilment of this aim, which now seems more realizable than at the time when Thomsen was writing.

The series of articles by Choquet [24] may also be consulted for an axiomatic development of plane geometry in a related fashion.

Further examples

(1) What is the meaning of **lP** = **Pm**? (*l is parallel to m and P is midway between the two.*)

(2) What is the meaning of **Pl** = **lQ**? (*l is the perpendicular bisector of PQ.*)

(3) What is the meaning of **Pl** = **mQ**? (*The line PQ is perpendicular to l and m, and distance Pl = distance mQ.*)

(4) What is the resultant of two glide-reflections? Do two glide-reflections commute?

(5) Show that for any three line-reflections **a, b, c, ab.bc.ca** = **I.** Interpret this as a theorem on rotations about the vertices of a triangle. Generalize and illustrate.

(6) Show that reflection in three given lines may be carried out by combining reflection in a suitable point with reflection in a suitable line.

(7) Show that **lPl** denotes the half-turn about the point which is the reflection of *P* in *l*. What does **PlP** denote? Make up further examples of this type. (*These are examples of similarity transformations.*)

(8) If **PQRS** = **I** and **RSTV** = **I** prove that **QPTV** = **I.** Give two proofs of this, by commonsense geometrical interpretation, and by algebra alone.

(**PQRS** = **I** ⇔ **P.PQRS** = **PI** ⇔ **QRS** = **P** ⇔ **RS** = **QP**, etc.)

(9) **ASBSCS** = **I** means that *S* is the centre of gravity of *A*, *B* and *C*.

(10) If **ASBS** = **BTCT** = **AM(TM)**^{2} = **I**, then **CM(SM)**^{2} = **I.** Essentially this is the theorem that the medians of a triangle are concurrent. (*Solved in* [120]. *See also* p. 306.)

(11) A term is said to be *involutory* if its square is equal to **I.** What is the geometrical significance of the following terms being involutory? **ab, Ab, abc, AbC, ABc.**

If the term $\alpha\beta$ is involutory (where α and β denote either half-turns or line-reflections) show that α and β commute, and conversely. Previous work shows that the incidence relations of ordinary geometry are intimately bound up with commutative relations in this algebra.

The notation $\alpha|\beta$ can be used to indicate that α and β commute, and this can be extended to relations of the form $\alpha, \beta|\gamma, \delta$ which means that each of $\boldsymbol{\alpha}$ and $\boldsymbol{\beta}$ commutes with each of $\boldsymbol{\gamma}$ and $\boldsymbol{\delta}$. Show that $\mathbf{l}|\mathbf{A}, \mathbf{B}$ if and only if l is the line AB. Show that $\mathbf{l}|\mathbf{AB}$ if and only if l is parallel to AB.

(12) Consider the propositions:

(i) If $\mathbf{l}, \mathbf{m}, \mathbf{n}|\mathbf{P}$ then there exists a line p such that $\mathbf{lmn} = \mathbf{p}$.

(ii) If $\mathbf{l}, \mathbf{m}, \mathbf{n}|\mathbf{g}$ then there exists a line p such that $\mathbf{lmn} = \mathbf{p}$.

What is the geometrical interpretation of these relations? It does not seem possible to deduce them from any of our previous relations, and Bachmann takes them as axioms. Their axiomatic nature is concealed in the intuitive approach which we have adopted in the text.

Using (ii) prove that $\mathbf{AbC}$ is a line if, and only if, there is a $\mathbf{v}$ with the property $\mathbf{A}, \mathbf{b}, \mathbf{C}|\mathbf{v}$. Prove that this is equivalent to saying that b is perpendicular to AC if and only if $\mathbf{bAC} = \mathbf{CAb}$.

Deduce that, if $\mathbf{aa}' = \mathbf{A}$, $\mathbf{cc}' = \mathbf{C}$ and $\mathbf{abc} = \mathbf{d}$ then $\mathbf{a}'\mathbf{bc}'$ is a line if, and only if, there is a $\mathbf{v}$ with the property $\mathbf{A}, \mathbf{d}, \mathbf{C}|\mathbf{v}$. *This theorem is of particular importance in the further development of the geometry of the plane, and was called the* fundamental theorem *by Hjelmslev.*

(13) If $\mathbf{ba}'\mathbf{c} = \mathbf{a}''$, $\mathbf{cb}'\mathbf{a} = \mathbf{b}''$, $\mathbf{ac}'\mathbf{b} = \mathbf{c}''$ and $\mathbf{a}'\mathbf{b}'\mathbf{c}'$ is involutory then $\mathbf{a}''\mathbf{b}''\mathbf{c}''$ is involutory. This introduces the notion of isogonal conjugates in the geometry of the triangle.

(14) For a treatment of the concurrence of the bisectors of the angles of a triangle, and the perpendicular bisectors of the sides, see [90]. The concurrence of the altitudes can be reduced to the concurrence of the perpendicular bisectors of the sides by considering a triangle with sides twice as big, or by using the ideas of example (13).

(15) The sequence of problems in Yaglom [134] provides a worthwhile study.

(16) An amusing finale is provided by another example from [10]. $\mathbf{abc}$ is, in general, a glide-reflection; hence its square $(\mathbf{abc})^2$ is a translation. Any two translations commute, so

$$(\mathbf{abc})^2\,(\mathbf{bca})^2 = (\mathbf{bca})^2\,(\mathbf{abc})^2.$$

Rearranging, and using $\mathbf{a}^2 = \mathbf{b}^2 = \mathbf{c}^2 = \mathbf{I}$,

$$\mathbf{abcabcbcabcacbacbcbacb} = \mathbf{I}.$$

This is an identity holding for any three reflections $\mathbf{a}$, $\mathbf{b}$ and $\mathbf{c}$. Illustrate!

Another point of view

In this type of geometry we are concerned all the time with the relative displacements of two planes; we move a sheet of tracing paper over a stationary sheet of drawing paper. As with all relative movement there are two points of view. The diagrams which illustrate this section have been drawn on the understanding that the points and lines in which the reflections take place are fixed (on the drawing paper), and that, as the operations are performed, other elements (on the tracing paper) move.

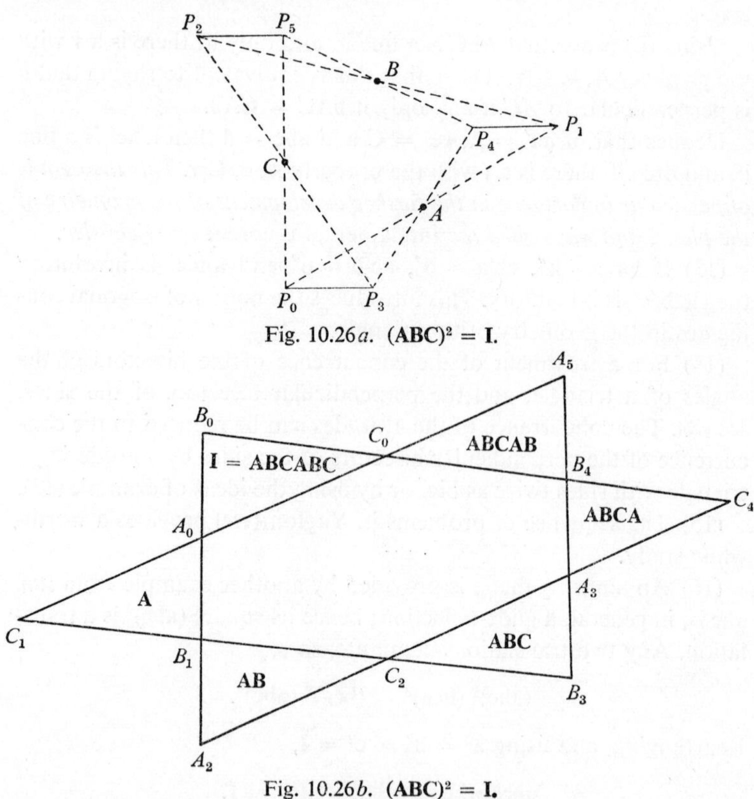

Fig. 10.26*a*. $(\mathbf{ABC})^2 = \mathbf{I}$.

Fig. 10.26*b*. $(\mathbf{ABC})^2 = \mathbf{I}$.

The other point of view interchanges the role of the two planes and regards the points and lines in which the reflections take place as moving (being as it were on a piece of card). This leads to diagrams drawn on a different convention, and some of these have a very pleasing appearance. The identity $(\mathbf{ABC})^2 = \mathbf{I}$ is then illustrated by Fig. 10.26 (*b*). The general principle underlying diagrams of this type is as follows. The relation $\mathbf{ABC}.\mathbf{ABC} = \mathbf{I}$ involves three points, A, B and C, so the basic configuration is a triangle. Six letters appear on the left-hand side of the identity and so the figure is formed by six copies of the triangle suitably fitted together.

The diagram to which reference has already been made [29, p. 348] is drawn on this convention, and very attractive figures result when illustrating such examples as (9) and (16).

4. DILATATIONS

In §3 attention was focused exclusively on isometries—shape- and size-preserving transformations. However, in Euclidean geometry—and in the real world—we are much concerned with similar, as well as with congruent figures; and this is the theme of this section.

We saw in §1 how 'tilings' or tessellations suggest ways of making precise and explicit intuitive ideas about 'same shape, different size'. There the treatment was static (unless we see that any pattern is in a sense dynamic). Here we sketch, in merest outline, a dynamic treatment, with the idea that the physical explorations of the early work will provide a store of ideas that can be used in the later stages.

The pantograph

Making and using a simple pantograph is a natural way of introducing 11–13-year-olds to the dynamic aspect of similarity, and giving them intuitive experience of the transformation $OP' = k.OP$. Older pupils on the other hand might be presented with the instrument and asked to explain how it works—a good exercise in abstracting the essential geometrical relationships from their concrete embodiment.

A pantograph can be made from strips of stout card and paper fasteners, or from Meccano, or from strips of peg-board (Fig. 10.27). P and P' are small holes, one for viewing the curve to be followed, the other for the pencil point. A, B, C, D are pivoted, so that $ABCD$ remains a parallelogram. The instrument is fixed to the drawing board with a drawing pin at O. The lengths of AD and BC can be varied so that the ratio OB/OA can be varied while OPP' remains a straight line.

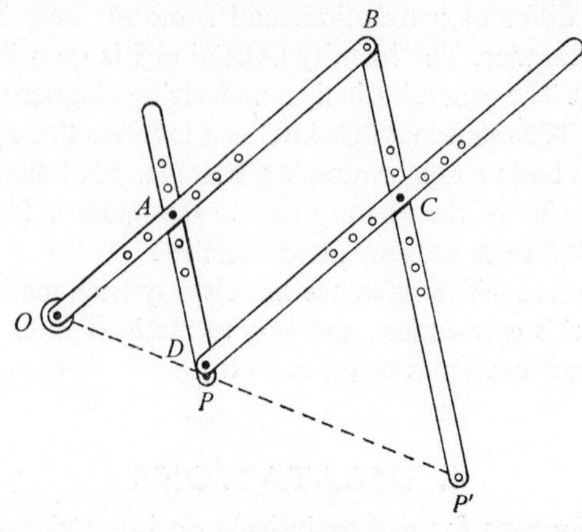

Fig. 10.27

Early lessons

(1) Pupils explore the use of the instrument, with the minimum of direction. They enlarge drawings and pictures of their own design or choice. They meet and solve the problem of altering the degree of enlargement, and may think out, or hit upon, a method of reducing instead of enlarging. If a pupil discovers that he can achieve a reversal by interchanging the roles of P and O, so much the better.

(2) Just what does the instrument do? The answer, 'It makes things bigger', has to be clarified and made precise. How much bigger? 'Twice as big'. Twice the length or twice the area? In what direction(s)? The figure is moved as well as enlarged; is it rotated at all? What difference does it make if we change the position of O? Can we put the image just where we want it on the page? Here is a drawing and its enlargement: can you find where the point O must have been?

(3) Just how does it work? What is the geometry of it? We are working towards the generalizations $OP' = k.OP$, $P'Q' = k.PQ$; if the class has a background of experience of 'tilings', as suggested in §1, the manipulation of the pantograph over a tiling drawing is a fruitful experience.

The dynamic treatment of similarity, in which the attention is not on the figures which are similar but on the transformation of the plane which assigns to every point P an image P' such that $OP' = k.OP$, is outlined in many traditional texts and in [30]; such a transformation, denoted perhaps by (O, k), is called a *dilatation* or *homothety*. Jeger's short book [66] outlines a school course for senior pupils in which the transformations of §3 (rotation, translation, reflection and glide-reflection), dilatations and 'dilatations, with rotation or reflection' are the central theme, with the group structure explicitly in view. The interested reader should also see the Russian text of Kutuzov [73], which is available in an English translation.

We conclude with an example of the classroom treatment of a famous theorem.

Lesson on the nine-point circle

Background knowledge assumed:

(i) The figure consisting of two non-intersecting circles with their common tangents has been explored from the dilatation or homothety viewpoint; the pupils are aware that the two circles can be related by a homothety in more than one way.

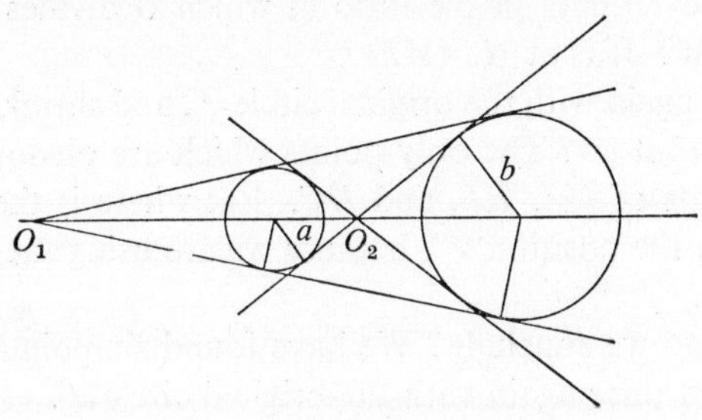

Fig. 10.28

The homothetic centres are O_1 and O_2, and the homothetic ratios are $\pm b/a$, where a and b measure the radii of the circles.

(ii) The Euler line property is known: that is, $OG:GH = 1:2$, where O, G and H are circumcentre, centroid, orthocentre. Here again this will have been regarded as a dilatation, either a 'shrink' $(G, -\frac{1}{2})$ or a 'stretch' $(G, -2)$.

We now consider Fig. 10.29 and find the two homothetic triangles ABC, $A'B'C'$.

How are they related? A shrink of $-\frac{1}{2}$ about G sends ABC to $A'B'C'$. What happens to $\mathscr{S}$, the circumcircle of ABC? It becomes $\mathscr{S}'$, the circumcircle of $A'B'C'$. What is its radius? Half what it was. Where is the centre of the transformed circle? We examine a 'detail'

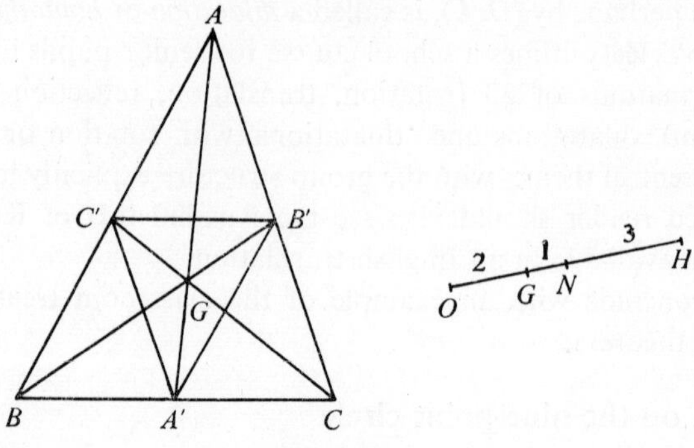

Fig. 10.29

of the figure, and locate it at N, the mid-point of OH; because $GN = -\frac{1}{2}GO$.

Where is the second homothetic centre of the two circles? It divides ON externally in the ratio in which G divides it internally. Where is that? It is at H. (*Why?*)

Now start again with the original circle $\mathscr{S}$, and shrink it by $\frac{1}{2}$ about H. Where will it go? The only points which are obviously on it are the mid-points of HA, HB, and HC. But where is the circle now? It is again in the position $\mathscr{S}'$, because we are using the other homothetic centre.

So what can we conclude? We have found six points on $\mathscr{S}'$.

Let us start once again with the original circle $\mathscr{S}$.

First reflect it in BC. Draw in the image-circle $\mathscr{S}_1$, noting particularly the position of A_1, the image of A in BC.

Again, transform $\mathscr{S}$ by a half-turn about A', getting a circle $\mathscr{S}_2$. Where is $\mathscr{S}_2$? It coincides with $\mathscr{S}_1$. (*Why?*) Note the position of A_2, the image of A under this transformation. $\mathscr{S}_1$ contains B, C, A_1 and A_2.

Now shrink $\mathscr{S}_1$ by $\frac{1}{2}$ with A as the homothetic centre. Its image is again 'half-size', and what points does it contain? The mid-points of AB, AC, AA_1 and AA_2; which are C', B', A' and the foot of the perpendicular from A to BC. So the circle is once more in the position $\mathscr{S}'$, and we have identified one more point on it. The two remaining points may be identified similarly.

Thus a sequence of transformations has established the traditional property of the nine points, and further properties of the figure may be explored by the same techniques.

5. STRIP PATTERNS

We give here a note on 'the seven essentially distinct ways to repeat a pattern on a strip of ribbon'. What is meant by 'essentially distinct' will be discussed below. Colour-fig. 15 lists the seven, giving the code symbol which Coxeter uses [30], but using for the pattern the hook form of symbol employed in §3.5 on symmetry groups.

A strip pattern is produced by translating a unit repeatedly along an axis. The unit may possess internal symmetries but such symmetries are only taken into account if they are symmetries of the pattern as a whole, and not simply of the unit.

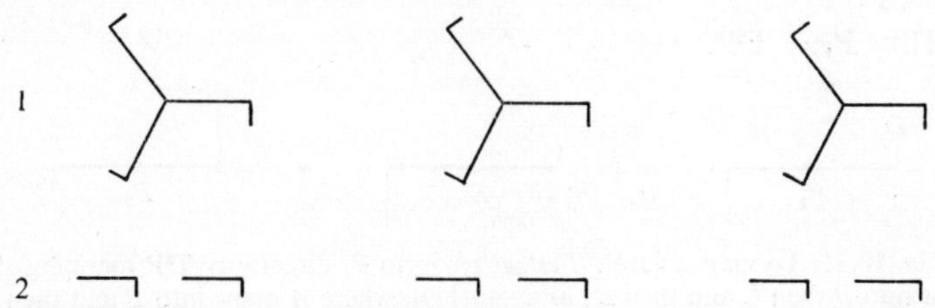

Fig 10.30. Two patterns not qualifying as extensions of the list in Colour-fig. 15.

Thus the internal symmetry group C_3 (see chapter 3) of the units of pattern 1 of Fig. 10.30 is disregarded, since the 120° rotation is not a symmetry of the whole pattern; this type is regarded as not essentially different from Type 1 of Colour-fig. 15. Likewise pattern 2 of Fig. 10.30 is not essentially different from Type 1 of Colour-fig. 15. We are concerned only with the plane symmetry groups which contain one translation. The group properties imply that all repetitions of the translations are also included. The problem consists of discovering what other symmetries of the plane will combine to form a group with this translation.

We count first the case in which there is no other symmetry. The group has as elements the translation $\mathbf{T}$ and its powers $\mathbf{T}^2$, $\mathbf{T}^3$, ..., $\mathbf{T}^{-1}$, $\mathbf{T}^{-2}$, ..., along with the identity $\mathbf{I}$. (See Colour-fig. 15, Type 1.) We shall be sure of having considered all the other cases by being able to identify transformations, other than translations, which preserve the identity of the pattern. We may refer to §3 where we discovered the group of isometric transformations. Coxeter states that 'any direct isometry is either a translation or a rotation; any opposite isometry is either a reflection or a glide reflection', and our §10.3 was concerned with exploring this situation. So other groups of transformations consist of the translation group C_∞ together with rotations, reflections or glide-reflections.

The only rotation which, combined with **T**, does not produce a translation in a new direction, is the half-turn or point-reflection. The pattern in Fig. 10.31 can be generated by the group of transformations which includes **T** and half-turns. When patterns are being explored for these relationships the tracing paper and cellulose acetate mentioned in §10.1 are invaluable. Using the symbolism developed in §10.3 the half turn **P** added to C_∞ allows us to deduce further half-turns $\mathbf{P}_1, \mathbf{P}_2, \ldots$.

Thus $\mathbf{P}_1 = \mathbf{T}^2\mathbf{P}$.

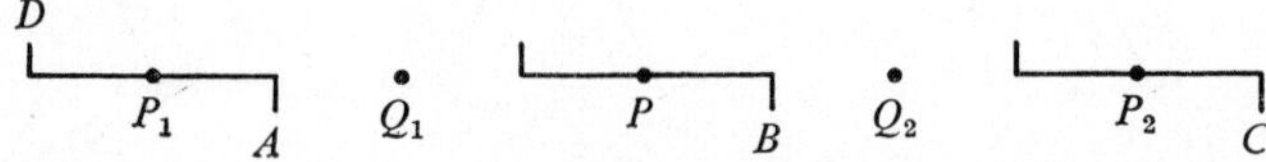

Fig. 10.31. To map A into D: *either* perform $\mathbf{P}_1$ directly *or* $\mathbf{T}^2\mathbf{P}$ mapping A first into B then C and then D; *or* again $\mathbf{TQ}_1$, where A maps into B and then D.

That is, in the Fig. 10.31 a half-turn of the pattern about point P_1 gives the same result as the combination of two translations, **T**, and a half-turn about the point P. Also half-turn symmetries exist now about the intermediate points $Q_1, Q_2, \ldots$, and $\mathbf{Q}_1 = \mathbf{TP}$. This group has $\mathbf{P}^2 = (\mathbf{PT})^2 = \mathbf{I}$; it is an example of the group D_∞, so called again by analogy with D_n of chapter 3. (See Colour-fig. 15, Type 2.)

Another pattern can be found which is generated by adding to C_∞ a reflection; this must be either parallel or perpendicular to the translation. The perpendicular case proves to be closely analogous to the rotational case above; a doubly infinite set of reflections is generated, with similar relations: $\mathbf{l}^2 = (\mathbf{lT})^2 = \mathbf{I}$, where the line l is the axis of reflection. This group is therefore also an example of D_∞ (Type 3). If a pattern exists where the reflection, **r**, is parallel to the translation, no further symmetries are generated; in this case **r** and **T** commute, and $(\mathbf{rT})^2 = \mathbf{rTrT} = \mathbf{TrrT} = \mathbf{T}^2$. This group may be symbolized as $D_1 \times C_\infty$, its elements being $\mathbf{T}^n$ and $\mathbf{rT}^n$, for any n (Type 4).

Patterns given by combinations of the above types are possible. Any two of the last three symmetry types together imply the third (they are the three types of the 'rectangle' group D_2). The result is a group of type $D_1 \times D_\infty$ (Type 5).

Finally, patterns can be made when a glide-reflection, **g**, is combined with C_∞. A repeated glide-reflection is a translation, which must be **T** in order for the combinations with C_∞ to be satisfied. Thus $\mathbf{g}^2 = \mathbf{T}$, and the elements of the group are **I** and all the powers of **g**. This is also a C_∞ (Type 6).

The only previous symmetry which combines with this one to give any new pattern is Type 3; the result of the combination is Type 7. The matrix approach to these problems is outlined in §12.3. The investigation of strip patterns and the reduction to the seven types is an interesting exploration treated at the stage B level. Further investigation of group properties can be found by extending the work fully to two dimensions. Crystallography uses the symmetry groups obtained by extending the work into three dimensions [127, 128].

Work like this can be followed up in many ways. It has been said that the first examples of higher mathematics in the history of the world were the symmetric repeating patterns designed by the ancient Egyptians. Some children may like to collect ornaments and patterns. Weyl's fascinating book [128] or the designs of the Dutch artist Escher will give them a good start.

The regular solids are standard mathematical exhibits, although their makers may not always have their symmetry groups [75] in mind. Can we suggest more mathematical exhibitions displaying symmetric repeating patterns?

11

VECTORS

The idea of a vector is introduced by considering displacements. This elementary notion is then extended to enable some traditional sixth-form coordinate geometry to be simplified and generalized through a vector approach.

An alternative introduction is illustrated in §3 where vectors are considered from the start as number pairs (x, y), number triples (x, y, z), etc., without any reference to geometry. This approach is powerful and general since all quantities involving a magnitude and direction can be handled, and so also can many other concepts.

In §4 vector algebra is applied to some simple Euclidean geometry involving parallelism, while in §5 the scalar product is introduced to provide an algebraic counterpart to perpendicularity.

Finally, in §6 the general idea of a *vector space* is introduced.

1. INTRODUCTION TO DISPLACEMENTS

Q. If you started at A (Fig. 11.1) and moved through two vertical bands to the right and, at the same time, upwards through two horizontal bands, where would you be?

Repeat with different starting-points and different displacements.

Q. Describe what movement or displacement has taken place in going from P to Q, also from Q to P.

Q. What would be a shorthand way of describing this last displacement?

A. $(1, -2)$.

Q. What can you say about the two displacements from P to Q and from R to S?

A. Some children would perhaps say they were the same, i.e. $(-1, +2)$; others would say they were not, since there are different starting-points and finishing-points.

If one wishes to specify the location of a displacement then the term 'located vector' *is used. One refers to the displacement as a* 'free vector' *and the notation* **PQ** *is used. Thus* **PQ** = **RS** = (−1, +2). *Other notations, such as* $\overrightarrow{PQ}$, *may be used to denote vectors.*

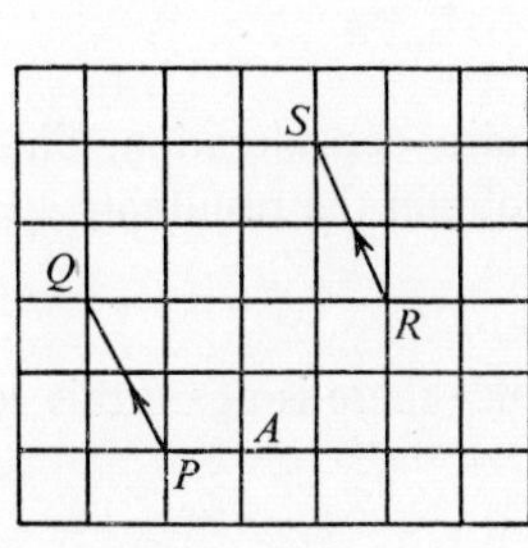

Fig. 11.1

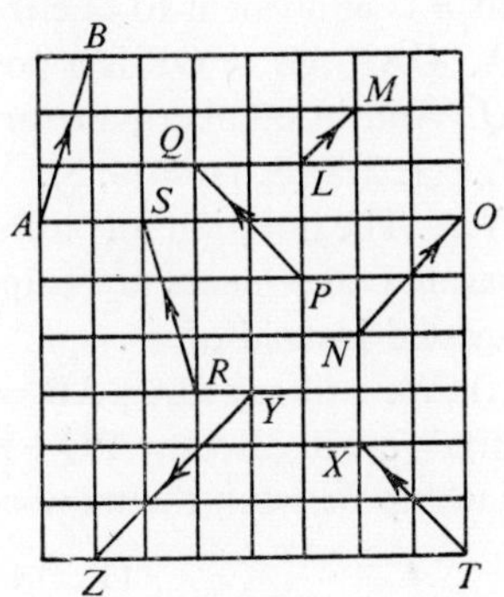

Fig. 11.2

Exercises

(To be done on squared paper Fig. 11.2.)

(1) Draw three located vectors with the same displacement as **AB**.

(2) What is **AB** in number pair form?

(3) Which free vectors are equal?

(4) Which located vectors are parallel?

(5) What conditions must the number pairs satisfy if the associated located vectors are parallel? *Here a tessellation of a plane by right-angled triangles helps, as in Fig. 11.6.*

(6) What conditions must be satisfied by two number pairs if their associated located vectors are in the same direction?

Henceforth the term *direction* (or *length*) of **PQ** will mean the direction (or length) of all associated located vectors. Also *vector* in future will imply *free vector*.

Addition of displacements

Q. Start at any intersection on your squared paper. Call it *O*. Make a displacement (1, 2) to a point *A* and follow it by a displacement (3, 1) to a point *B*. What displacement would have taken you directly from *O* to *B* (Fig. 11.3)?

A. (4, 3). *The principle of adding the respective components of the separate displacements to obtain the components of the resultant displacement will emerge from examples such as this.*

Q. What about the written representation? How could we put in shorthand that the displacement from O to A followed by one from A to B is equivalent to one from O to B?

A. $\mathbf{OA}\oplus\mathbf{AB} = \mathbf{OB}$ is a possibility ($\oplus$ *meaning followed by*).

Q. And in number pair form?

A. $(1, 2)\oplus(3, 1) = (4, 3)$.

Note. The notation for 'followed by' requires consideration. Since separate components are being added to obtain the resultant, + is suggested instead of $\oplus$.

(i) We will say $\mathbf{OA}+\mathbf{AB} = \mathbf{OB}$.

(ii) We will also say $\mathbf{PQ}+\mathbf{RS} = \mathbf{LM}$ since there is again this relationship between the number pairs

$$(1, 2)+(3, 1) = (4, 3).$$

(iii) Geometrically speaking, $\mathbf{PQ}+\mathbf{RS} = \mathbf{LM}$ if a triangle OAB exists where $\mathbf{OA} = \mathbf{PQ}$, $\mathbf{AB} = \mathbf{RS}$ and $\mathbf{OB} = \mathbf{LM}$.

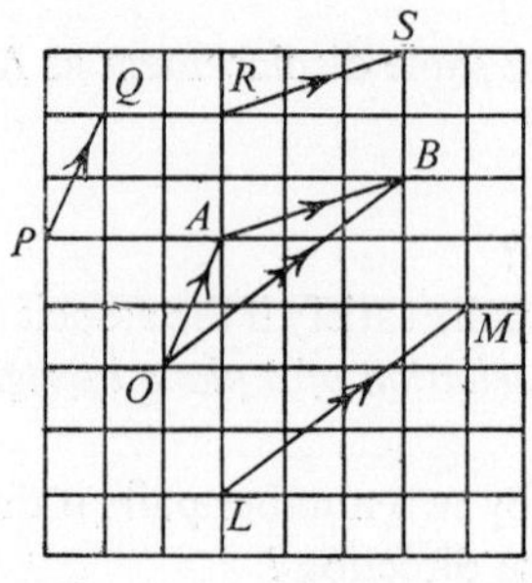

Fig. 11.3

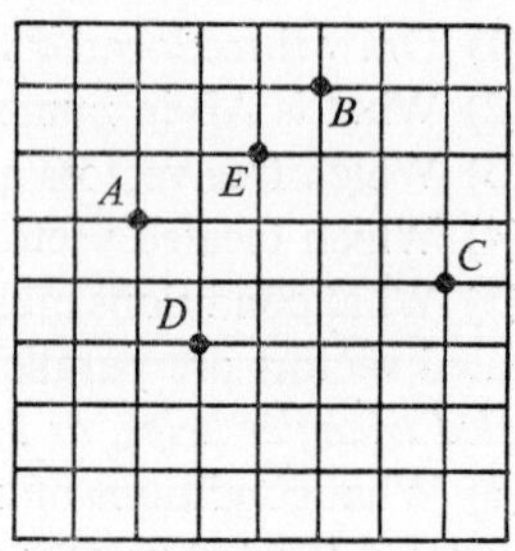

Fig. 11.4

Exercises (See Fig. 11.4)

(1) Put in number pair form the equation $\mathbf{AB}+\mathbf{BC} = \mathbf{AC}$.

(2) Find $\mathbf{AB}+\mathbf{BC}$ and $\mathbf{BC}+\mathbf{AB}$ geometrically, and also in number pair form.

(3) Find

$$[(1, 2)+(2, 3)]+(-1, -2) \text{ and also } (1, 2)+[(2, 3)+(-1, -2)].$$

Draw diagrams to illustrate.

(4) By drawing, find (**AB**+**DC**)+**EB** and **AB**+(**DC**+**EB**). What are the number pair forms of the corresponding equations?

(5) Find (**AE**+**EB**)+**BC** and also (**EB**+**BC**)+**AE**.

(6) Without drawing, find the displacement which is equivalent to (2, 3)+(3, 4)+(−5, −7). How is the answer to be interpreted? *The answer (0, 0) implies that one is back where one started. (0, 0) is called the* 'zero vector'.

(7) If you make a displacement (1, 2) what displacement will bring you back to the original position? *The answer (−1, −2) is called the* 'inverse' of *(1, 2). The inverse of* **OA** *is* **AO**.

From exercises such as these one can develop an awareness of the group structure of vectors. Speaking informally, the order of addition is immaterial, and whatever displacements are made one can always find another to take one back to the starting-point.

A formal summary of the structure

(i) For all vectors $\mathbf{p}$ and $\mathbf{q}$, $\mathbf{p}+\mathbf{q} = \mathbf{q}+\mathbf{p}$.

(ii) There is a $\mathbf{0}$ such that for all $\mathbf{p}$, $\mathbf{p}+\mathbf{0} = \mathbf{p}$.

(iii) For all $\mathbf{p}$ there is an element $-\mathbf{p}$ (the inverse of $\mathbf{p}$) such that $\mathbf{p}+(-\mathbf{p}) = \mathbf{0}$.

(iv) For all $\mathbf{p}$, $\mathbf{q}$, $\mathbf{r}$, $(\mathbf{p}+\mathbf{q})+\mathbf{r} = \mathbf{p}+(\mathbf{q}+\mathbf{r})$.

Vectors in a plane thus form a *commutative group* under the law of addition (see §3.6).

Further development

Some teachers would prefer to use a parallelogram law rather than a triangle law for addition of vectors. In any case the equivalence can be demonstrated.

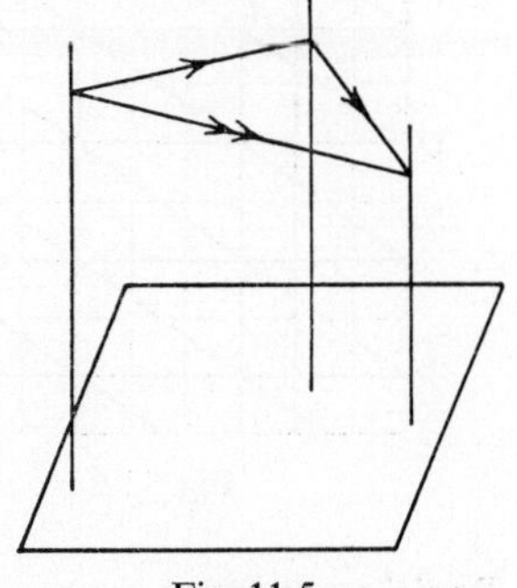

Fig. 11.5

A column form for vectors is sometimes desirable rather than the row form used here, one advantage being that the way in which components are added is made particularly clear.

The work in two dimensions can be extended to three dimensions. Here, dowelling with an appropriately marked scale can be placed in pegboard, and thread used to represent the vectors (see Fig. 11.5).

2. POSITION VECTORS

Introductory stages

The ability to interpret vectors both algebraically and geometrically is very important, and after the addition and subtraction of vectors the multiplication of vectors by scalars can be developed. The notation **PQ** = 3**OA** is used to indicate that PQ is in the same direction as OA and is three times as long. A set of equally spaced parallel lines on a squared background (Fig. 11.6) shows that if, for example,

$$\mathbf{PQ} = 3\mathbf{OA} \quad \text{where} \quad \mathbf{PQ} = (a, b) \quad \text{and} \quad \mathbf{OA} = (l, m),$$

$$\text{then } (a, b) = (3l, 3m).$$

Problems leading to vector equations involving addition, subtraction and multiplication can follow. For example,

$$2(3, 4)+\mathbf{x} = (4, 6)$$
$$(6, 8)+\mathbf{x} = (4, 6)$$
$$\mathbf{x} = (4, 6)-(6, 8)$$
$$= (-2, -2).$$

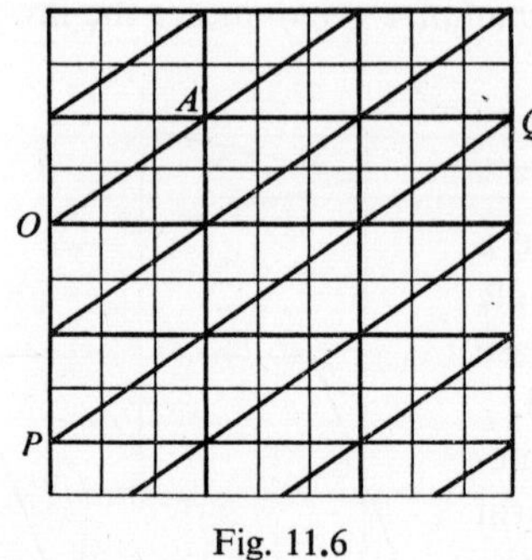

Fig. 11.6

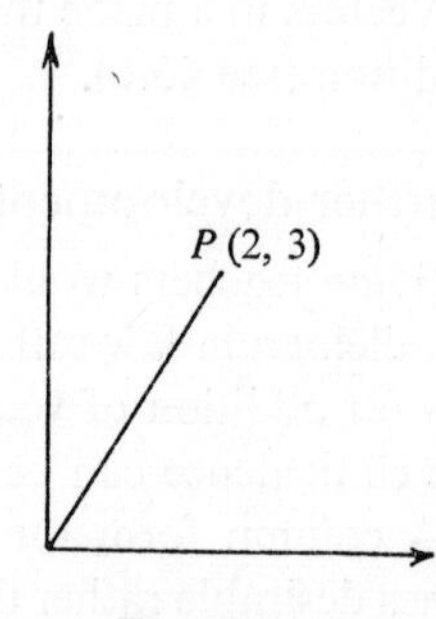

Fig. 11.7

Position vectors

With agreed axes and origin the position of any point P is completely specified by the vector **OP**. This vector is referred to as the *position vector* of the point, and may be denoted by a lower-case letter (e.g. **p**).

POSITION VECTORS

The co-ordinates of a point may be defined as the components of the position vector. Thus when we label a point P (2, 3) we may think of the 2 and 3 separately as coordinates, or we may think of (2, 3) as the position vector of P (Fig. 11.7).

Problem

In coordinate geometry one frequently wishes to specify a point which divides a line in a given ratio. The calculations and diagrams may become cumbersome, particularly when working in three dimensions. A vector approach is more concise, easier to visualize, and will apply to both two and three dimensions.

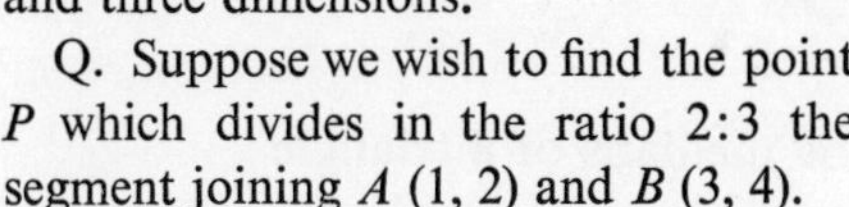

Q. Suppose we wish to find the point P which divides in the ratio 2:3 the segment joining A (1, 2) and B (3, 4).

First of all, what vectors can we specify?

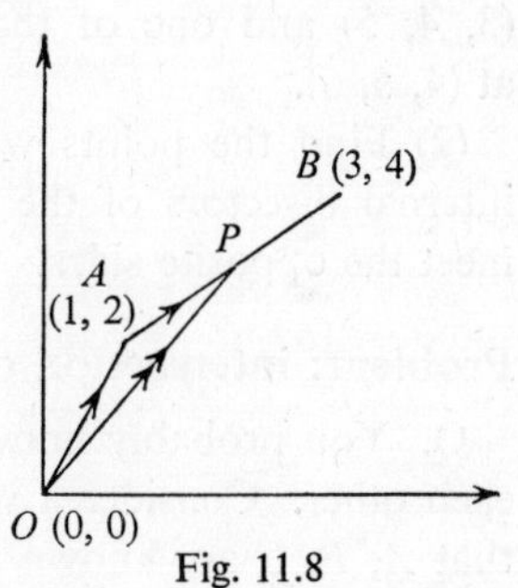

Fig. 11.8

A. $\mathbf{OA} = (1, 2)$, $\mathbf{OB} = (3, 4)$, $\mathbf{AB} = (2, 2)$, etc.

Q. We require the vector **OP**. Can we express it in terms of known vectors and thus obtain the required coordinates?

A.
$$\begin{aligned}\mathbf{OP} &= \mathbf{OA}+\mathbf{AP},\\ &= \mathbf{OA}+\tfrac{2}{5}\mathbf{AB},\\ &= (1, 2)+\tfrac{2}{5}(2, 2),\\ &= (1, 2)+(\tfrac{4}{5}, \tfrac{4}{5}),\\ &= (\tfrac{9}{5}, \tfrac{14}{5}).\end{aligned}$$

Three-dimensional problems may be handled in the same way as two-dimensional ones. After considering external division, the general formula for the position vector of a point P dividing a line AB in the ratio $\lambda:\mu$ can be evolved:

$$\mathbf{p} = \frac{\lambda\mathbf{b}+\mu\mathbf{a}}{\lambda+\mu}.$$

This formula applies to both two and three dimensions.

Proof.

$$\mathbf{p} = \mathbf{a}+\frac{\lambda}{\lambda+\mu}\,\mathbf{AB},$$

$$= \mathbf{a}+\frac{\lambda}{\lambda+\mu}\,(\mathbf{b}-\mathbf{a}),$$

$$= \frac{\lambda\mathbf{b}+\mu\mathbf{a}}{\lambda+\mu}.$$

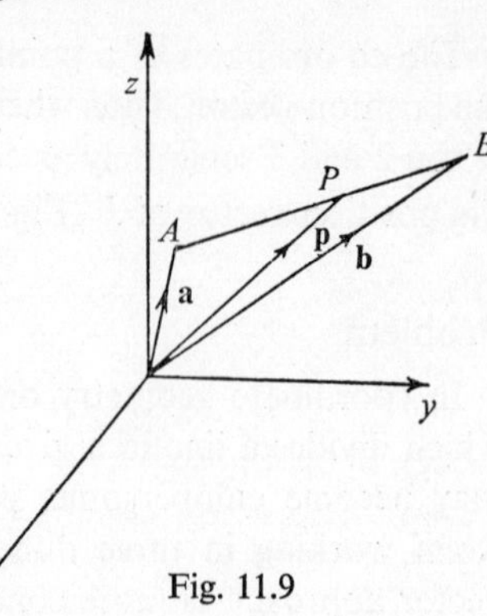

Fig. 11.9

Exercises

(1) Find the centre of gravity of two masses, one of two units at (3, 4, 5) and one of three units at (4, 5, 6).

(2) Find the points where the internal bisectors of the triangle *ABO*, *A* (3, 4), *B* (3, 0), *O* (0, 0), meet the opposite sides.

Problem: intersection of the medians of a triangle

Q. You probably know that the medians of a triangle all trisect each other. Consider a vector approach to this problem. Suppose that *A*, *B*, *C* are known points with position vectors **a**, **b**, **c**, and that *F* is the mid-point of *AB*. Can you find the position vector of the point *D*, which divides *CF* in the ratio of 2:1, in terms of **a**, **b**, **c**?

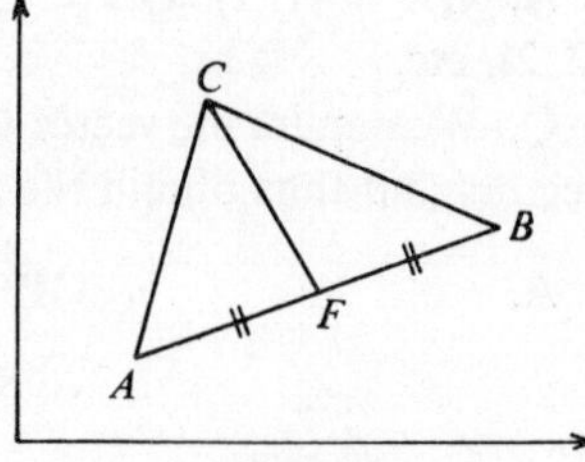

Fig. 11.10

A. It is uniquely determined by **a**, **b**, **c** so it ought to be possible.

$$\mathbf{f} = \frac{\mathbf{a}+\mathbf{b}}{2},$$

$$\mathbf{d} = \frac{2\mathbf{f}+1\mathbf{c}}{2+1},$$

$$= \frac{\mathbf{a}+\mathbf{b}+\mathbf{c}}{3}.$$

Q. Would you have obtained the same result for the position vector of a corresponding point of trisection of another median?

A. The expression is symmetrical, so whatever median we choose we obtain the same result. The points of trisection of medians further from the vertices coincide, thus proving that the medians of a triangle trisect each other.

Note. By considering number pairs or triples the coordinates of the centre of gravity of a triangular lamina can be obtained, in the form

$$X = \tfrac{1}{3}(x_1+x_2+x_3);\ Y = \tfrac{1}{3}(y_1+y_2+y_3);\ Z = \tfrac{1}{3}(z_1+z_2+z_3).$$

Exercises

(1) Find the centre of gravity of the triangular lamina ABC, $A\,(1, 2, 3)$, $B\,(-4, -6, 7)$, $C\,(4, 2, -3)$.

(2) Show that the joins of the mid-points of the opposite edges of a tetrahedron bisect each other.

(3) Show that the lines joining the vertices of a tetrahedron to the centroids of the opposite faces intersect.

(4) Show that the four mid-points of the sides of a skew quadrilateral in space are the vertices of a parallelogram.

These problems can be solved by first specifying the minimum of information required to determine the situation (in these cases the position vectors of the vertices), then obtaining all remaining vectors in terms of them.

3. NON-GEOMETRICAL VECTORS

Vectors occur in many branches of mathematics. They can be used as measures of quantities involving magnitude and direction, such as forces, velocities, accelerations, displacements; but, speaking more generally, provided only that entities possess the structure outlined in §11.6, then these entities are vectors, even if they do not directly involve a magnitude and direction.

The next lesson for younger children illustrates this idea.

A non-geometrical application

Q. Mrs Smith wants to buy 6 apples and 3 pears. Mrs Jones doesn't want any apples but she wants 12 pears and 4 bananas. Each lady wants 7 oranges and Mrs Smith wants 10 bananas. Can you arrange this confused information on the blackboard into some systematic order?

A. We might have

	Apples	Bananas	Oranges	Pears
Mrs Smith	6	10	7	3
Mrs Jones	0	4	7	12

or

	Mrs Smith	Mrs Jones
Apples	6	0
Bananas	10	4
Oranges	7	7
Pears	3	12

Various possibilities arise. Suppose the convention is adopted of arranging the amounts of fruit, in alphabetical order, in a horizontal row. Mrs Smith's list can be expressed simply as (6, 10, 7, 3) and referred to as **s**; Mrs Jones's as (0, 4, 7, 12) and referred to as **j**.

Q. Suppose Mrs Carter went shopping for both ladies what would her list be?

A. $\mathbf{c} = \mathbf{s}+\mathbf{j}$ = (6, 14, 14, 15). *We notice here that the method for addition is the same as it was for displacements. Sets of four numbers are being added instead of two and three.*

Q. If Mrs Smith's list is the same each week what would be her shopping list for a month? Suppose Mrs Jones wants two more oranges and two more bananas, what must be added to the previous list? Express this in the form adopted previously.

To what does $4\mathbf{s}+4\mathbf{j}$ correspond? *Expressions such as* $\mathbf{s}-\mathbf{j}$, $-\mathbf{j}$ and $\mathbf{0}$ *have interpretations but they can become rather forced and artificial. However, questions such as those indicated above do serve to emphasize the similarity in structure between 'list mathematics' and 'displacement mathematics'.*

Q. If we are to calculate the bill for each list what do we need to know? *The price of each item of fruit is required. If expressed in pence then the prices could be put in the same form as the lists.* $\mathbf{p}$ = *(3, $2\frac{1}{2}$, 6, 4) conveys the information required.*

Q. How do we now find the cost of Mrs Smith's weekly shopping list? Does Mrs Jones spend more or less?

A. Mrs Smith's bill is $6\times 3+10\times 2\frac{1}{2}+7\times 6+3\times 4$, etc. *Consideration should be given to the precise nature of this calculation. There are some similarities with ordinary multiplication so the term* 'scalar product', *and the notation* $\mathbf{c} = \mathbf{s}.\mathbf{p}$ *can be introduced. Expressed alternatively,*

$$(a, b, c, d).(k, l, m, n) = ak+bl+cm+dn.$$

Some exercises with only two fruits could be given and the situation represented by points on graph paper. This could lead on to further work on the scalar product as in §11.5.

Reference

Kemeny, Snell and Thompson [70].

4. VECTORS IN THE GEOMETRY COURSE

After a stage A or early stage B treatment of the parallelogram, vector methods may well be applied when appropriate to geometrical problems. Two lessons based on the mid-points of the sides of a quadrilateral and of a hexagon are now considered.

Lesson 1

Present a familiar figure—a quadrilateral with the mid-points of sides joined in order—and run over the usual Euclidean proof that *PQRS* is a parallelogram.

Q. Can we translate this into vector terms? How shall we say that *P* and *S* are mid-points of *AB* and *AD*?

A. $\mathbf{AP} = \frac{1}{2}\mathbf{AB}$, $\mathbf{AS} = \frac{1}{2}\mathbf{AD}$.

Q. And from this, what about **PS**?

A. $\mathbf{PS} = \mathbf{AS} - \mathbf{AP} = \frac{1}{2}\mathbf{AB} - \frac{1}{2}\mathbf{AD}$
$= \frac{1}{2}(\mathbf{AB} - \mathbf{AD}) = \frac{1}{2}\mathbf{BD}$.

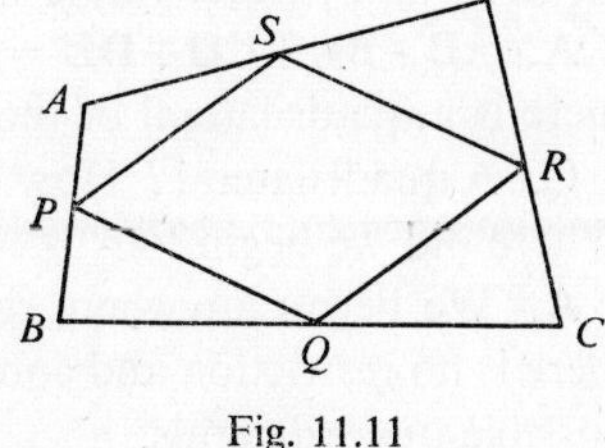

Fig. 11.11

Q. And what about **QR**?

A. $\mathbf{QR} = \frac{1}{2}\mathbf{BD}$ (similarly), so $\mathbf{PS} = \mathbf{QR}$, and this tells us *PQRS* is a parallelogram.

Q. This was not very original. Let us go back to the beginning, forgetting about the 'old' proof. *ABCD* is a quadrilateral. Say that in vectors.

A. $\mathbf{AB} + \mathbf{BC} + \mathbf{CD} + \mathbf{DA} = \mathbf{0}$.

Q. The first two terms centre on *B*, the second two on *D*; but we must bring in these mid-points. So what is the next step?

A. (*After discussion.*) $(2\mathbf{PB} + 2\mathbf{BQ}) + (2\mathbf{RD} + 2\mathbf{DS}) = \mathbf{0}$.

Q. And eventually this leads to ... what?

A. **PQ**+**RS** = **0**, that is **PQ** = **SR**, so *PQRS* is a parallelogram.

Q. What other final equation would be just as good?

A. **PS** = **QR**.

Help the class to organize an argument which will lead to this, by getting them to work 'backwards' *from*

PS = **QR** *to* **BA**+**AD**+**DC**+**CB** = **0**.

Class exercise: aim to get **RS** = **QP** *as the conclusion.*

Q. Since we have been working backwards, does this mean we have proved a *converse?* What is the converse of the theorem we have just proved? What data would we have?

A. *PQRS* is a parallelogram: in vectors, **PQ** = **SR**.

Q. And what are we going to investigate?

A. Is there a quadrilateral *ABCD* with *P*, *Q*, *R* and *S* the mid-points of its sides?

Q. Well, starting with any point *A*, we can contrive (*drawing the figure*) that *P* bisects *AB*, *Q* bisects *BC*, *R* bisects *CD*; and *S*...?

A. We don't know that it bisects *DA*, but (*with a hint*!) we can mark a point *E* so that it bisects *DE*.

Q. Now starting with **PQ** = **SR**, and using the 'reversed' argument we had before, where will we get (eventually)?

A. **AB**+**BC**+**CD**+**DE** = **0**, that is **AE** = **0**, so *E* falls on *A*, so there is a quadrilateral of the kind we were looking for.

Q. A quadrilateral? How did we draw this notable quadrilateral? Where did we begin?

A. We began anywhere so we could have begun anywhere else; there is no restriction and consequently there is an infinite number of possible quadrilaterals.

As a supporting drawing exercise at this point—on quarter-inch squared paper, mark P, Q, R, S as the vertices of the parallelogram; choose any point A, and, using the grid, mark, in succession, B, C, D, ... and A.

Lesson 2

The next stage is to explore the hexagon, postponing the pentagon on the grounds that since we took the sides in pairs, like (**AB**+**BC**), it will be more akin to what we have just done if we have an even number of sides.

As above, the argument 'forwards' yields

$$(\mathbf{AB}+\mathbf{BC})+(\mathbf{CD}+\mathbf{DE})+(\mathbf{EF}+\mathbf{FA}) = \mathbf{0}$$
$$\mathbf{PQ} \quad + \quad \mathbf{RS} \quad + \quad \mathbf{TU} \quad = \mathbf{0}$$

(*P being the mid-point of AB, etc.*)

Q. What does it mean when the sum of three vectors is zero? It is a pity the lines representing the vectors in the figure are so far apart. Could we translate?

A. These vectors would make a closed triangle.

Q. Suppose we had begun $(\mathbf{BC}+\mathbf{CD})+(\mathbf{DE}+\ldots)$. What would we have reached?

A. $\mathbf{QR}+\mathbf{ST}+\mathbf{UP} = \mathbf{0}$ implying another closed triangle!

Q. Is this a new result really? Should we have known?

A. Of course! $\mathbf{PQ}+\mathbf{QR}+\mathbf{RS}+\mathbf{ST}+\mathbf{TU}+\mathbf{UP}$ had to be zero.

Q. What about a converse? What do we start with?

A. Points *P, Q, R, S, T, U* so disposed that $\mathbf{PQ}+\mathbf{RS}+\mathbf{TU} = \mathbf{0}$.

Q. Can we reverse the argument; and if so...?

A. It's like the quadrilateral. If you can do it at all, you can start at any point *A* and finish up with a closed hexagon with each side bisected. But what about the pentagon?

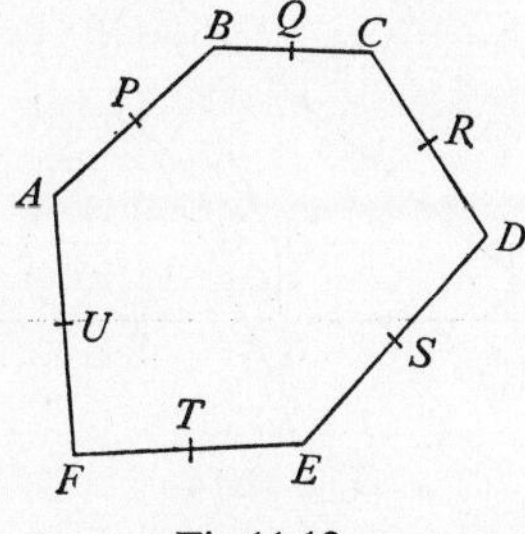

Fig.11.12

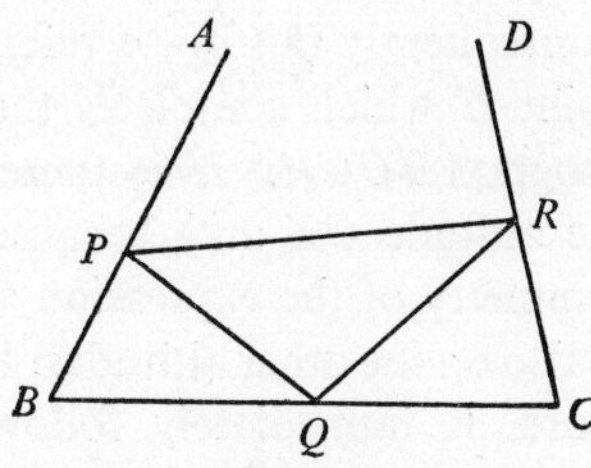

Fig. 11.13

Q. All right, but what about the triangle? Given *P, Q, R* can we construct a triangle with its sides bisected at *P, Q, R*?

Again, the figure is familiar from 'mid-point theorem' work. We 'see' that there is just one position for *A* in this case. Let us study this more closely in vector terms.

Certainly we want $\mathbf{AD} = \mathbf{0}$, i.e.

$$\mathbf{AB}+\mathbf{BC}+\mathbf{CA} = \mathbf{0}, \text{ so } 2\mathbf{PQ}+\mathbf{CA} = \mathbf{0}.$$

So we need a line through R parallel to PQ. The pentagon requirement similarly is found to be

$$\mathbf{AB}+\mathbf{BC}+\mathbf{CD}+\mathbf{DE}+\mathbf{EA} = \mathbf{0}, \quad \text{i.e. } 2\mathbf{PQ}+2\mathbf{RS}+2\mathbf{TA} = \mathbf{0}.$$

So given five points P, Q, R, S, T there is just one point A from which we can start, to get a closed pentagon with the given points as the mid-points of its sides; and this last equation tells us enough to find it.

Q. We have considered polygons with 3, 4, 5, 6 sides. Can we generalize our conclusions? *A related problem is discussed in* §*10.3.*

Having considered a vector approach to some elementary problems in Euclidean geometry it is worth noting that vector methods afford elegant proofs of some of the well-known geometrical theorems. Two examples are given below.

The altitudes of a triangle are concurrent

Let the circumcentre O of the triangle ABC be the origin of vectors. The position vector of G is $\frac{1}{3}(\mathbf{a}+\mathbf{b}+\mathbf{c})$. What point has a position vector $\mathbf{a}+\mathbf{b}+\mathbf{c}$?

$\mathbf{a}+\mathbf{b}$ corresponds to a point T on the perpendicular bisector of AB. Consequently $(\mathbf{a}+\mathbf{b})+\mathbf{c}$ refers to the point H which completes the parallelogram $TOCH$. H is therefore on the altitude CX and, because of the symmetry of the expression $\mathbf{a}+\mathbf{b}+\mathbf{c}$, is also on the other altitudes. Furthermore, it immediately follows that, since $\mathbf{OH} = 3\mathbf{OG}$, the point G divides OH in the ratio 1:2. *Refer to the nine-point circle lesson in* §*10.4.*

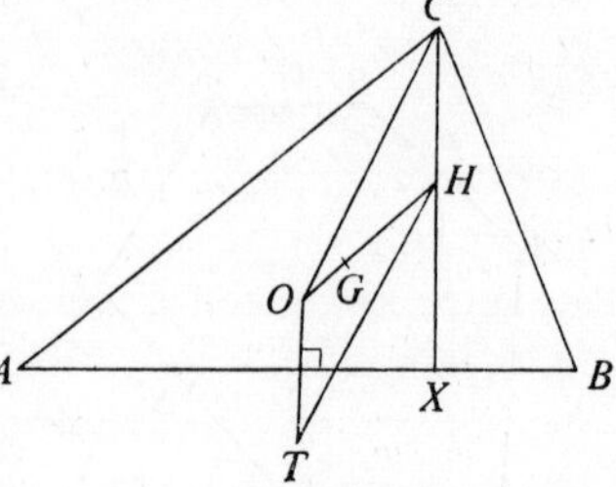

Fig. 11.14

Ceva's theorem

Given any triangle ABC, with D, E and F points on its sides such that AD, BE and CF are concurrent at a point O, then

$$\frac{AF}{FB}\cdot\frac{BD}{DC}\cdot\frac{CE}{EA} = 1.$$

Let O be the origin of vectors. Since **a** can be expressed as a linear sum of **b** and **c**, there exists an expression of the form

$$l\mathbf{a}+m\mathbf{b}+n\mathbf{c} = 0$$

connecting them.

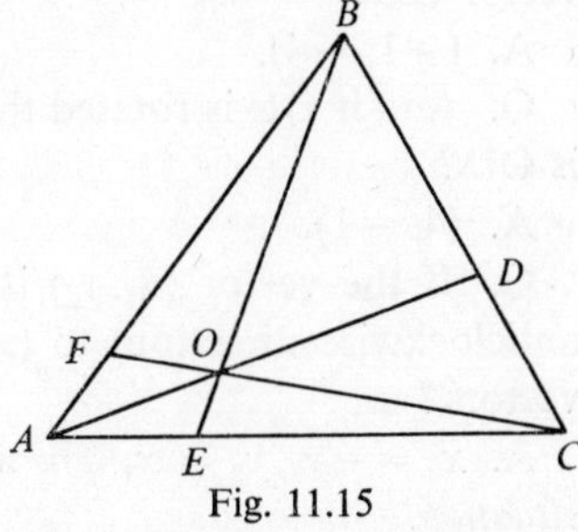

Fig. 11.15

Now

$$\mathbf{d} = k\mathbf{a}$$

$$= \frac{m\mathbf{b}+n\mathbf{c}}{-l/k},$$

so by comparison with the expression

$$\frac{\lambda\mathbf{c}+\mu\mathbf{b}}{\lambda+\mu},$$

it can be seen that D divides BC in the ratio $n:m$. Similarly E divides CA in the ratio $l:n$ and F divides AB in the ratio $m:l$. Consequently

$$\frac{AF}{FB}\cdot\frac{BD}{DC}\cdot\frac{CE}{EA} = 1.$$

5. THE SCALAR PRODUCT

The scalar product was introduced in §11.3, where it arose in the problem of determining a bill from a quantity vector and a cost vector. Here we are concerned more especially with the geometrical aspects of the scalar product. These could be approached graphically by drawing various pairs of vectors and calculating the corresponding scalar products and trying to see how they are related to the geometry of the situation, but it might be better to prepare the ground by making certain that the notion of rotation through a right angle, and the accompanying ideas of the symmetries of a square are fully understood. This may be done by means of a lesson on the square on the lines of the lesson on the rectangle in §10.1. If the work on regular polygons in §3.7 is familiar, then one could refer back and recapitulate the rotational symmetries of a square, the group C_4; but it is almost as easy to start from scratch.

Lesson

Q. If **OA** is the vector (1, 4), and **OB** is obtained by rotating **OA** through a right angle, in an anti-clockwise direction, what is the vector **OB**?

A. From Fig. 11.16 **OB** is $(-4, 1)$.

Q. If **OB** is rotated through another right angle to **OC** what is the vector **OC**?

A. $(-1, -4)$.

Q. And if **OC** is rotated through another right angle to **OD** what is **OD**?

A. $(4, -1)$.

Q. If the vector (x_1, y_1) is rotated through a right angle (in an anticlockwise direction) to (x_2, y_2) what equations connect the two vectors?

A. $x_2 = -y_1, y_2 = x_1$. *The work in § 12.3 bears closely on the present situation.*

Q. This being so what is the scalar product of (x_1, y_1) and (x_2, y_2)?

A. $x_1x_2+y_1y_2 = x_1(-y_1)+y_1x_1 = 0$.

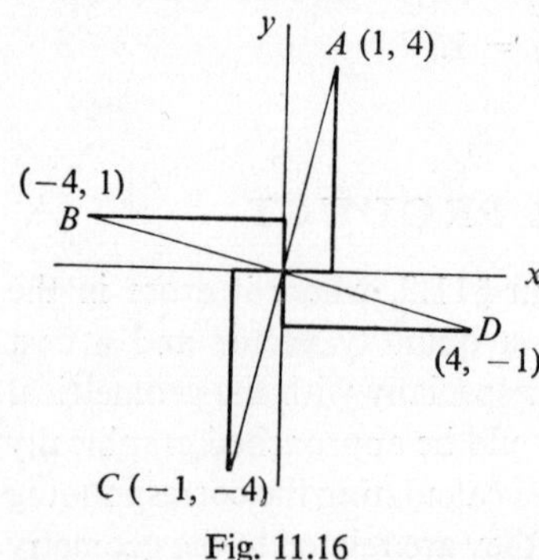

Fig. 11.16

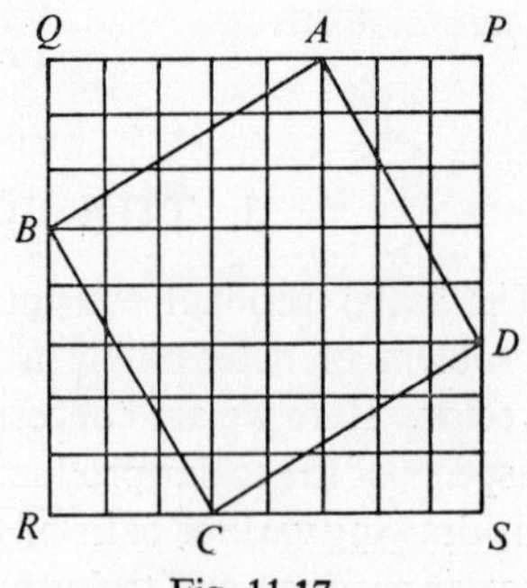

Fig. 11.17

Fig. 11.16 is closely related to Bhaskara's figure (Fig. 11.17), which may be used to give a proof of Pythagoras' Theorem from a consideration of the areas involved. If we wish to pursue this we might continue:

Q. Look at Fig. 11.17. The sides of *PQRS* are 8 units and the points *ABCD* are all 3 units along from the respective vertices *PQRS*. What can be said about the angles *ABC*, *BCD*, etc.?

A. Because of the rotational symmetry, the angles are all right angles and the sides of *ABCD* are all equal. *ABCD* is a square.

Q. What are the vectors **AB, BC, CD** and **DA** in number pair form?

A. $(-5, -3)$, $(3, -5)$, $(5, 3)$, $(-3, 5)$.

Q. What relations do we find between the perpendicular vectors?

A. As we have seen before, the scalar products are zero.

Q. So far we have only considered perpendicular vectors which are of the same length. If (x_2, y_2) is perpendicular to (x_1, y_1), but not necessarily of the same length, what can we say about it?

A. $(x_2, y_2) = (-ky_1, kx_1)$, where k may be either positive or negative.

Q. What is the value of the scalar product in this case?

A. It is still zero.

Q. Now if two vectors are not perpendicular, to what does the scalar product refer? Thinking geometrically, upon what could the scalar product depend? *This may require discussion; it is perhaps not too much to hope for the conjecture that it may depend in some way on the lengths of the vectors involved and the angle between them.*

Q. If we have two free vectors **PQ** (x_1, y_1) and **RS** (x_2, y_2), what are their lengths?

A. $\sqrt{(x_1^2+y_1^2)}$ and $\sqrt{(x_2^2+y_2^2)}$.

Q. And how can we find the angle between them?

A. We can consider the two corresponding located vectors passing through the origin, and find the angle between them (Fig. 11.18).

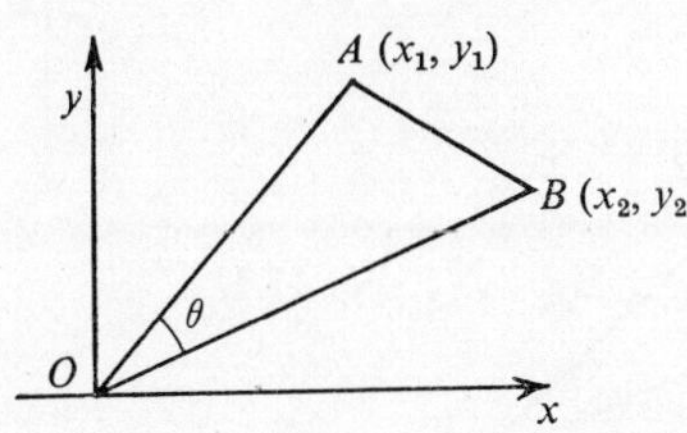

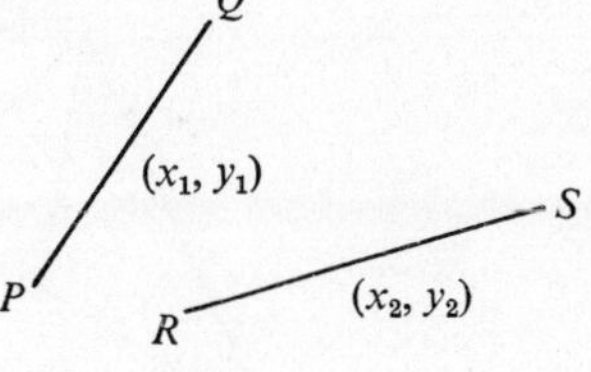

Fig. 11.18

From the cosine formula $AB^2 = OA^2+OB^2-2OA.OB\cos\theta$,

$$(y_2-y_1)^2+(x_2-x_1)^2 = x_1^2+y_1^2+x_2^2+y_2^2-2OA.OB\cos\theta,$$

$$-2x_1x_2-2y_1y_2 = -2OA.OB\cos\theta, \qquad (1)$$

$$\frac{x_1x_2+y_1y_2}{\sqrt{[(x_1^2+y_1^2)(x_2^2+y_2^2)]}} = \cos\theta.$$

Q. Now the answer we are seeking should be apparent. How is $x_1x_2+y_1y_2$ to be interpreted geometrically?

A. From equation (1), $x_1x_2+y_1y_2 = OA.OB\cos\theta$, so the scalar product is to be interpreted as the product of the length of the vectors and the cosine of their included angle. *This result is of great value. It means that if we have two known vectors, wherever they are located, we can find the angle between them. Some exercises may be given and then the three-dimensional case may be considered.*

Q. How is the scalar product of **OA** (x_1, y_1, z_1) and **OB** (x_2, y_2, z_2) to be defined?

A. $x_1x_2+y_1y_2+z_1z_2$.

Q. What is the geometrical interpretation of this expression?

A. Presumably it is $OA.OB\cos\theta$, where θ is the angle between **OA** and **OB**. *This result can be verified in the same way as it was for two dimensions. Exercises can then be given. The structural aspect of the scalar product needs investigation. Is it commutative? Is it distributive over addition?*

The power of the scalar product notion is now illustrated in two examples.

Problem

What is the angle θ between **OA** *and* **BC** *in Fig. 11.19?*

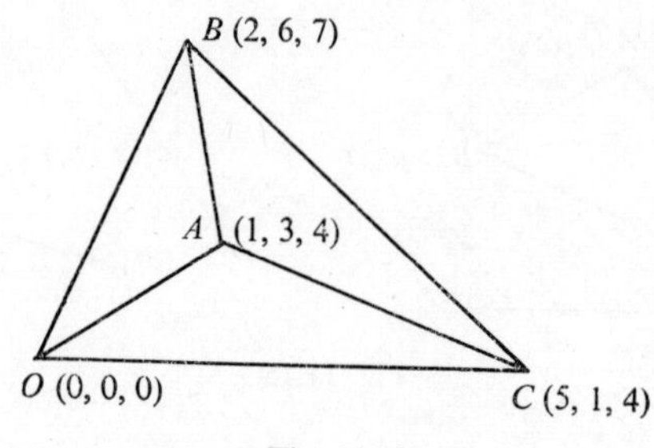

Fig. 11.19

$$\mathbf{OA}.\mathbf{BC} = (1, 3, 4).(3, -5, -3),$$

$$\sqrt{26}\sqrt{43}\cos\theta = 3-15-12,$$

$$\cos\theta = \frac{-24}{\sqrt{26}\sqrt{43}},$$

$$\theta = \text{arc cos}\left(\frac{-24}{\sqrt{26}\sqrt{43}}\right).$$

Problem

Find the great circle distance between two points P and Q on the earth's surface with latitude and longitude (θ_1, ϕ_1) *and* (θ_2, ϕ_2).

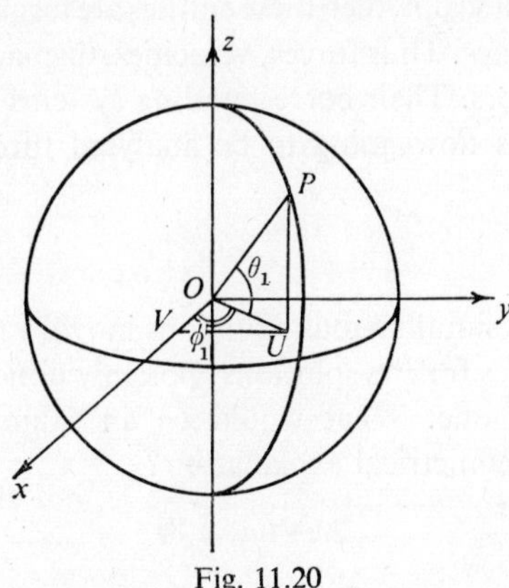

Fig. 11.20

The essence of the problem is the calculation of the angle between **OP** and **OQ**. We need to know **OP** and **OQ** in number triple form.

$$\mathbf{OP} = (OV, VU, VP)$$
$$= (R\cos\theta_1\cos\phi_1,\quad R\cos\theta_1\sin\phi_1,\quad R\sin\theta_1),$$

where R is the radius of the earth. A similar expression holds for **OQ**. But

$$\mathbf{OP}.\mathbf{OQ} = OP.OQ\cos\psi,$$

where ψ is the required angle. Hence,

$$(R\cos\theta_1\cos\phi_1, R\cos\theta_1\sin\phi_1, R\sin\theta_1).(R\cos\theta_2\cos\phi_2, R\cos\theta_2\sin\phi_2, R\sin\theta_2) = R^2\cos\psi,$$

$$\cos\theta_1\cos\theta_2\cos\phi_1\cos\phi_2+\cos\theta_1\sin\phi_1\cos\theta_2\sin\phi_2 +\sin\theta_1\sin\theta_2 = \cos\psi,$$

and so

$$\cos\theta_1\cos\theta_2\cos(\phi_1-\phi_2)+\sin\theta_1\sin\theta_2 = \cos\psi.$$

The required distance is therefore

$$R\arccos[\cos\theta_1\cos\theta_2\cos(\phi_2-\phi_1)+\sin\theta_1\sin\theta_2].$$

Reference

S.M.S.G. [111].

6. TOWARDS VECTOR SPACES

(*A sixth-form approach*)

If any system of entities is closed under the operations of addition and multiplication by a scalar, then these entities are termed *vectors*, and the system a *vector space*. Thus forces, velocities, displacements, accelerations are all vectors. Their corresponding systems have an identical structure which is now going to be analysed through a particular example.

Introduction

Consider linear simultaneous equations in two variables. There are various possibilities for the solutions; possibly none at all, an infinite number, or only one. What would be an example of each type together with a geometrical explanation?

(*a*) *No solutions*

$$2\lambda+6\mu = 8,$$

$$\lambda+3\mu = 7.$$

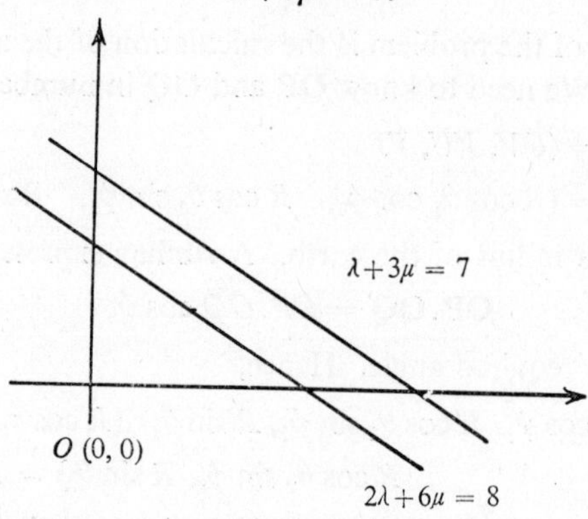

Fig. 11.21

The sets of points whose coordinates satisfy the equations are disjoint.

(*b*) *An infinite number of solutions*

$$2\lambda+6\mu = 8,$$

$$\lambda+3\mu = 4.$$

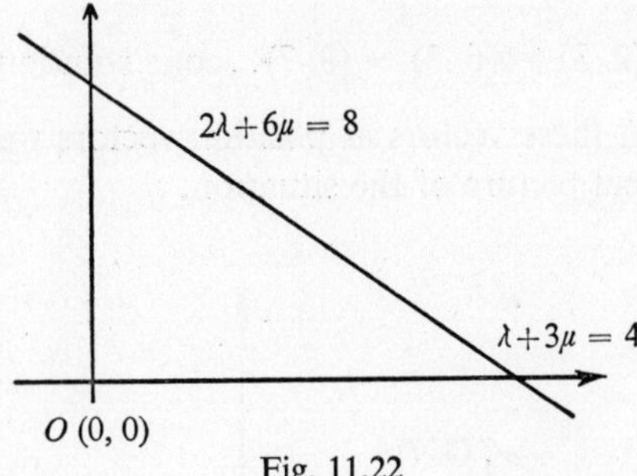

Fig. 11.22

The sets of points whose coordinates satisfy the equations are identical.

(*c*) *One solution only*

$$2\lambda+6\mu = 8,$$
$$3\lambda+3\mu = 7.$$

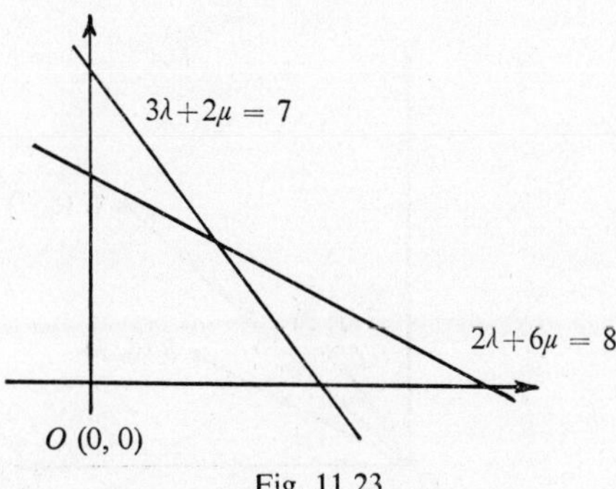

Fig. 11.23

Here the two sets of points whose coordinates satisfy the equations have only one element in common.

Let us look at the situation from another viewpoint. Each pair of simultaneous equations can be expressed as a single vector equation.

Case (*a*)

$$\lambda(2, 1)+\mu(6, 3) = (8, 7)\ldots\text{no solutions.}$$

Case (*b*)

$$\lambda(2, 1)+\mu(6, 3) = (8, 4)\ldots\text{an infinite number of solutions.}$$

Case (c)

$$\lambda(2, 3)+\mu(6, 3) = (8, 7)\ldots\text{one solution.}$$

If we interpret these vectors as position vectors we have an alternative geometrical picture of the situation.

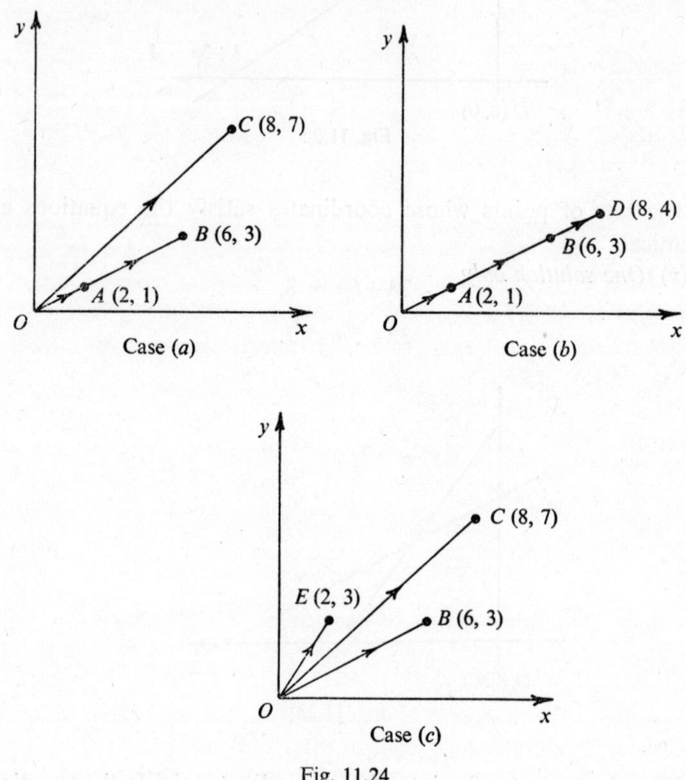

Fig. 11.24

The explanations are now as follows:

Case (a): O, A, B are collinear, so whatever multiples of **a** and **b** are formed their sum will always be the position vector of a point on OAB.

Case (b): Whatever multiple λ of **a** is formed a value of μ can be found such that $\lambda\mathbf{a}+\mu\mathbf{b} = \mathbf{d}$. There are therefore an infinite number of solutions.

Case (*c*):

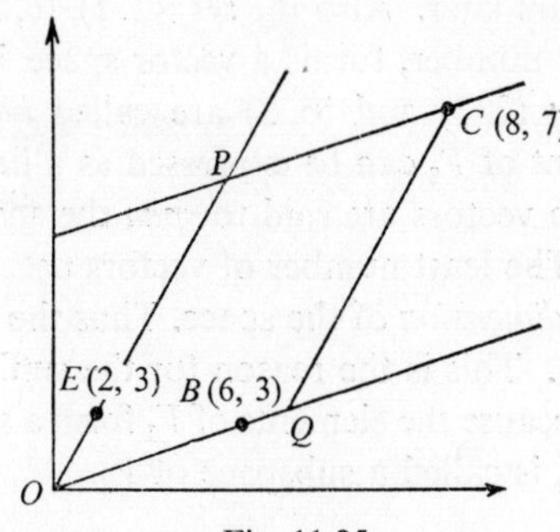

Fig. 11.25

OBE are not collinear so a parallelogram *OQCP* may be drawn with sides *OQ* and *OP* passing through *B* and *E* respectively. The values of λ and μ are given by OQ/OB and OP/OE.

Simultaneous equations may be solved geometrically by this method and exercises can be given.

Recapitulation

$$a_1x+b_1y = c_1,$$

$$a_2x+b_2y = c_2,$$

$$(a_1, a_2) \neq k(b_1, b_2)\ldots\text{one solution.}$$

$$(a_1, a_2) = k(b_1, b_2) \neq k'(c_1, c_2)\ldots\text{no solutions.}$$

$$(a_1, a_2) = k(b_1, b_2) = k'(c_1, c_2)\ldots\text{an infinite number of solutions.}$$

The geometrical viewpoint of a similar situation with three equations and three variables also applies in three dimensions.

Vector spaces

Some of the main features of vector spaces may be illustrated by the above consideration of simultaneous equations.

(1) *Linear combination*: A *linear combination* of a set of vectors **a**, **b**, **c**, etc., is a vector of the type $\lambda\mathbf{a}+\mu\mathbf{b}+\nu\mathbf{c}$, etc.

(2) *Vector space*: If a set of vectors satisfies the condition, that whatever linear combination is formed the resultant is still within the set, then this set is termed a *vector space*.

Thus the set of vectors $\{(x, y)\}$ forms a vector space V_2. The reason for the suffix appears later. Also the set $\{(2, 1), (6, 3), (8, 4), (2\rho, \rho)\}$, where ρ is any real number, forms a vector space V_1.

(3) *Base vectors*: (2, 3) and (6, 3) are called *base vectors* of V_2 because any element of V_2 can be expressed as a linear combination of them. Any such vectors are said to *span* the space.

(4) *Dimension*: The least number of vectors necessary to span the space is called the *dimension* of the space. Thus the dimensions of V_2 and V_1 are 2 and 1. This is the reason for the suffixes.

(5) *Subspace*: Because the elements of V_1 form a space and are also contained in V_2, V_1 is called a subspace of V_2.

Further development

These notions can be developed by considering the three-dimensional situation.

Vectors of the type (x, y, z) form a vector space V_3 with dimension 3.

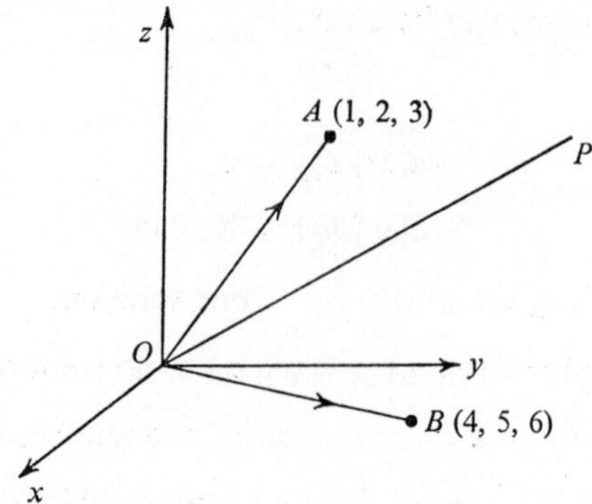

Fig. 11.26

Vectors of the type $(x, y, z) = \lambda(1, 2, 3)+\mu(4, 5, 6)$ form a subspace of V_3 with dimension 2.

Furthermore, there is a geometrical representation consisting of the set of position vectors **OP**, where P lies in the plane OAB.

Incidentally, the parametric equations of the plane OAB are seen to be:

$$x = \lambda+4\mu, \quad y = 2\lambda+5\mu, \quad z = 3\lambda+6\mu,$$

and from these the cartesian equation may easily be derived.

This work is of value in coordinate geometry, analysis of linear simultaneous equations and linear programming. Further examples of vector spaces occur in chapter 2; in these spaces the only numbers appearing are 0 and 1, for they are spaces over the *ground field* GF(2). In these spaces the ideas of linear dependence, subspace and dimension all apply, save that some of the spaces consist of vectors with an infinite number of components, and their dimension is infinite.

References

Cohn [25, 26]; O.E.E.C. [90]; S.M.S.G. [111]; Sawyer [109].

12

MATRICES

1. MATRIX CODES

This lesson is a possible extension of the earlier work on coding, in the chapter on finite arithmetics (p. 48). It also requires a knowledge of simultaneous equations. It is elementary and motivates the study of 2×2 matrices.

Lesson

Suppose the letters of the alphabet are numbered 1 to 26 inclusive and 27, 28 and 0 are reserved for comma, full-stop and space. In the earlier lesson, one symbol was coded at a time. Here we try to make the code harder to crack by taking two adjacent symbols from the message at a time. If x and y are two such adjacent symbols, let the coding equations be

$$\left.\begin{aligned} x' &\equiv 2x+y \\ y' &\equiv 3x+2y \end{aligned}\right\} \pmod{29}.$$

To encode the message YES. we replace the symbols Y, E, S and full-stop by 25, 5, 19 and 28. Next we take the two adjacent numbers 25 and 5 as the values of x and y in the encoding formulae.
Thus

$$\left.\begin{aligned} x' &\equiv 2(25)+5 \equiv 55 \equiv 26 \\ y' &\equiv 3(25)+2(5) \equiv 85 \equiv 27 \end{aligned}\right\} \pmod{29}.$$

and 26, 27 are now replaced by Z and comma. In the same way, S and full-stop are replaced by H and Z. The coded message now reads 'Z,HZ'.

Notice that the original message should be made up to an even number of symbols since we have to take two at a time and that a 'space' is needed for the decoding process. But can we decode? To do this we must solve the coding equations for x and y. The solutions are

$$\left.\begin{aligned} x &\equiv 2x'-y' \\ y &\equiv -3x'+2y' \end{aligned}\right\} \pmod{29}.$$

As an exercise, decode the message Z,HZ.

The coding formulae can themselves be written down in an abbreviated form. Whatever the values of x and y, we always take the same combinations to form x' and y'. Thus only a note of the combinations is needed. One way of doing this is to write down the coefficients of x and y in a row, one row for each equation. Thus

$$\begin{matrix} 2, 1 \\ 3, 2 \end{matrix}$$

A rectangular array of numbers like this (in this case square) is called a *matrix*, and is usually provided with brackets to group the numbers together which, with spacing, allows the commas to be omitted:

$$\begin{bmatrix} 2 & 1 \\ 3 & 2 \end{bmatrix}.$$

The matrix for the decoding equations is

$$\begin{bmatrix} 2 & -1 \\ -3 & 2 \end{bmatrix}.$$

Since the decoding matrix restores the initial symbols, it is natural to call it the 'inverse' of the coding matrix. Further, the coding matrix is the inverse of the decoding matrix. The inverse of the square matrix $\mathbf{A}$ will be written $\mathbf{A}^{\text{inv}}$. (Other notations are also common.)

Make up some more examples.

It will be found that, in this arithmetic, fractions are no problem. (*Why?*)

Not all coding matrices are feasible, because unambiguous decoding may be impossible. This occurs when the coding equations cannot be solved for x and y, and the coding matrix is then said to have no inverse. Such matrices are called singular. What happens if we use the following matrix to encode?

$$\begin{bmatrix} 1 & 2 \\ 3 & 6 \end{bmatrix}.$$

We find, for example, that *both* pairs of symbols (D, A) and (space, C) are coded as (F, R). This happens throughout, and so decoding is impossible, because we cannot decide unambiguously what the original letters were. *This is an example of a mapping which is many–one; coding demands a mapping which is one–one.* Investigate what happens if we try to decode by solving the encoding equations for x and y.

Consider the suggestion that a better code can be obtained by coding with a matrix two symbols at a time, and then coding the result *again*; for example

$$\left.\begin{aligned} x' &\equiv 3x+4y \\ y' &\equiv 2x+3y \end{aligned}\right\} \pmod{29},$$

followed by

$$\left.\begin{aligned} x'' &\equiv 5x'+2y' \\ y'' &\equiv 7x'+3y' \end{aligned}\right\} \pmod{29}.$$

What happens if, in the second coding, we replace x' and y' by the values assigned in the first coding?

$$\left.\begin{aligned} x'' &\equiv 5(3x+4y)+2(2x+3y) \\ y'' &\equiv 7(3x+4y)+3(2x+3y) \end{aligned}\right\} \pmod{29},$$

that is

$$\left.\begin{aligned} x'' &\equiv (5\times 3+2\times 2)\,x+(5\times 4+2\times 3)\,y \\ y'' &\equiv (7\times 3+3\times 2)\,x+(7\times 4+3\times 3)\,y \end{aligned}\right\} \pmod{29},$$

or

$$\left.\begin{aligned} x'' &\equiv 19x+26y \\ y'' &\equiv 27x+8y \end{aligned}\right\} \pmod{29}$$

We now observe that the coding formulae for x'' and y'' can be expressed directly in terms of the original x and y with coding matrix:

$$\begin{bmatrix} 19 & 26 \\ 27 & 8 \end{bmatrix} \pmod{29}.$$

Moreover, the final coding is again based on two symbols at a time, and we see that if you code and then code again, then, of course, the whole process is equivalent to coding once only. The point of interest is how the resultant matrix is related to the other two matrices. In the example we have just worked out, at one stage we obtained:

$$\left.\begin{aligned} x'' &\equiv (5\times \underline{3}+2\times \underline{2})\,x+(5\times \underline{4}+2\times \underline{3})\,y \\ y'' &\equiv (7\times \underline{3}+3\times \underline{2})\,x+(7\times \underline{4}+3\times \underline{3})\,y \end{aligned}\right\} \pmod{29}.$$

The new coding matrix can be written:

$$\begin{bmatrix} 5\times\underline{3}+2\times\underline{2} & 5\times\underline{4}+2\times\underline{3} \\ 7\times\underline{3}+3\times\underline{2} & 7\times\underline{4}+3\times\underline{3} \end{bmatrix} \pmod{29}.$$

This resultant matrix has a certain pattern (look at the figures underlined). Next write down the individual coding matrices next to each other, the one which coded first on the right:

$$\begin{bmatrix} 5 & 2 \\ 7 & 3 \end{bmatrix} \quad \begin{bmatrix} 3 & 4 \\ 2 & 3 \end{bmatrix}.$$

With a little effort you should be able to say how the rows of the left-hand matrix operate on the columns of the right-hand matrix to produce the numbers in the resultant matrix.

Consider the number $5\times 3+2\times 2$ in the *first* row and *first* column of the resultant. Looking at the row and column in isolation:

$$[5 \quad 2] \qquad \begin{bmatrix} 3 \\ 2 \end{bmatrix}.$$

Now rotating [5 2] in a clockwise direction so that it stands shoulder to shoulder with $\begin{bmatrix} 3 \\ 2 \end{bmatrix}$ we obtain

$$\leftarrow\!\!\begin{bmatrix} 5 \\ 2 \end{bmatrix}\!\!-\!\!\begin{bmatrix} 3 \\ 2 \end{bmatrix}\!\!\rightarrow$$

The lines drawn show the numbers that correspond to each other. Corresponding numbers are multiplied together and then the resulting products added together. In this case $5\times 3+2\times 2$.

Can you obtain the other numbers in the resultant matrix in a similar way? This sort of 'multiplication' is called 'row and column multiplication'.

We shall, in fact, now *define* the resultant coding matrix as the *product* of the individual coding matrices, writing

$$\begin{bmatrix} 19 & 26 \\ 27 & 8 \end{bmatrix} \equiv \begin{bmatrix} 5 & 2 \\ 7 & 3 \end{bmatrix} \begin{bmatrix} 3 & 4 \\ 2 & 3 \end{bmatrix} \pmod{29},$$

where, by product, we understand the multiplication to be by row and column. But beware! Does

$$\begin{bmatrix} 3 & 4 \\ 2 & 3 \end{bmatrix} \begin{bmatrix} 5 & 2 \\ 7 & 3 \end{bmatrix}$$

give the same resultant? The order of coding may matter.

The idea of multiplying two matrices together can be extended to any two matrices **A** and **B** to give a product **AB**, provided the number of columns of **A** is the same as the number of rows in **B**. As a special case, we now have a neat way of writing our original coding equations at the point where the word 'matrix' was introduced. We may put

$$\begin{bmatrix} x' \\ y' \end{bmatrix} \equiv \begin{bmatrix} 2 & 1 \\ 3 & 2 \end{bmatrix} \begin{bmatrix} x \\ y \end{bmatrix}.$$

In this notation we check that the two matrices used in the double coding problem appear as a product in a natural way, for

$$\begin{bmatrix} x' \\ y' \end{bmatrix} \equiv \begin{bmatrix} 3 & 4 \\ 2 & 3 \end{bmatrix} \begin{bmatrix} x \\ y \end{bmatrix},$$

hence $$\begin{bmatrix} x'' \\ y'' \end{bmatrix} \equiv \begin{bmatrix} 5 & 2 \\ 7 & 3 \end{bmatrix} \begin{bmatrix} x' \\ y' \end{bmatrix} \equiv \begin{bmatrix} 5 & 2 \\ 7 & 3 \end{bmatrix} \begin{bmatrix} 3 & 4 \\ 2 & 3 \end{bmatrix} \begin{bmatrix} x \\ y \end{bmatrix}.$$

Is it possible to find a matrix which will also decode as well as code? To do this it is necessary to find a matrix **A** such that $\mathbf{A}^{\text{inv}} = \mathbf{A}$, or $\mathbf{A}^2 = \mathbf{I}$ (mod 29), where **I** is the matrix $\begin{bmatrix} 1 & 0 \\ 0 & 1 \end{bmatrix}$. **I** is called a *unit matrix* and it has the property of leaving all the letters unchanged if it is used for coding. The general solution of this problem is too cumbersome to be worth while, but we start as follows:

$$\begin{bmatrix} a & b \\ c & d \end{bmatrix} \begin{bmatrix} a & b \\ c & d \end{bmatrix} \equiv \begin{bmatrix} a^2+bc & ab+bd \\ ca+dc & bc+d^2 \end{bmatrix} \equiv \begin{bmatrix} 1 & 0 \\ 0 & 1 \end{bmatrix} \pmod{29}.$$

Without seeking general formulae some particular solutions are easily found. For example, it will do if we have $a+d \equiv 0$ (mod 29), so we may choose $a = 3$, $d = 26$; and then we need $bc \equiv 21$ (mod 29), so we may choose $b = 3$, $c = 7$. Therefore

$$\begin{bmatrix} 3 & 3 \\ 7 & 26 \end{bmatrix}$$

is its own inverse modulo 29, and the same matrix will decode as well as code.

Lessons on this topic have been developed by Dr G. Matthews [141], and these have, to some extent, turned our thoughts in this direction. Further work on these lines may be found in Peck [94].

2. MATRICES AND NETWORKS

Matrices have two most important properties: they act as a compact store of information, and they allow assemblies of information to be handled simultaneously instead of one item at a time. We later consider them as operators; for example, the product of the matrices for two rotations in the plane is the matrix which represents the resultant rotation; but now we look at them partly as operators and partly as counting mechanisms.

We first consider some properties of the interrelations between the nodes, branches and regions of networks, continuing the work in §8.3. We shall allow more than one branch between any pair of nodes, but any branch will have different nodes at each end (i.e. there will be no loops).

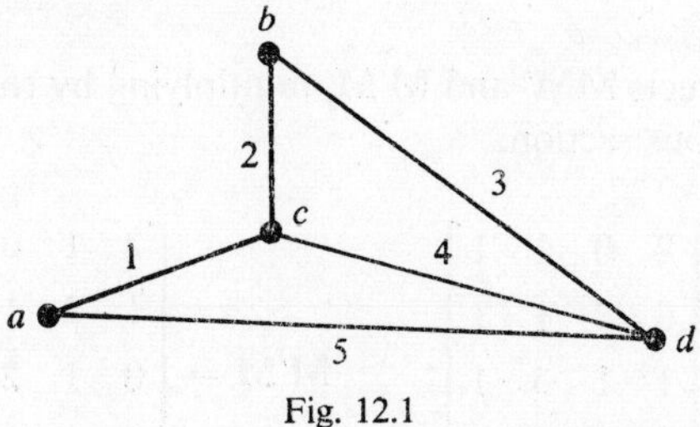

Fig. 12.1

The network in Fig. 12.1 consists of the nodes *a*, *b*, *c*, *d* and branches 1, 2, 3, 4, 5. We construct a table, in which rows represent nodes and columns represent branches. In the table we record a 1 if a node lies on a particular branch and 0 otherwise. We may regard this table as a matrix, **M**, with 4 rows and 5 columns. An entry in the table is called an element of the matrix. It is convenient to denote the element in the *i*th row and *j*th column of the matrix **A** by a_{ij}.

M =

	1	2	3	4	5
a	1	0	0	0	1
b	0	1	1	0	0
c	1	1	0	1	0
d	0	0	1	1	1

Here the matrix is a store of information.

It is worth noting that the statement 'node *a* lies on branch 5' is equivalent to the statement 'branch 5 contains node *a*', so that it does not matter which set of elements is used for the rows. Also a 1 which appears in the matrix, not only tells that the event has happened, but also that it occurs 'once'. The *transpose* of a matrix is another matrix whose *k*th column is the *k*th row of the original matrix. Thus the transpose of **M**, written **M′**, is

M′ =	*a*	*b*	*c*	*d*
1	1	0	1	0
2	0	1	1	0
3	0	1	0	1
4	0	0	1	1
5	1	0	0	1

Form the products **MM′** and **M′M**, multiplying by row and column as in the previous section.

$$\mathbf{MM'} = \begin{bmatrix} 2 & 0 & 1 & 1 \\ 0 & 2 & 1 & 1 \\ 1 & 1 & 3 & 1 \\ 1 & 1 & 1 & 3 \end{bmatrix}; \qquad \mathbf{M'M} = \begin{bmatrix} 2 & 1 & 0 & 1 & 1 \\ 1 & 2 & 1 & 1 & 0 \\ 0 & 1 & 2 & 1 & 1 \\ 1 & 1 & 1 & 2 & 1 \\ 1 & 0 & 1 & 1 & 2 \end{bmatrix}.$$

Are **MM′** and **M′M** the same matrix? What happens in either matrix if you exchange rows and columns? The operation produces the same matrix again. Such matrices are said to be *symmetric*. What information is given to us by the elements of **MM′**? Consider the number in row 3 and column 4 of **MM′**. It is obtained by row and column multiplication of the third row of **M** and the fourth column of **M′**. This is $1\times0+1\times0+0\times1+1\times1+0\times1 = 1$, and we note that the only non-zero contributions to the left-hand side arise where the corresponding elements from the row and column concerned are both 1. But if there is a 1 in row 3 and column *j* of **M** *and* a 1 in row *j* and column 4 of **M′**, this means that node *c* lies on branch *j*, and branch *j* also passes through node *d*. In this case there is one branch from node *c* to node *d*.

The elements of **MM′** which are not on the principal diagonal (the diagonal from the top-left to the bottom-right corners) count the number of paths between the nodes in question. The elements on the principal diagonal are a little different. What do they count? They count the number of branches associated with each node.

What information comes from the elements of **M′M**?

Matrices such as **M** are often called *incidence matrices.*

Exercise

Calculate the matrix **MM′** for the network in Fig. 12.2.

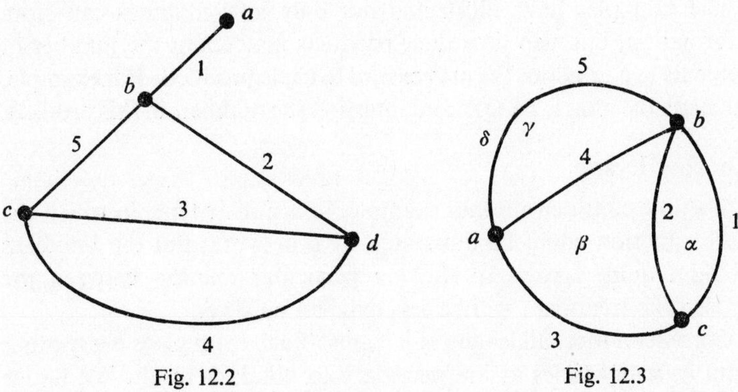

Fig. 12.2 Fig. 12.3

We will now go a stage further and take into account the regions into which the network divides the plane.

In Fig. 12.3 the regions are α, β, γ and δ, δ being the infinite region. The incidence matrices are

M =	1	2	3	4	5
a	0	0	1	1	1
b	1	1	0	1	1
c	1	1	1	0	0

P =	α	β	γ	δ
1	1	0	0	1
2	1	1	0	0
3	0	1	0	1
4	0	1	1	0
5	0	0	1	1

Q =	α	β	γ	δ
a	0	1	1	1
b	1	1	1	1
c	1	1	0	1

Calculate and interpret the matrices **MM′**, **M′M**, **PP′**, **P′P**, **QQ′** and **Q′Q**.

No doubt you are asking 'why three incidence matrices? Surely **M** and **P** contain all the information required to construct **Q**.' For example: from **M**, node a lies on branches 3, 4, 5 and from **P**, branch 3 is in region β, δ, branch 4 in β, γ, branch 5 in γ, δ. Hence node a is in regions β, γ, δ. We would expect the matrices to be connected, so let us see what the connection is.

Consider the matrix product **MP**. The element in row i and column j of this matrix evidently tells us the number of branches passing through node i which are boundaries of region j. But this number must be 2 or 0, according as node i is or is not a boundary point of region j. So $\mathbf{MP} = 2\mathbf{Q}$.

The examples have illustrated not only how matrices can store information, but also how their products may count the number of elements in a relation. We may extend to triple products. For example, calculate the matrix $\mathbf{M'QP'}$ and interpret the numbers in this product.

Another lesson

We now consider another example, of a similar type, in which the basic situation could be illustrated by a network, but the branches would require arrows to show a particular relation between the nodes. The treatment will be less detailed however.

Consider a football league of n teams. Each team plays every other team as many times as are necessary to obtain a result. We notice straight away that no team can play itself and that no draws are allowed. It is convenient to record the final outcomes in an $n \times n$ matrix **B**, in which the element in row i and column j, b_{ij}, is 1 if team i beats team j and is zero otherwise. By this definition $b_{ii} = 0$ and if $b_{ij} = 1$ then $b_{ji} = 0$. The sum of the units in row i is the total number of games won by team i. The relationship is not transitive, that is, if $b_{ij} = 1$, $b_{jk} = 1$ then b_{ik} may, or may not, equal 1. (If $b_{ik} = 0$ then an 'upset of form' has occurred.) If $b_{ij} = 1$ we shall say that team i is *directly superior* to team j. If $b_{ij} = 1$, $b_{jk} = 1$ we shall say that team i is *indirectly superior* to team k.

How can we discover the number of different ways in which team i is indirectly superior to team j?

Consider the element $b_{ij}^{(2)}$ of matrix **BB**, written $\mathbf{B}^2$.

By the definition of matrix multiplication

$$b_{ij}^{(2)} = \sum_k b_{ik} b_{kj}.$$

Any product $b_{ik}b_{kj}$ is non-zero if, and only if, b_{ik} and b_{kj} are both 1, that is if and only if team i is indirectly superior to team j through team k. Hence $b_{ij}^{(2)}$ is the required number of ways. Further work on this kind of dominance matrix may be found in [70].

3. MATRICES AND GEOMETRICAL TRANSFORMATIONS

We now see the use of matrices at a different level in the course. If matrices are already familiar in some other context, such as the one provided in the previous section, the knowledge can be applied to the geometrical problems discussed here, as matrices are a convenient algebraic notation with which to handle geometrical transformations in the plane. Alternatively, the ideas here can serve to introduce matrices to pupils who are already familiar with reflections, rotations and translations as they are treated in chapter 10, and have some knowledge of Cartesian coordinates and trigonometry.

Lesson

We will use P_0 with coordinates (x_0, y_0) to represent the position of a point in a plane before an operation is applied to it, and P_1 with coordinates (x_1, y_1) to represent the position of the point after the operation has been applied. If the operation **X** means *reflect in the x-axis* then **X** sends any point P_0 to P_1.

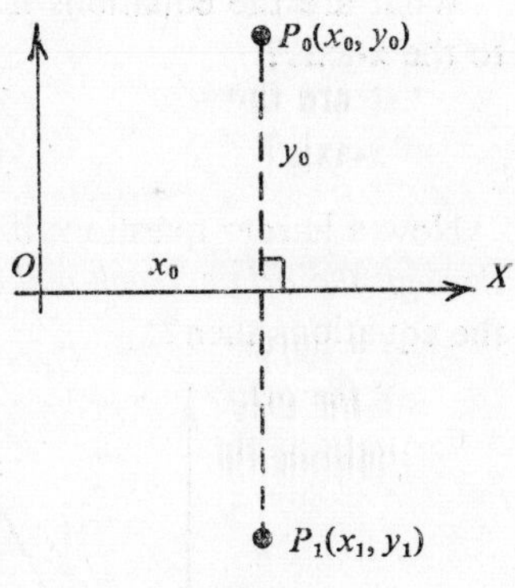

Fig. 12.4

Can we express the result of this operation algebraically?

$$\mathbf{X}\begin{cases} x_1 = x_0, \\ y_1 = -y_0. \end{cases}$$

These linear equations completely specify the operation **X**; that is, they tell us what happens to any point $P_0(x_0, y_0)$ when **X** is applied. If the operation **Y** means *reflect in the y-axis* can we write down the two linear equations which completely specify the operation **Y**?

$$\mathbf{Y}\begin{cases} x_1 = -x_0, \\ y_1 = y_0. \end{cases}$$

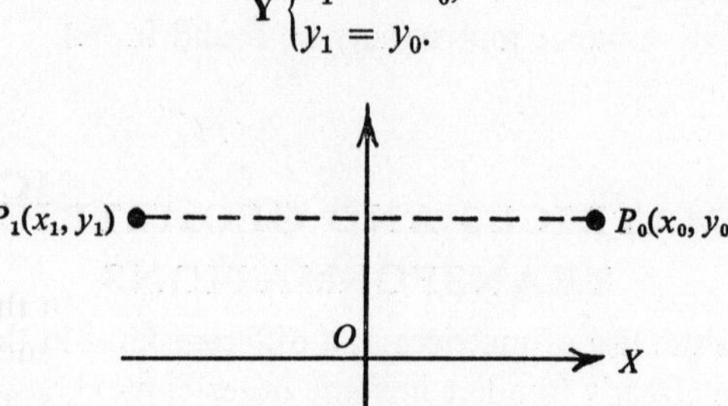

Fig. 12.5

What are the equations if we reflect in the line inclined at 45° to the x-axis?

$$x_1 = y_0,$$

$$y_1 = x_0.$$

What are the equations if we reflect in the line inclined at 135° to the x-axis?

$$x_1 = -y_0,$$

$$y_1 = -x_0.$$

Now a harder question; if the operation **L** means *reflect in a line through the origin which makes an angle* α *with the x-axis*, what are the equations then?

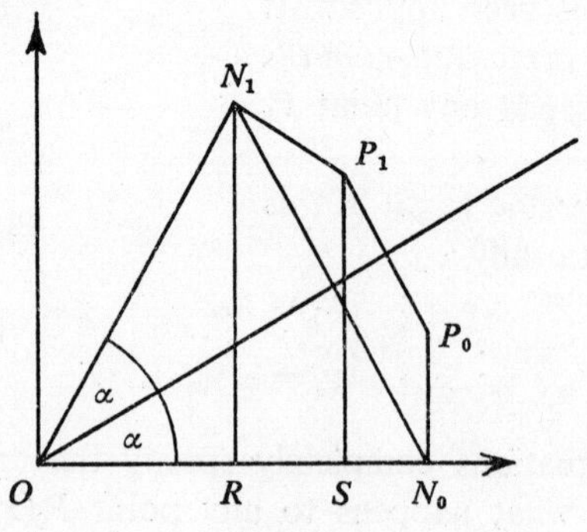

Fig. 12.6

From Fig. 12.6
$$\begin{aligned} OS &= OR+RS \\ &= ON_1 \cos 2\alpha + N_1P_1 \sin 2\alpha \\ &= ON_0 \cos 2\alpha + N_0P_0 \sin 2\alpha \\ &= x_0 \cos 2\alpha + y_0 \sin 2\alpha. \end{aligned}$$

Similarly,
$$\begin{aligned} SP_1 &= ON_1 \sin 2\alpha - N_1P_1 \cos 2\alpha \\ &= x_0 \sin 2\alpha - y_0 \cos 2\alpha. \end{aligned}$$

Hence
$$\mathbf{L}\begin{cases} x_1 = x_0 \cos 2\alpha + y_0 \sin 2\alpha, \\ y_1 = x_0 \sin 2\alpha - y_0 \cos 2\alpha. \end{cases}$$

Now consider the operation $\mathbf{R}_\pi$, which means *rotate through an angle of 180° about the origin.* Can you write down the equations which completely specify the operation $\mathbf{R}_\pi$?

$$\mathbf{R}_\pi\begin{cases} x_1 = -x_0, \\ y_1 = -y_0. \end{cases}$$

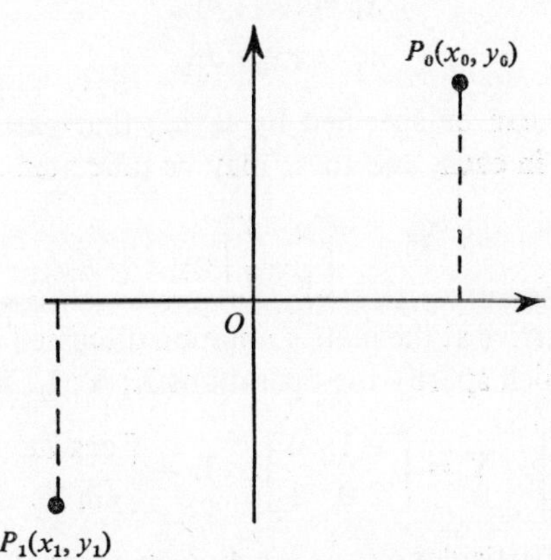

Fig. 12.7

Can we generalize this result to give the equations which completely specify the operation $\mathbf{R}_\theta$, which means *rotate through an angle θ about the origin?* From Fig. 12.8.

$$\mathbf{R}_\theta\begin{cases} x_1 = x_0 \cos \theta - y_0 \sin \theta, \\ y_1 = x_0 \sin \theta + y_0 \cos \theta. \end{cases}$$

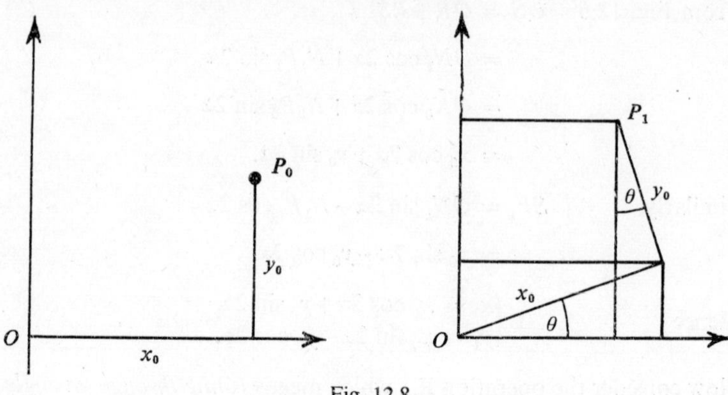

Fig. 12.8

On looking back at all of these results what do we notice? In every case x_1 and y_1 are given by linear expressions in x_0 and y_0, and so we have in each case equations of the form

$$x_1 = ax_0 + by_0,$$
$$y_1 = cx_0 + dy_0.$$

The operations can be specified by saying that particular numbers a, b, c, d occur in each, and these may be tabulated

$$\begin{bmatrix} a & b \\ c & d \end{bmatrix}.$$

In this way we arrive at the matrix notation discussed in §1. What are the matrices which specify the operations $\mathbf{X}$; $\mathbf{Y}$; $\mathbf{L}$; $\mathbf{R}_\pi$; $\mathbf{R}_\theta$?

$$\mathbf{X} = \begin{bmatrix} 1 & 0 \\ 0 & -1 \end{bmatrix}, \quad \mathbf{Y} = \begin{bmatrix} -1 & 0 \\ 0 & 1 \end{bmatrix}, \quad \mathbf{L} = \begin{bmatrix} \cos 2\alpha & \sin 2\alpha \\ \sin 2\alpha & -\cos 2\alpha \end{bmatrix}, \text{ etc.}$$

To show that a particular matrix is acting on a point with coordinates (x_0, y_0) we write our equations in tabular form:

$$\begin{bmatrix} x_1 \\ y_1 \end{bmatrix} = \begin{bmatrix} a & b \\ c & d \end{bmatrix} \begin{bmatrix} x_0 \\ y_0 \end{bmatrix},$$

or

$$\mathbf{P}_1 = \mathbf{A}\mathbf{P}_0,$$

where $\mathbf{A} = \begin{bmatrix} a & b \\ c & d \end{bmatrix}$, $\mathbf{P}_0 = \begin{bmatrix} x_0 \\ y_0 \end{bmatrix}$ and $\mathbf{P}_1 = \begin{bmatrix} x_1 \\ y_1 \end{bmatrix}$.

If we consider any point P_0 our earlier operation $\mathbf{X}$ sends P_0 to P_1, where

$$\mathbf{P}_1 = \mathbf{X}\mathbf{P}_0.$$

If now operation $\mathbf{Y}$ is performed it will send P_1 to P_2, where

$$\mathbf{P}_2 = \mathbf{Y}\mathbf{P}_1 = \mathbf{Y}\mathbf{X}\mathbf{P}_0.$$

N.B. *The resultant of two matrices* **AB** *represents the result of applying the operations* **B** *and* **A** *in succession. Operators in mathematics are written in different conventions. Sometimes they apply from right to left, as here, and sometimes from left to right, as in §3.6 and chapter 10. Mathematicians employ both conventions, and the student must get used to both, sooner or later. For that reason we do not attempt to put our operators in the same order as elsewhere, but this could be done by representing points by row vectors instead of column vectors.*

What can we deduce from this?

$$\mathbf{P}_1 = \mathbf{X}\mathbf{P}_0$$

means $\begin{cases} x_1 = x_0 \\ y_1 = -y_0 \end{cases}$ or $\begin{bmatrix} x_1 \\ y_1 \end{bmatrix} = \begin{bmatrix} 1 & 0 \\ 0 & -1 \end{bmatrix} \begin{bmatrix} x_0 \\ y_0 \end{bmatrix}$

and $$\mathbf{P}_2 = \mathbf{Y}\mathbf{P}_1$$

means $\begin{cases} x_2 = -x_1 \\ y_2 = y_1 \end{cases}$ or $\begin{bmatrix} x_2 \\ y_2 \end{bmatrix} = \begin{bmatrix} -1 & 0 \\ 0 & 1 \end{bmatrix} \begin{bmatrix} x_1 \\ y_1 \end{bmatrix}.$

Thus $\begin{cases} x_2 = -x_0 \\ y_2 = -y_0 \end{cases}$ or $\begin{bmatrix} x_2 \\ y_2 \end{bmatrix} = \begin{bmatrix} -1 & 0 \\ 0 & -1 \end{bmatrix} \begin{bmatrix} x_0 \\ y_0 \end{bmatrix}.$

Also $$\mathbf{P}_2 = \mathbf{Y}\mathbf{X}\mathbf{P}_0$$

means $\begin{bmatrix} x_2 \\ y_2 \end{bmatrix} = \begin{bmatrix} -1 & 0 \\ 0 & 1 \end{bmatrix} \begin{bmatrix} 1 & 0 \\ 0 & -1 \end{bmatrix} \begin{bmatrix} x_0 \\ y_0 \end{bmatrix}.$

But we know that $\mathbf{X}$ followed by $\mathbf{Y}$ results in $\mathbf{R}_\pi$; or $\mathbf{Y}\mathbf{X} = \mathbf{R}_\pi$, hence

$$\begin{bmatrix} -1 & 0 \\ 0 & 1 \end{bmatrix} \begin{bmatrix} 1 & 0 \\ 0 & -1 \end{bmatrix} = \begin{bmatrix} -1 & 0 \\ 0 & -1 \end{bmatrix}.$$

In this special case the two matrices combine in this way. Now if we have any two matrices, **A** and **B**, can we find the matrix **C**, which is equivalent to applying the operations **B** and **A** in succession; that is, can we find the matrix **C** such that

$$\mathbf{C} = \mathbf{A}\mathbf{B}?$$

Let $$\mathbf{A} = \begin{bmatrix} a & b \\ c & d \end{bmatrix} \quad \text{and} \quad \mathbf{B} = \begin{bmatrix} p & q \\ r & s \end{bmatrix},$$

then operation **B** sends P_0 to P_1, where

$$\mathbf{P}_1 = \mathbf{BP}_0,$$

i.e. $$\left.\begin{aligned} x_1 &= px_0+qy_0 \\ y_1 &= rx_0+sy_0 \end{aligned}\right\} \quad \text{or} \quad \begin{bmatrix} x_1 \\ y_1 \end{bmatrix} = \begin{bmatrix} p & q \\ r & s \end{bmatrix} \begin{bmatrix} x_0 \\ y_0 \end{bmatrix}$$

and the operation **A** sends P_1 to P_2, where

$$\mathbf{P}_2 = \mathbf{AP}_1,$$

i.e. $$\left.\begin{aligned} x_2 &= ax_1+by_1 \\ y_2 &= cx_1+dy_1 \end{aligned}\right\} \quad \text{or} \quad \begin{bmatrix} x_2 \\ y_2 \end{bmatrix} = \begin{bmatrix} a & b \\ c & d \end{bmatrix} \begin{bmatrix} x_1 \\ y_1 \end{bmatrix}.$$

Substituting, as on p.   , we find that

$$\begin{bmatrix} a & b \\ c & d \end{bmatrix} \begin{bmatrix} p & q \\ r & s \end{bmatrix} = \begin{bmatrix} ap+br & aq+bs \\ cp+dr & cq+ds \end{bmatrix}.$$

This product is formed by 'row and column multiplication' as described earlier. If we put an arrow against any row of the first matrix and an arrow against any column of the second matrix, the terms so singled out produce a term in the product matrix. The place where it appears is shown by putting two arrows at the side of the product matrix in positions corresponding to the arrows on the left-hand side of the equation. Thus, the numbers from the top row and the right-hand column appear in the position in the product where the top row meets the right-hand column.

$$\begin{matrix} \rightarrow \\ \end{matrix}\begin{bmatrix} a & b \\ & \end{bmatrix} \begin{bmatrix} & q \\ & s \end{bmatrix} = \begin{matrix} \rightarrow \\ \end{matrix}\begin{bmatrix} & aq+bs \\ & \end{bmatrix}.$$
$$\qquad\qquad\qquad\quad\ \uparrow \qquad\qquad\qquad\qquad\qquad \uparrow$$

A numerical example will emphasize that in this type of multiplication the order of the matrices is very important. In view of the geometrical interpretation of many of our matrices is this surprising?

$$\begin{bmatrix} 2 & 1 \\ 0 & -1 \end{bmatrix} \begin{bmatrix} 1 & 3 \\ -2 & 1 \end{bmatrix} = \begin{bmatrix} 2\times1+1\times(-2) & 2\times3+1\times1 \\ 0+(-1)\times(-2) & 0+(-1)\times1 \end{bmatrix}$$
$$= \begin{bmatrix} 0 & 7 \\ 2 & -1 \end{bmatrix}.$$

But what happens if we multiply the same two matrices the other way round?

$$\begin{bmatrix} 1 & 3 \\ -2 & 1 \end{bmatrix}\begin{bmatrix} 2 & 1 \\ 0 & -1 \end{bmatrix} = \begin{bmatrix} 2 & -2 \\ -4 & -3 \end{bmatrix}.$$

In cases where two matrices give the same product in whichever order they are taken we say that they *commute*. Can you think of any two of the geometrical operations with which we have been dealing which do not commute? Can you think of any two which *do* commute? What happens, in each case, when the corresponding matrices are multiplied together?

Let us now consider the operation of leaving things as they are. As before, this operation will be denoted by **I**. What is the matrix representing **I**?

$$\mathbf{I} = \begin{bmatrix} 1 & 0 \\ 0 & 1 \end{bmatrix}.$$

To reflect twice in a line sends any point P_0 to ...?

Let us investigate further by seeing what the matrix $\mathbf{X}^2$ represents, remembering that $\mathbf{X}^2$ denotes a double reflection in the x-axis.

$$\mathbf{X}^2 = \begin{bmatrix} 1 & 0 \\ 0 & -1 \end{bmatrix}\begin{bmatrix} 1 & 0 \\ 0 & -1 \end{bmatrix} = \begin{bmatrix} 1 & 0 \\ 0 & 1 \end{bmatrix}.$$

This checks with our previous knowledge that reflecting twice in the x-axis sends any point P_0 back to its original position. Is this true for a double reflection in any line through the origin? Is $\mathbf{L}^2 = \mathbf{I}$? If we rotate twice through an angle θ about the origin do we get back to an original position? In general

$$\mathbf{R}_\theta^2 = \begin{bmatrix} \cos\theta & -\sin\theta \\ \sin\theta & \cos\theta \end{bmatrix}\begin{bmatrix} \cos\theta & -\sin\theta \\ \sin\theta & \cos\theta \end{bmatrix}$$

$$= \begin{bmatrix} \cos^2\theta - \sin^2\theta & -2\sin\theta\cos\theta \\ 2\sin\theta\cos\theta & -\sin^2\theta + \cos^2\theta \end{bmatrix},$$

i.e.

$$\mathbf{R}_\theta^2 = \begin{bmatrix} \cos 2\theta & -\sin 2\theta \\ \sin 2\theta & \cos 2\theta \end{bmatrix}$$

and so

$$\mathbf{R}_\theta^2 = \mathbf{R}_{2\theta}.$$

Again, this checks with our previous geometrical knowledge. In particular $\mathbf{R}_\pi^2 = \mathbf{I}$.

What must $\mathbf{R}_\theta$ be multiplied by in order to bring us back to our original position? In other words, what is the operation inverse to $\mathbf{R}_\theta$? Can you write down the matrices representing the operations inverse to $\mathbf{X}$, $\mathbf{Y}$, $\mathbf{L}$, $\mathbf{R}_\pi$, $\mathbf{R}_\theta$? Check the products algebraically. Are the products always $\mathbf{I}$?

Further numerical examples may be worked, and also other examples illustrating the ideas on pp. 337–9, as far as this is possible at this stage. Dilatations about the origin are easily done, and dilatations and rotations may be combined. As a more sophisticated exercise a class which is familiar with the groups C_n and D_n (see p. 57) can work out sets of matrices which describe the operations of the groups.

So far we have dealt with the matrices which represent the geometrical operations of reflection and rotation. Let us now introduce the operations $\mathbf{T}_{1,0}$ and $\mathbf{T}_{0,1}$, which are translations through one unit in the x and y directions respectively. The operation $\mathbf{T}_{1,0}$ means *translate through one unit in the direction of the x-axis*, and so it sends any point P_0 to P_1, where

$$\mathbf{T}_{1,0}\begin{cases} x_1 = x_0+1, \\ y_1 = y_0. \end{cases}$$

Observe that the value of x_1 now involves more than just the values of x_0 and y_0, so the matrices we have been using are no longer big enough to store all the necessary information. Can we manage by using matrices with two rows and three columns? For the moment let us put

$$\mathbf{T}_{1,0} = \begin{bmatrix} 1 & 0 & 1 \\ 0 & 1 & 0 \end{bmatrix}.$$

Now a translation through one unit in the x-direction done twice is equivalent to a translation through two units in the x-direction. Algebraically we can represent this by

$$x_1 = x_0+1,$$

$$y_1 = y_0$$

and

$$x_2 = x_1+1,$$

$$y_2 = y_1,$$

which gives

$$x_2 = x_0+2,$$

$$y_2 = y_0.$$

Thus we need to have

$$[\mathbf{T}_{1,0}]^2 = \begin{bmatrix} 1 & 0 & 1 \\ 0 & 1 & 0 \end{bmatrix} \begin{bmatrix} 1 & 0 & 1 \\ 0 & 1 & 0 \end{bmatrix} = \begin{bmatrix} 1 & 0 & 2 \\ 0 & 1 & 0 \end{bmatrix}.$$

However, our rule of multiplication of matrices breaks down here; because, for example, on selecting the first row of the first matrix and the first column of the second matrix the corresponding numbers are

$$[1 \quad 0 \quad 1] \quad \begin{bmatrix} 1 \\ 0 \\ * \end{bmatrix}.$$

There is no number at * in the column to make a product with the third element in the row, and it is similar in every other case. If we are to be able to multiply we need a third row to specify the asterisk, and, as we know what the first and second rows should be in our product matrix, we can see that we require:

$$\mathbf{T}_{1,0} = \begin{bmatrix} 1 & 0 & 1 \\ 0 & 1 & 0 \\ 0 & 0 & 1 \end{bmatrix}.$$

Similarly,

$$\mathbf{T}_{0,1} = \begin{bmatrix} 1 & 0 & 0 \\ 0 & 1 & 1 \\ 0 & 0 & 1 \end{bmatrix}.$$

Now what was the operation $\mathbf{T}_{1,0}$? And what will happen if we repeat it n times? We see that

$$[\mathbf{T}_{1,0}]^n = \begin{bmatrix} 1 & 0 & n \\ 0 & 1 & 0 \\ 0 & 0 & 1 \end{bmatrix} \quad \text{and} \quad [\mathbf{T}_{0,1}]^n = \begin{bmatrix} 1 & 0 & 0 \\ 0 & 1 & n \\ 0 & 0 & 1 \end{bmatrix}.$$

These results come from our geometrical knowledge, but are they also clear from the point of view of matrix multiplication? If there is any doubt they will provide an exercise on induction. What matrix corresponds to a translation of a in the x-direction? ...of b in the y-direction? ...a translation with components (a, b)? Verify that

$$\mathbf{T}_{a,o}\mathbf{T}_{o,b} = \mathbf{T}_{o,b}\mathbf{T}_{a,o} = \mathbf{T}_{a,b}$$

and that

$$\mathbf{T}_{a,b}\mathbf{T}_{c,d} = \mathbf{T}_{a+c,b+d}.$$

When dealing with operations involving reflections, rotations and translations we must augment our existing matrices as follows:

$$\mathbf{X} = \begin{bmatrix} 1 & 0 & 0 \\ 0 & -1 & 0 \\ 0 & 0 & 1 \end{bmatrix}, \quad \mathbf{R}_\theta = \begin{bmatrix} \cos\theta & -\sin\theta & 0 \\ \sin\theta & \cos\theta & 0 \\ 0 & 0 & 1 \end{bmatrix},$$

$$\mathbf{P}_0 = \begin{bmatrix} x_0 \\ y_0 \\ 1 \end{bmatrix}, \quad \mathbf{P}_1 = \begin{bmatrix} x_1 \\ y_1 \\ 1 \end{bmatrix},$$

in order to comply with the rule of multiplication of matrices.

Exercises

It is now possible to express the results developed in §10.5 in matrix terminology, and numerical exercises on these lines are easy to construct, once the principles have been understood.

Thus the matrices which describe the symmetries of the strip ornaments of Type 4 in Fig. 10.31 are all of the form

$$\begin{bmatrix} 1 & 0 & n \\ 0 & \pm 1 & 0 \\ 0 & 0 & 1 \end{bmatrix}.$$

Those describing Type 2 are of the form

$$\begin{bmatrix} \pm 1 & 0 & n \\ 0 & \pm 1 & 0 \\ 0 & 0 & 1 \end{bmatrix},$$

where the two $\pm$ signs are the same, and those describing Type 7 are of the form

$$\begin{bmatrix} \pm 1 & 0 & n \\ 0 & 1 & 0 \\ 0 & 0 & 1 \end{bmatrix} \quad \text{or} \quad \begin{bmatrix} \pm 1 & 0 & n \pm \frac{1}{2} \\ 0 & -1 & 0 \\ 0 & 0 & 1 \end{bmatrix}.$$

The other cases may be worked out as an exercise.

To complete the comparison with chapter 10 we may, much later in the course, obtain the matrix which reflects the plane in the line $y = mx + c$. The resulting formula is of no major importance in itself, although it occurs, not infrequently, as an examination question. It is convenient to think of a piece of paper, carrying the line, and placed on a table. The table has a fixed coordinate system on it, and the aim is to turn the paper over, about the line.

The line $y = mx+c$ cuts the y-axis at the point $(0, c)$, and is inclined to the x-axis at an angle θ, where $\tan\theta = m$. Therefore we may reflect in the line by carrying out a sequence of five operations for which we already know the matrix representation.

If we first translate the paper in the y-direction through $-c$ units, the line on the paper passes through the origin on the table. If we then rotate about the origin through an angle $-\theta$, where $\tan\theta = m$, the line becomes coincident with the x-axis. We can then reflect in the x-axis, rotate about the origin through an angle θ and translate in the y-direction through c units to get the required result. What is this sequence of operations in matrix notation?

$$\mathbf{P}_1 = \mathbf{T}_{0,c}\,\mathbf{R}_\theta\,\mathbf{X}\mathbf{R}_{-\theta}\,\mathbf{T}_{0,-c}\,\mathbf{P}_0,$$

where $\tan\theta = m$. (*Be careful over the order!*) So

$$\begin{bmatrix} x_1 \\ y_1 \\ 1 \end{bmatrix} = \begin{bmatrix} 1 & 0 & 0 \\ 0 & 1 & c \\ 0 & 0 & 1 \end{bmatrix} \begin{bmatrix} \cos\theta & -\sin\theta & 0 \\ \sin\theta & \cos\theta & 0 \\ 0 & 0 & 1 \end{bmatrix} \begin{bmatrix} 1 & 0 & 0 \\ 0 & -1 & 0 \\ 0 & 0 & 1 \end{bmatrix}$$

$$\begin{bmatrix} \cos\theta & \sin\theta & 0 \\ -\sin\theta & \cos\theta & 0 \\ 0 & 0 & 1 \end{bmatrix} \begin{bmatrix} 1 & 0 & 0 \\ 0 & 1 & -c \\ 0 & 0 & 1 \end{bmatrix} \begin{bmatrix} x_0 \\ y_0 \\ 1 \end{bmatrix}.$$

Therefore

$$\begin{bmatrix} x_1 \\ y_1 \\ 1 \end{bmatrix} = \begin{bmatrix} \cos^2\theta-\sin^2\theta & 2\sin\theta\cos\theta & -2c\sin\theta\cos\theta \\ 2\sin\theta\cos\theta & \sin^2\theta-\cos^2\theta & c(\cos^2\theta-\sin^2\theta+1) \\ 0 & 0 & 1 \end{bmatrix} \begin{bmatrix} x_0 \\ y_0 \\ 1 \end{bmatrix}.$$

But $\tan\theta = m$, and so $\sin\theta = \dfrac{m}{\sqrt{(1+m^2)}}$ and $\cos\theta = \dfrac{1}{\sqrt{(1+m^2)}}$.

Hence
$$\begin{bmatrix} x_1 \\ y_1 \\ 1 \end{bmatrix} = \begin{bmatrix} \dfrac{1-m^2}{1+m^2} & \dfrac{2m}{1+m^2} & \dfrac{-2mc}{1+m^2} \\ \dfrac{2m}{1+m^2} & \dfrac{m^2-1}{1+m^2} & \dfrac{2c}{1+m^2} \\ 0 & 0 & 1 \end{bmatrix} \begin{bmatrix} x_0 \\ y_0 \\ 1 \end{bmatrix},$$

or
$$x_1 = \frac{(1-m^2)\,x_0+2m\,y_0-2mc}{1+m^2},$$

$$y_1 = \frac{2mx_0+(m^2-1)\,y_0+2c}{1+m^2}.$$

Reflections in lines parallel to the y-axis are not included in this formula; but these are easily written down separately.

Advanced notes

The method of moving the line to a more convenient position, carrying out the operation there, and then moving back, produces matrix products of the form $\mathbf{PAP}^{-1}$. If $\mathbf{B} = \mathbf{PAP}^{-1}$ the matrices $\mathbf{A}$ and $\mathbf{B}$ are said to be *similar*. The study of similar matrices is an important topic later on, and here we see the type of situation which raises the problem.

In a similar way, we can find the matrix which rotates the plane through an angle θ about the point (x, y). One has then expressed all the isometries of the Euclidean plane in matrix form. That is, we have constructed a matrix representation of the Euclidean group of isometries which was built up in chapter 10.

4. FOUR-TERMINAL NETWORKS

A lesson on 2×2 matrices may be given around the topic of four-terminal networks. The lesson presupposes a knowledge of electricity as far as Ohm's law. We see a technological application of matrices, and the students are faced with the problem of formalizing a physical situation in mathematical terms, and of interpreting their calculations in terms of something in the real world. This ability may not matter to any potential research pure mathematicians in the class, but it is at all times a vital lesson for those boys who will later apply mathematics in industry and in the sciences. Routine, manipulative exercises occur in the course of the investigation.

In order to keep the electrical problems as simple as possible we restrict attention to the easiest cases; and in these examples two of the leads in our circuits are always, in effect, earthed, although the theory is quite capable of handling situations which are more general than this. Also, in our examples, all of the circuit boxes are connected in series, but there are further developments of the theory in which other interconnections are possible, and some of these may be found in [27]. Hyperbolic functions occur in one of the examples, but they are not required for the main discussion.

Lesson

Let us consider an electrical problem of a type which can well occur in practice. We have a 'black box' with four terminals. We do not know what is inside the box, but two terminals are labelled 'input' and two are labelled 'output'. As is often the case with electrical equipment the terminals are marked + and –, and to keep the problem simple, I will tell you that the – terminals are always to be connected to an earth line.

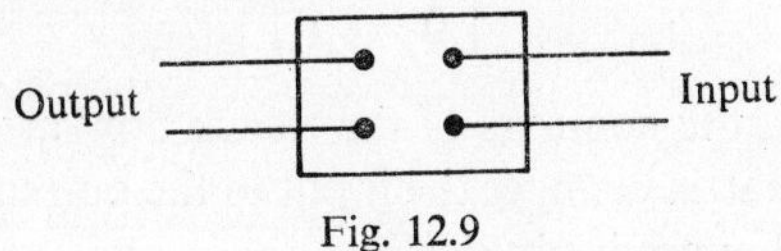

Fig. 12.9

Now if we do not know what is inside the box, can we find out by experiment? *This is a genuine question, and the physicists in the class can profitably devote time to discussing it; although here we shall not get anywhere near to a complete solution.*

One method might be to feed a current into the input, and measure it; and see what comes out of the output. This would lead to the scheme:

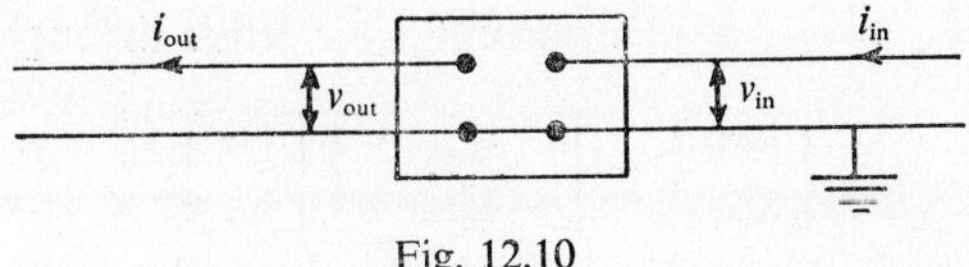

Fig. 12.10

The currents and voltages are to be measured by ammeters and voltmeters, and we may imagine that we might find various relations between them experimentally.

A start could be made on the problem by considering what relations arise in a few especially simple cases. Suppose the box contains

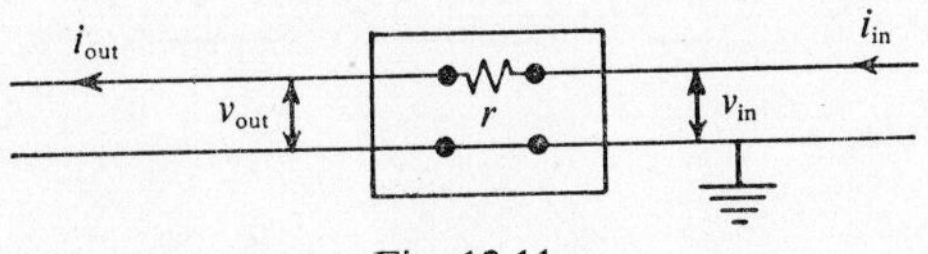

Fig. 12.11

where r is a resistance.

How are v_{out} and i_{out} related to v_{in} and i_{in}?

It is easy to see that $i_{out} = i_{in}$, and the voltage is altered by the drop across the resistance. Therefore

$$v_{out} = v_{in} - r\,i_{in},$$

$$i_{out} = i_{in}.$$

How can this be written in matrix notation?

$$\begin{bmatrix} v_{out} \\ i_{in} \end{bmatrix} = \begin{bmatrix} 1 & -r \\ 0 & 1 \end{bmatrix} \begin{bmatrix} v_{in} \\ i_{in} \end{bmatrix}.$$

In other words, the contents of the 'black box' are described by a matrix, and the matrix changes the input to the output (algebraically) just as the box changes it physically.

How does this box behave?

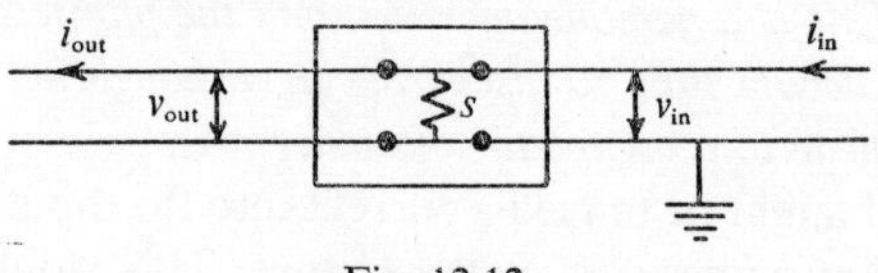

Fig. 12.12

This time $v_{out} = v_{in}$, and a current v_{in}/s leaks through the resistance, so

$$\begin{bmatrix} v_{out} \\ i_{out} \end{bmatrix} = \begin{bmatrix} 1 & 0 \\ -1/s & 1 \end{bmatrix} \begin{bmatrix} v_{in} \\ i_{in} \end{bmatrix}.$$

Again the box is described by a matrix. Now what happens if we connect a number of boxes in series? At each stage the output of a box becomes the input of the next box, and it is then multiplied by the corresponding matrix. So we may imagine the following scheme.

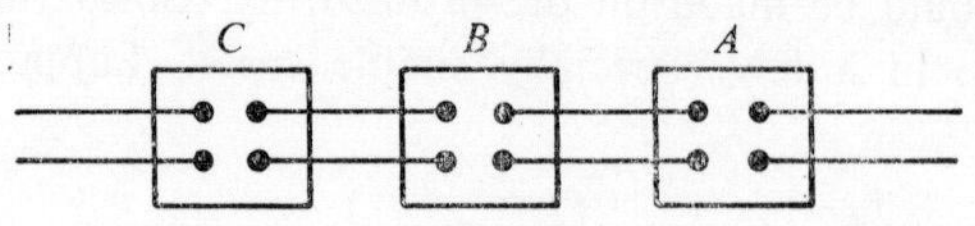

Fig. 12.13

$$\begin{bmatrix} v_{out} \\ i_{out} \end{bmatrix} = \mathbf{CBA} \begin{bmatrix} v_{in} \\ i_{in} \end{bmatrix}.$$

The matrices appear as an algebraic representation of the boxes, and the algebraic equation is almost a diagram of the circuit. If

$$\mathbf{R} = \begin{bmatrix} 1 & -r \\ 0 & 1 \end{bmatrix} \quad \text{and} \quad \mathbf{S} = \begin{bmatrix} 1 & 0 \\ -1/s & 1 \end{bmatrix},$$

draw the circuits corresponding to **RSR** and **SRS** and evaluate the corresponding matrix products. Further examples may easily be constructed.

Matrices may be found corresponding to other electrical components, but it is important to remember that this theory is restricted to components which are *linear*, that is, where v_{out} and i_{out} are related to v_{in} and i_{in} by equations which are of the first degree and, in this case, without constant terms. In spite of these limitations the theory is adequate for a number of tasks, and it can be extended to alternating currents by introducing complex impedances.

As a harder example, what is the matrix for the network below, where there are n resistances of each type?

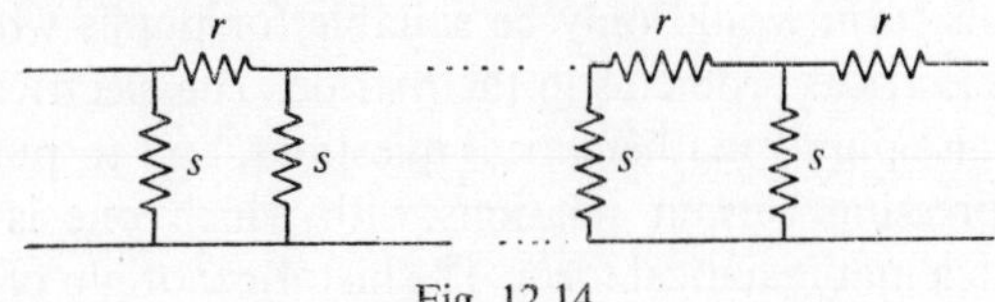

Fig. 12.14

If we put

$$\mathbf{A} = \mathbf{SR} = \begin{bmatrix} 1 & 0 \\ -1/s & 1 \end{bmatrix} \begin{bmatrix} 1 & -r \\ 0 & 1 \end{bmatrix} = \begin{bmatrix} 1 & -r \\ -1/s & 1+r/s \end{bmatrix},$$

then we have to find $\mathbf{A}^n$. When we take powers the matrix soon becomes complicated; but a convenient trick has been discovered [131] in the substitution

$$\mu = \sqrt{rs}, \quad \cosh u = 1+r/2s.$$

Then
$$\sinh \tfrac{1}{2}u = \tfrac{1}{2}\sqrt{(r/s)}$$

and
$$\mathbf{A} = \begin{bmatrix} 1 & -2\mu \sinh \frac{1}{2}u \\ -(2/\mu) \sinh \frac{1}{2}u & 2\cosh u - 1 \end{bmatrix}$$

$$= \frac{1}{\cosh \frac{1}{2}u} \begin{bmatrix} \cosh \frac{1}{2}u & -\mu \sinh u \\ -(1/\mu) \sinh u & \cosh \frac{3}{2}u \end{bmatrix}.$$

It is now an exercise on induction to prove that

$$\mathbf{A}^n = \frac{1}{\cosh \frac{1}{2}u}\begin{bmatrix} \cosh (n-\frac{1}{2})\, u & -\mu \sinh nu \\ -(1/\mu) \sinh nu & \cosh (n+\frac{1}{2})\, u \end{bmatrix}.$$

Circuits like this are important in the design of electrical filters, and they provide examples in recurrence relations, continued fractions, and summation of trigonometric series.

5. LESSONS ON THE WHEATSTONE BRIDGE

In this section we study the Wheatstone bridge, a piece of electrical apparatus which is traditionally part of the school A-level physics course and which will therefore be familiar to many boys. In seeking to analyse this structure we will find the opportunity to combine many of the ideas in previous chapters and the doors will be opened to a number of branches of modern algebra and topology as well as to a particular branch of technology.

Some of the points in this section are definitely difficult and the lesson in this form would only be suitable for pupils who can take the purely electrical problems in their stride. The electrical situation is used to raise purely mathematical questions, and to put the problem of expressing certain relations with which one is intuitively familiar into a mathematical form. The historical origin of the branch of topology called homology is to be found in Kirchhoff's study of electrical networks.

Lesson

A Wheatstone bridge is an electrical network arranged as in Fig. 12.15. *On the blackboard this figure would be built up step by step, as the discussion develops.*

Any circuit is composed of *nodes* connected by *branches*. In this case there are four nodes and six branches, which we may letter and number. If the bridge is already known from physics lessons we ask the class how they proceed; in which case, before long somebody will say that we have to specify a certain number of *loop currents*, and this is our starting-point. In the network problems of sixth-form physics it is customary to pick fundamental loops by guesswork. How can we make this guesswork systematic? The first notion to make clear is that with any network we may associate a *tree*. In general, there are many associated trees, and we produce a tree by imagining branches removed from the circuit one at a time, until the removal of one more branch is bound to disconnect one or more of the nodes entirely. It is easy to see that this inevitably involves reducing the number of

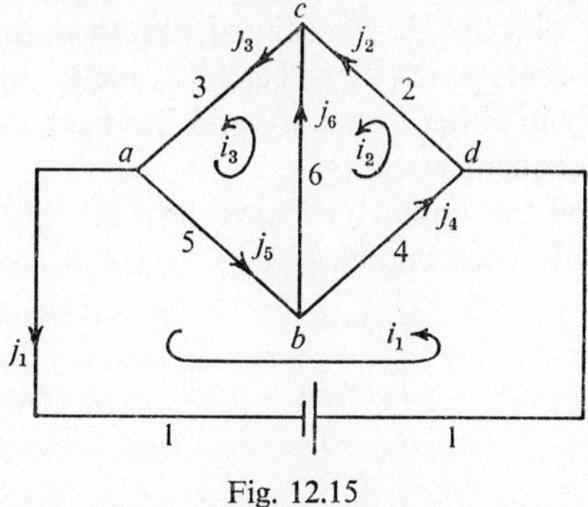

Fig. 12.15

branches to one less than the total number of nodes. Some of the various ways of deriving a tree from the circuit of Fig. 12.15 are shown in Fig. 12.16. The branches remaining we call *tree-branches*, and the branches which we imagine removed we call *links*. Replacing links one at a time produces closed loops. *This situation calls for thorough discussion by the class.*

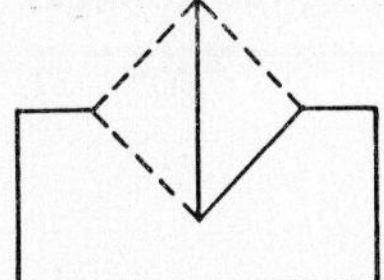 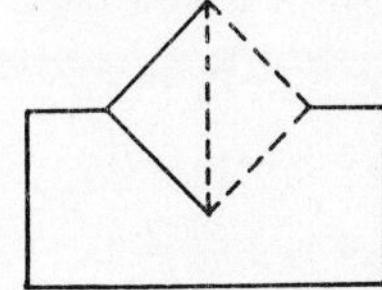 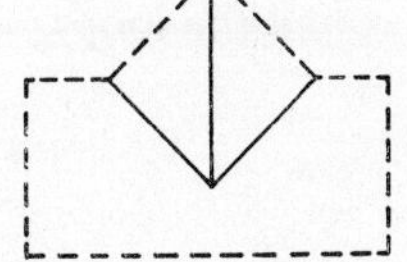

Fig. 12.16

We will now proceed to analyse the circuit by using the third tree, but any other tree would do. It is advisable to use any branches in which there are batteries as links, and it is useful, but not essential, to number the links of the circuit before the tree-branches, and this we have done in Fig. 12.15.

We will denote the currents flowing in the six branches of the circuit by $j_1, j_2, \ldots, j_6$ and we may draw them initially in either direction, but once they have been indicated the sense must be maintained. It is familiar practice in physics to express these in terms of the loop currents, and here we have three loop currents corresponding to the three links. If we imagine the tree and insert link 1, we form the closed loop containing branches 1, 4 and 5. If we start with the tree and insert link 2, we get the closed loop containing the branches 2, 4 and 6. Similarly with link 3. Associated with these three loops there are three loop currents which we will call i_1, i_2 and i_3. Again, these may be drawn initially in either direction, and the signs look after themselves in the subsequent working.

Having labelled the circuit elements, how are the i's and j's related? Summing in each branch gives

$$\begin{aligned} j_1 &= i_1, \\ j_2 &= i_2, \\ j_3 &= i_3, \\ j_4 &= -i_1+i_2, \\ j_5 &= -i_1+i_3, \\ j_6 &= -i_2+i_3. \end{aligned}$$

It is now time to introduce the voltages, and we may use $v_1, \ldots, v_6$ to denote the voltages across the branches of the circuit and e_1, e_2, e_3 to denote any applied voltages in the links. *In our case the only applied voltage is e_1 in link 1, but the general method should be clear.*

What equations connect the v's and e's? Summing the voltages round the loops

$$\begin{aligned} v_1-v_4-v_5 &= e_1, \\ v_2+v_4-v_6 &= 0, \\ v_3+v_5+v_6 &= 0. \end{aligned}$$

One should be quite clear that v_1 denotes the voltage-drop across any resistance which is in branch 1, and e_1 denotes the voltage supplied by the battery.

Can you express these two sets of equations in matrix form? Both involve the same matrix; although in one case the matrix is transposed; that is, the rows and columns are interchanged. Let us write down the matrix involved, denoting it by **T**, and its transpose by **T**$'$,

$$\mathbf{T} = \begin{bmatrix} 1 & 0 & 0 & -1 & -1 & 0 \\ 0 & 1 & 0 & 1 & 0 & -1 \\ 0 & 0 & 1 & 0 & 1 & 1 \end{bmatrix}.$$

The equations are then

$$\mathbf{j} = \mathbf{T'i} \quad \text{and} \quad \mathbf{Tv} = \mathbf{e}.$$

This matrix clearly has an importance quite apart from electrical considerations; what is this? Familiarity with the lesson on p. might suggest the answer—this is an incidence matrix, and it shows the relationships between the loops and the branches in the circuit; so it might suitably be labelled:

		Branch					
		1	2	3	4	5	6
Loop	1	1	0	0	−1	−1	0
	2	0	1	0	1	0	−1
	3	0	0	1	0	1	1

This matrix has the technical name of a *tie-set schedule*. Observe it carefully, and make yourself familiar with the sign conventions involved. This matrix is a way of describing the interconnectedness of the circuit, in fact it may be regarded as characterizing certain properties of a tetrahedron. (*Why?*)

This matrix has many properties. For example, loop *abdc* may be regarded in a certain sense as the 'sum' of loops 2 and 3. If we add rows 2 and 3 of the matrix we get the row vector (0, 1, 1, 1, 1, 0) which tells us correctly the branches in the new loop.

How may we express the loop *a*, battery, *d*, *c*? Describe other loops in a similar way. Describe the path obtained by taking loop 1 minus loop 2.

It should transpire from discussion that if different people work with different trees, then they will have different matrices **T**, but anybody will be able to express the rows of the other peoples' matrices by taking linear sums of the rows in his own matrix. Each person's three loops form a *basis*, in terms of which all possible closed loops in the circuit may be expressed. Changing from one tree to another involves changing the basis. *In an important way these relations need not be new; because empirical experiment with methods of solution which are familiar in the physics laboratory soon gives an instinctive appreciation that different starts to the problem lead to the same solution, and that the equations obtained by two methods of starting are mutually derivable from one another. The notion of a vector space with different bases is merely the sophisticated formulation of this idea.*

We have not yet mentioned the resistances in the different branches. How must these be introduced? If the resistances are $r_1, r_2, \ldots, r_6$ then

$$v_s = r_s j_s,$$

for $s = 1, \ldots, 6$. We may write in matrix notation

$$\mathbf{v} = \mathbf{Rj},$$

where **R** is a diagonal matrix; that is, a matrix in which all the terms which are not in the principal diagonal are zero.

Combining the equations

$$\mathbf{e} = \mathbf{Tv} = \mathbf{TRj} = \mathbf{TRT'i}.$$

This equation shows what voltages must be applied in the links to maintain specified currents in the loops. It may be regarded as the matrix equivalent of Ohm's law, the matrix **TRT′** being the 'resistance' of the network, relating the matrix current **i** and the matrix voltage **e**.

The teacher may note that we have here, inherent in the situation, a matrix transformation of the form $\mathbf{R} \rightarrow \mathbf{TRT'}$. *This is an example of a congruence transformation, the study of which is an established part of more advanced matrix theory. This example is unusual, from the matrix algebra point of view, because the matrix* **T** *is not square.*

The method, at the beginning of this lesson, of constructing trees and deciding the basic loop currents may seem unnecessarily involved, as the finite regions of the circuit diagram can be used as basic loops, and the class may see this. This is only the case, however, when the circuit has a planar *graph—that is when the circuit can be drawn 'flat' on the paper. The methods we have outlined apply even when the circuit can not.*

Further developments

This is perhaps as far as one might expect to go at school, but the situation contains further important ideas which we give here for the teacher, inviting him to experiment with them if a suitable occasion arises. The basic problem is topological rather than electrical, and it would seem that there is much to learn by attempting to analyse different networks for oneself on the lines which follow.

Inherent in this situation there is a certain duality. We develop a way of thinking in which voltage and current, short-circuit and open-circuit, loops and nodes are dual terms. The dual point of view may not seem quite so important physically in this particular example, because, whereas before we introduced voltages into the links of the circuit, we will now have to introduce currents at the tree-branch nodes, and this is a less common state of affairs in elementary electricity. *For practical purposes, using the present circuit, it is probably better to remove branch 1, and to consider the currents flowing in and out at a and d as coming from external sources, but for the sake of the example we continue with the circuit as it is.*

Previously we considered the various loops produced by open-circuiting all but one of the links at a time. Now we consider the *cut-sets* produced by short-circuiting all but one of the tree-branches at a time. Short-circuiting all but one of the tree-branches divides the nodes into two sets at different potentials. We may imagine the nodes to be little spheres held together by the branches, which are pieces of elastic (*there is an opportunity for a visual aid here*). We take the tree-branch in question and stretch it between our hands. The tree-branches which are short-circuited are each bunched up, the elastic being entirely in one hand. Each link in the circuit is then constrained to behave in a particular way—it may be bunched in either hand or it may be stretched between the two.

The set of branches which is stretched between the hands we call a *cut-set*; the name means simply that by cutting them we can separate the circuit into two parts. The aim at present is to identify sets of branches with this property, just as before the aim was to identify sets of branches which formed closed loops. Accordingly we now construct a cut-set schedule which shows which branches fall into which cut-sets. There is one basic cut-set corresponding to each tree-branch just as before there was one basic tie-set (closed loop) corresponding to each link. This gives the matrix **C**.

	Cut-set	Branch 1	2	3	4	5	6	Short-circuit all tree-branches except	'Picked-up' nodes
C	1	1	−1	0	1	0	0	4	*a*, *b*, *c*
	2	1	0	−1	0	1	0	5	*a*
	3	0	1	−1	0	0	1	6	*a*, *b*, *d*

Some features of this matrix still require explanation. When we pick up the particular tree-branch involved it is convenient to think of it as being held vertically, one hand above the other, with the arrow on the branch pointing downwards. The matrix records the branches which are stretched between the hands, with $+1$ if the arrow on the branch is pointing downwards, and -1 if it is pointing upwards.

Before considering the electrical relations pause to investigate how the matrix describes the interconnections of the circuit. It describes three ways of cutting the circuit in two, or more strictly of dividing the nodes into two disjoint sets. These three ways are 'basic' in the sense that other ways may be described in terms of them just as all of the closed loops of the circuit could be described in terms of the three loops.

Another way of cutting the circuit into two parts would be to sever branches 2, 3, 4 and 5. How can this be expressed in terms of our basic cut-sets? If we subtract row 2 of the matrix from row 1 we get the row vector $(0, -1, +1, +1, -1, 0)$. This shows the branches which have to be cut, and also the directions on the arrows if the cutting is done by picking up nodes b and c according to the convention. (*Study other cases of this.*)

The matrix $\mathbf{C}$ may also be used to express Kirchhoff's laws. Whereas previously there were certain basic loop currents, in terms of which all the branch currents could be expressed, there are now certain basic voltages (which we can call f_4, f_5 and f_6) appearing at the ends of the tree-branches, and all the branch voltages can be expressed in terms of these;

$$\mathbf{v} = \mathbf{C}'\mathbf{f} \quad \text{where} \quad \mathbf{f} = \begin{bmatrix} f_4 \\ f_5 \\ f_6 \end{bmatrix}.$$

To preserve the duality we must analyse the case where there are no sources of voltage in the bridge, but where currents k_4, k_5 and k_6 are introduced into the nodes at the ends of the tree-branches (the sense being the same as the conventional sense attached to the corresponding tree-branch currents earlier) and extracted from the root of the tree. The current equations are then

$$\mathbf{Cj} = \mathbf{k},$$

and for the currents in the individual branches

$$\mathbf{j} = \mathbf{Gv},$$

where **G** is a diagonal matrix giving the conductances (reciprocals of resistances) of the individual branches. Then,

$$\mathbf{k} = \mathbf{Cj} = \mathbf{CGv} = \mathbf{CGC'f}.$$

This shows the currents which must be supplied from external sources to produce assigned voltages across the branches. As exercises the analysis may be repeated using other trees, and other circuits may be studied in a similar way.

Note that in general **T** and **C** have different numbers of rows, and the total is equal to the number of columns. The scalar product of any row of **T** with any row of **C** is zero. This being so the vector spaces, which are generated by the rows of **T** and the rows of **C**, are called *orthogonal complements*.

We will end with two questions:

(1) How can further lessons on homology be developed from the starting-points here?

(2) By using a relation $\mathbf{i} \leqslant \mathbf{i}_0$ (that is, assigning an upper limit to the flow in any branch) how applicable is the above analysis to transportation problems in linear programming?

References

For the electrical questions involved see Guillemin [57], and for a formal approach, at university level, to the topological questions see Patterson [93]. Related topics may be found in Berge [13].

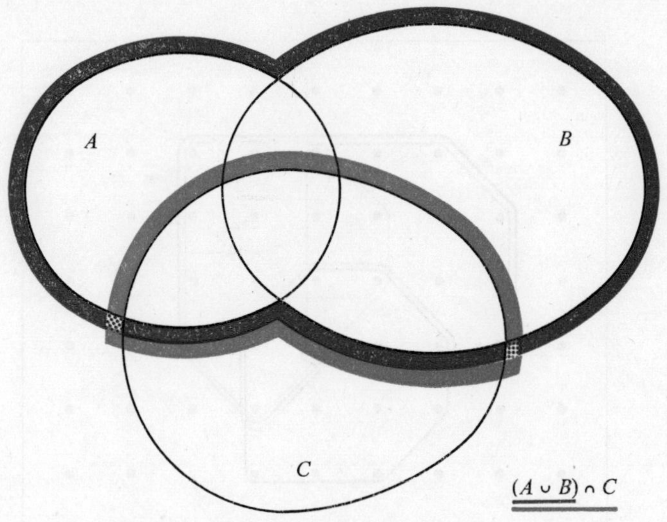

Colour-fig. 1. (See p. 131.)

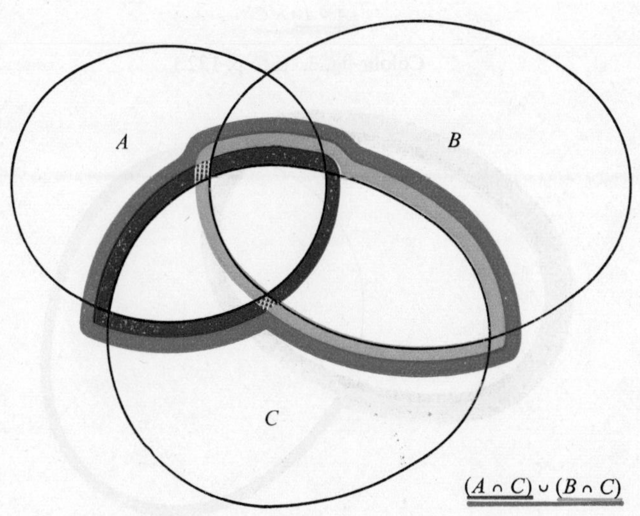

Colour-fig. 2. (See p. 131.)

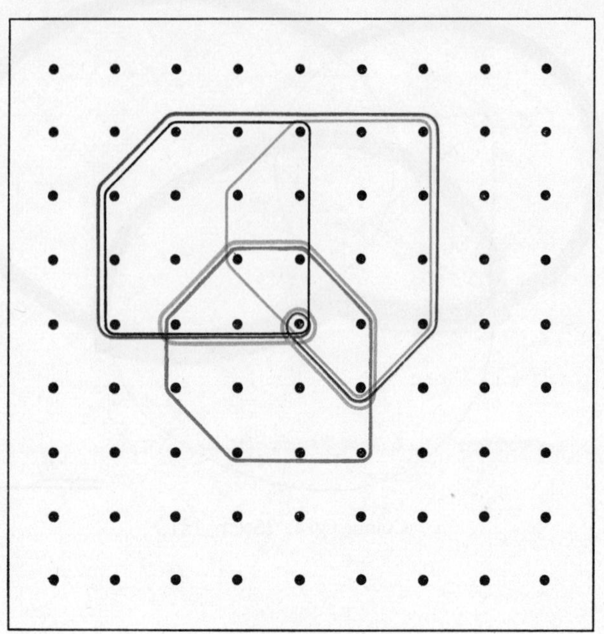

$\underline{(A \cup B)} \cap C$

Colour-fig. 3. (See p. 132.)

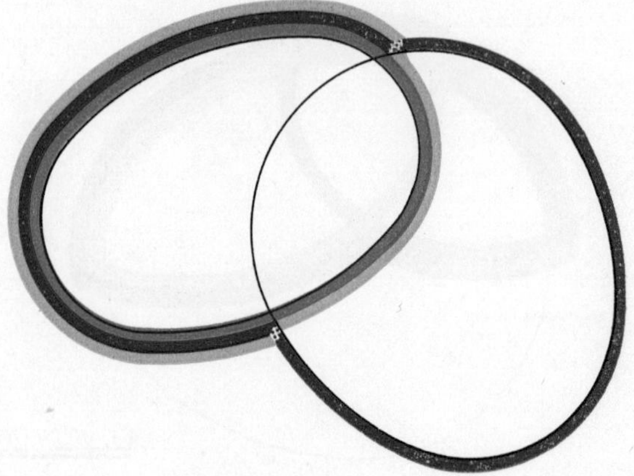

$\underline{A(A+B)} = \underline{A}$

Colour-fig. 4. (See p. 135.)

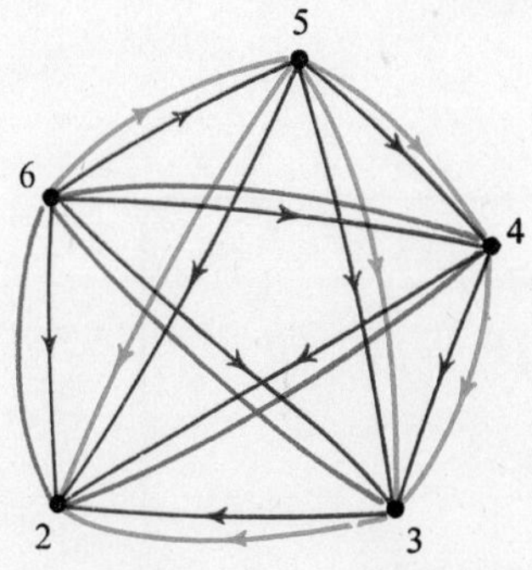

m n R: m is greater than n

m n S: m has a factor in common with n

m n T: there is a connected polygon $\left\{ {m \atop n} \right\}$

$$T = R \cap S'$$

Colour-fig. 5. (See p. 196.)

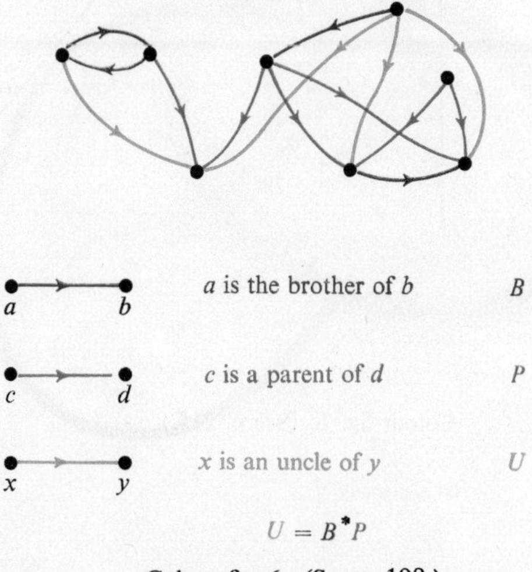

Colour-fig. 6. (See p. 198.)

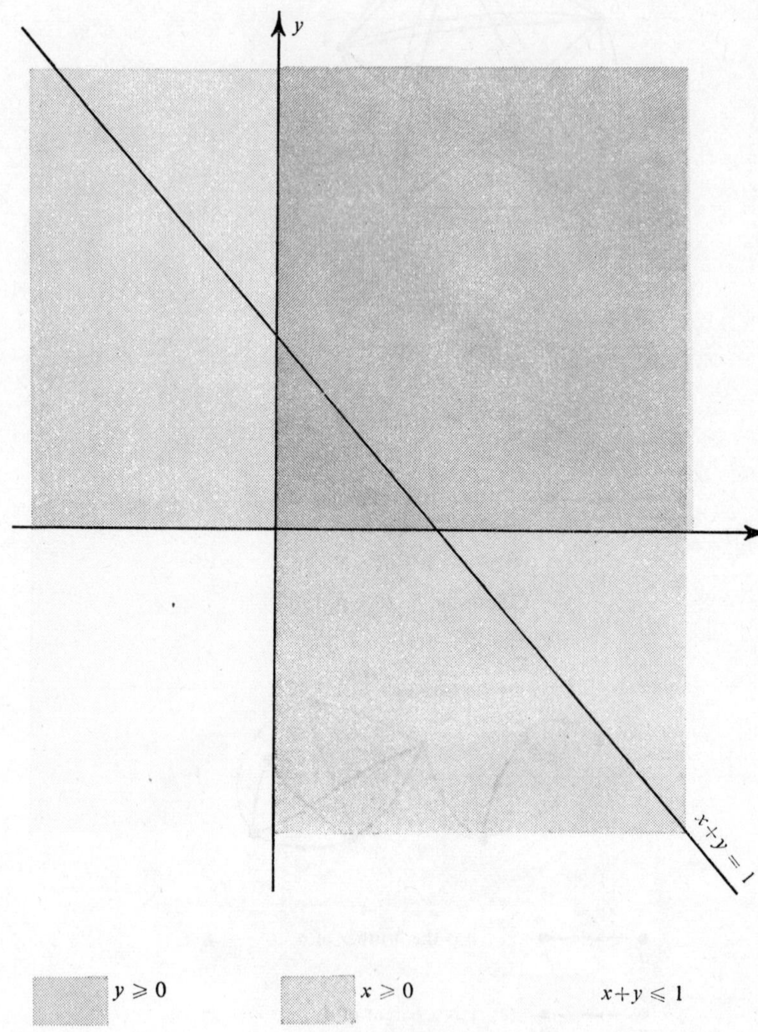

Colour-fig. 7. (See p. 215.)

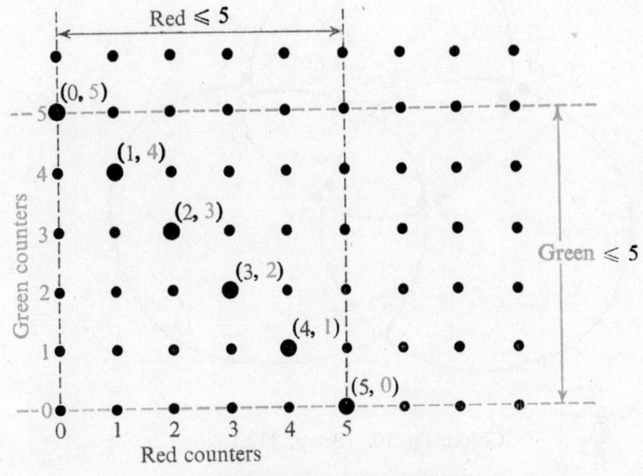

Colour-fig. 8. (See p. 211.)

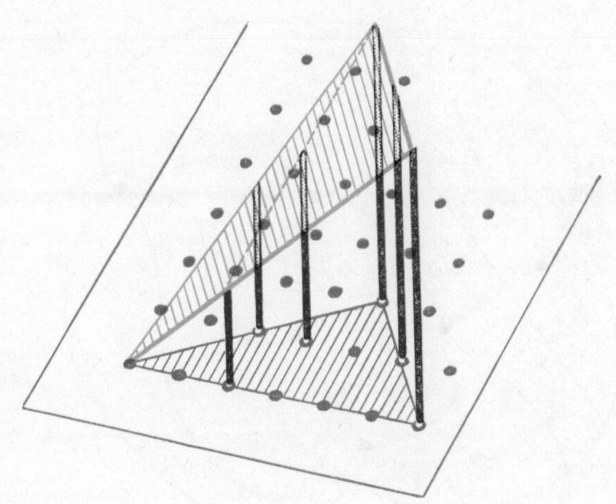

Colour-fig. 9. (See p. 218.)

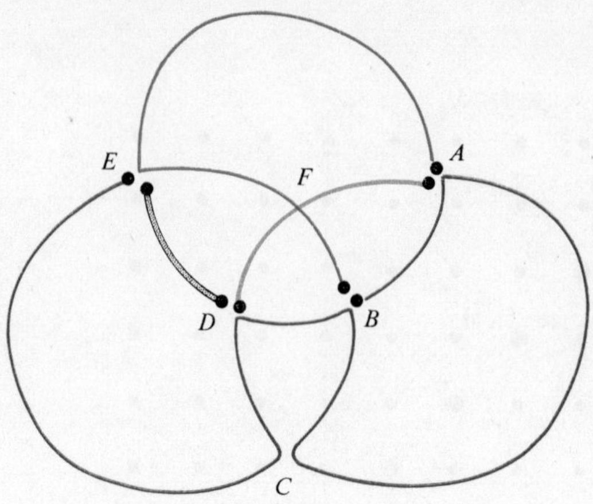

Colour-fig. 10. (See p. 232.)

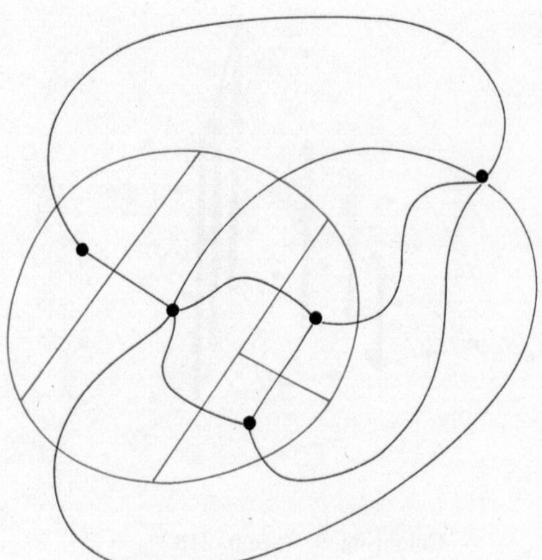

Colour-fig. 11. (See p. 243.)

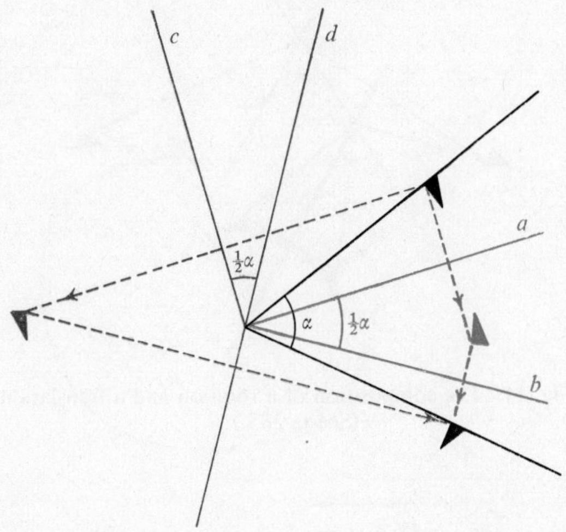

Colour-fig. 12. Equality of two pairs of reflections, **ab** = **cd.** (See p. 281.)

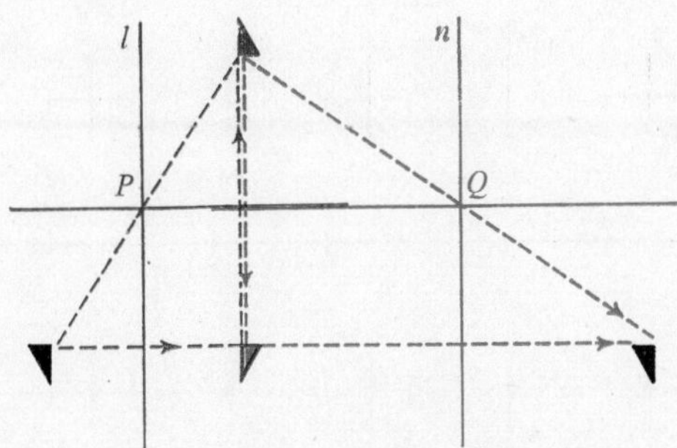

Colour-fig. 13. The composition of two half-turns, seen also as four reflections; **PQ** = **ln.** (See p. 282.)

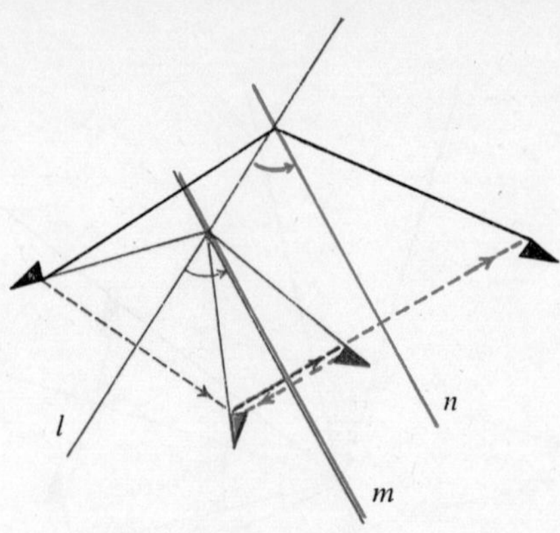

Colour-fig. 14. The composition of a rotation and a translation. (See p. 285.)

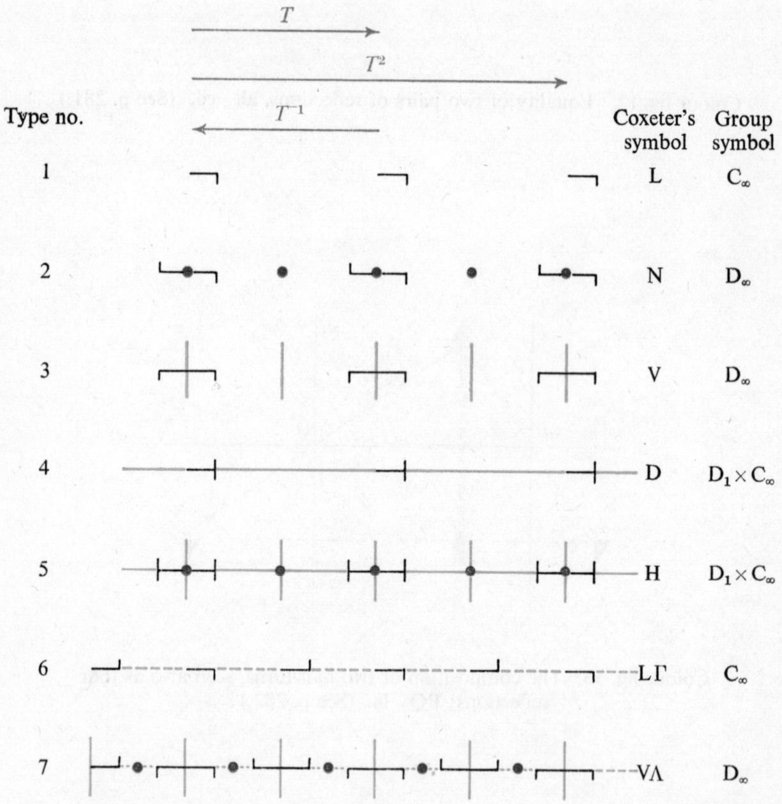

T

T^2

T^{-1}

Type no.	Coxeter's symbol	Group symbol
1	L	C_∞
2	N	D_∞
3	V	D_∞
4	D	$D_1 \times C_\infty$
5	H	$D_1 \times C_\infty$
6	LΓ	C_∞
7	VΛ	D_∞

Colour-fig. 15. The seven plane symmetry groups which contain one translation. The patterns are in black, with 'geometrical comment' in colour. (See p. 297.)

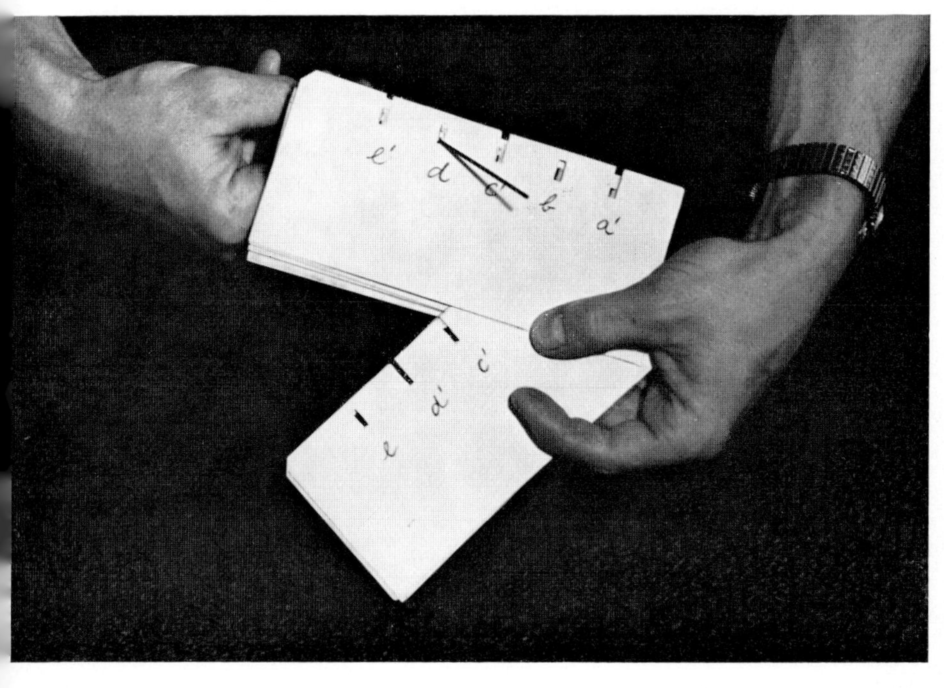

PLATE I. Punched cards. The set *d* is being lifted up on the needle.

PLATE II(*a*). A demonstration binary adder.

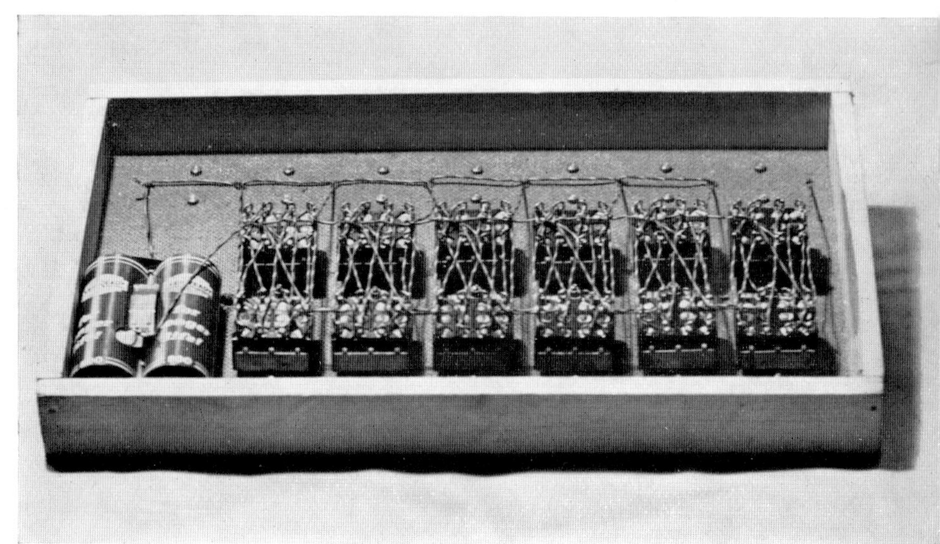

PLATE II(*b*). Underside of a binary adder.

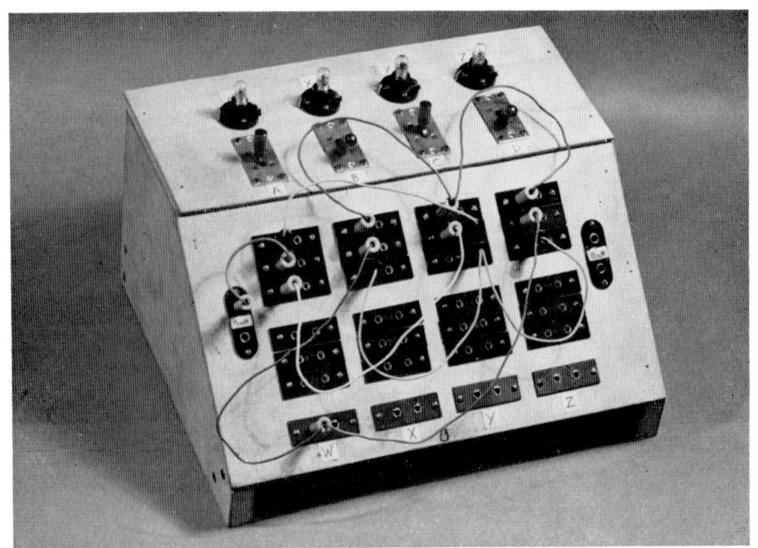

PLATE III(*a*). Front view of a logical circuit board, wired for the wolf, goat and cabbages problem.

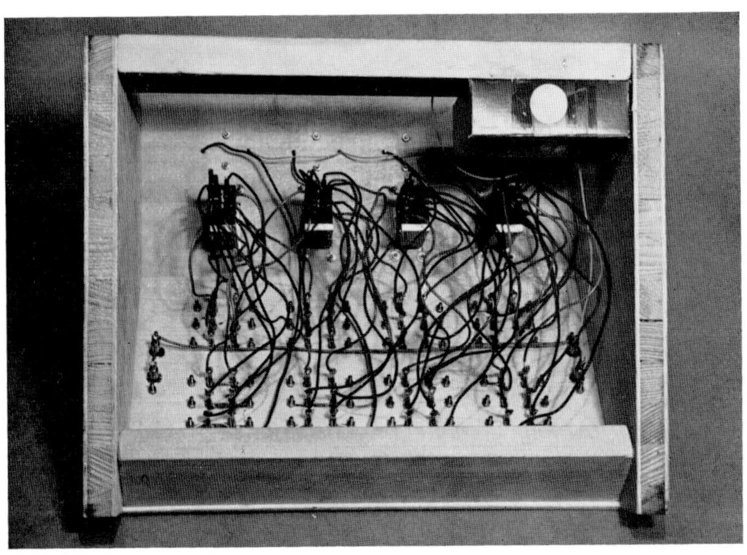

PLATE III(*b*). Underside of a logical circuit board, showing internal wiring.

BIBLIOGRAPHY

[1] Adam, P. P. *Didactica matematica euristica.* Madrid: Instituto Enseñanza Laboral, 1956.

[2] Adam, P. P. *La matematica y su enseñanza actual.* Madrid: Ministerio de Educacion Nacional, 1960.

[3] Adler, I. *The New Mathematics.* Dobson, 1959; paperback ed., Mentor Books, 1960.

[4] Adler, I. *Thinking Machines.* Dobson, 1961.

[5] Albert, A. *Fundamental Concepts of Higher Algebra.* Chicago: Univ. of Chicago Press, 1956.

[6] Alexandroff, P. S. *An Introduction to the Theory of Groups.* Blackie, 1959.

[7] Alexandroff, P. S. *Elementary Concepts of Topology.* New York: Dover, 1961.

[8] Andree, R. V. *Selections from Modern Abstract Algebra.* New York: Holt, Rinehart & Winston, 1961.

[9] Arnold, B. H. *Intuitive Concepts in Elementary Topology.* Englewood Cliffs: Prentice-Hall, 1962.

[10] Bachmann, F. *Aufbau der Geometrie aus dem Spiegelungsbegriff.* Berlin [etc.]: Springer, 1959.

[11] Beberman, M. *An Emerging Program of Secondary School Mathematics.* Cambridge (Mass.): Harvard Univ. Press; London: Oxford Univ. Press, 1959.

[12] Beberman, M. and Hart, A. New Approach to Mathematics. *Discovery* (March 1962), pp. 17–21.

[13] Berge, C. *The Theory of Graphs.* Methuen, 1962.

[14] Birkhoff, G. and MacLane, S. *A Survey of Modern Algebra.* Revised ed. New York: Macmillan, 1953.

[15] Blumenthal, L. M. *A Modern View of Geometry.* San Francisco, London: Freeman, 1961.

[16] Botsch, O. *Bewegungsgeometrie.* 2nd ed. Frankfurt-a-M.: Diesterweg, 1958.

[17] Bowden, B. V. (ed.) *Faster Than Thought.* Pitman, 1953.

[18] Bowman, M. E. *Romance in Arithmetic.* Univ. London Press, 1950.

[19] Bréard, C. *Mathématiques 5e –2e.* Paris: Editions de l'Ecole, 1961 (in progress).

[20] Brisac, R. *Exposé élémentaire des principes de la géométrie euclidienne.* Paris: Gauthier-Villars, 1955.

[21] Butler, R. and Kerr, E. *An Introduction to Numerical Methods.* Pitman, 1962.

[22] Cajori, F. *A History of Mathematics.* 2nd ed. New York: Macmillan, 1953.

[23] Carroll, L. Symbolic Logic; and The Game of Logic (*Mathematical Recreations*, vol. 1). New York: Dover, 1958.

[24] Choquet, G. A Pedagogically Satisfactory Axiomatic Basis for Elementary Geometry. *Maths Teaching*, no. 18 (spring 1962), pp. 45–50; no. 19 (summer 1962), pp. 62–5; no. 20 (autumn 1962), pp. 55–9; no. 21 (winter 1962), pp. 62–7.

[25] Cohn, P. M. *Linear Equations.* Routledge & Kegan Paul, 1958.

[26] Cohn, P. M. *Solid Geometry.* Routledge & Kegan Paul, 1961.

[27] Colebrook, F. M. and Head, J. W. *Basic Mathematics for Radio and Electronics.* 3rd ed. Iliffe, 1957.

[28] College Entrance Examination Board. Commission on Mathematics. *Program for College Preparatory Mathematics.* New York: C.E.E.B., 1959.

[29] Courant, R. and Robbins, H. *What is Mathematics*? New York: Oxford Univ. Press, 1941.

[30] Coxeter, H. S. M. *An Introduction to Geometry.* New York, London: Wiley, 1961.

[31] Crank, J. *Mathematics and Industry.* Oxford Univ. Press, 1962.

[32] Culbertson, J. T. *Mathematics and Logic for Digital Devices.* Princeton [etc.]: Van Nostrand, 1959.

[33] Cundy, H. M. and Rollett, A. P. *Mathematical Models.* 2nd ed. Oxford: Clarendon Press, 1961.

[34] Dantzig, T. *Number, the Language of Science.* 4th ed. Allen & Unwin, 1963.

[35] Davies, O. L. *The Design and Analysis of Industrial Experiments.* Oliver & Boyd, 1956.

[36] Dienes, Z. P. *Building up Mathematics.* Hutchinson Educational, 1960.

[37] D.S.I.R. *Automation. A Report on Technical Trends and their Impact on Management and Labour.* H.M.S.O., 1956.

[38] Eggleston, H. G. *Convexity.* Cambridge Univ. Press, 1958.

[39] Englebart, D. C. Games that Teach the Fundamentals of Computer Operation. *I.R.E. Trans. on Electronic Computers*, EC–10, no. 1 (March 1961), pp. 31–41.

[40] Félix, L. *Exposé moderne des mathématiques élémentaires.* Paris: Dunod, 1959.

[41] Félix, L. *Mathématiques modernes $\cap$ enseignement élémentaire.* Paris: Blanchard, 1960.

[42] Félix, L. *The Modern Aspect of Mathematics.* New York: Basic Books, 1960.

[43] Feller, W. *An Introduction to Probability Theory and its Applications.* Vol. 1, 2nd ed. New York: Wiley; London: Chapman & Hall, 1957.

[44] Finney, D. J. Statistical Science and Agricultural Research. *Mathematical Gazette*, 31, no. 293 (February 1947), pp. 21–30.

[45] Fletcher, T. J. Finite Geometry by Co-ordinate Methods. *Mathematical Gazette*, 37, no. 319 (February 1953), pp. 34–8.

[46] Fletcher, T. J. The New Gamesmanship. *Maths Teaching*, (July 1958), pp. 16–23.

[47] Fletcher, T. J. *0 and 1*. Solihull: Mathematical Pie, 1961.

[48] Fletcher, T. J. Teaching Aids and Logic. V: Punched Cards and Conclusions. *Maths Teaching*, no. 15 (March 1961), pp. 36–42.

[49] Forder, H. G. *Geometry*. Hutchinson Univ. Library, 1950.

[50] Fox, L. Practical Methods for the Solution of Linear Equations and the Inversion of Matrices. *J. R. Stat. Soc.* B (Methodological), 12, no. 1 (1950), pp. 120–36.

[51] Gardner, M. Mathematical Games. *Scientific American*, 206, no. 2 (February 1962), pp. 154–6.

[52] Gardner, M. *Logic Machines and Diagrams*. New York [etc.]: McGraw-Hill, 1958.

[53] Gattegno, C. *Modern Mathematics with Numbers in Colour*. Reading: Cuisenaire Co., 1959.

[54] Gattegno, C. (ed.). *Le matériel pour l'enseignement des mathématiques*. Neuchâtel, Paris: Delachaux & Niestlé, 1958.

[55] Good, I. J. Normal Recurring Decimals. *J. Lond. Math. Soc.* 21, no. 3 (May 1946), pp. 167–9.

[56] Goodstein, R. L. *Fundamental Concepts of Mathematics*. Oxford [etc.]: Pergamon Press, 1962.

[57] Guillemin, E. A. *The Mathematics of Circuit Analysis*. New York: Wiley, 1949.

[58] Hafstrom, J. E. *Basic Concepts in Modern Mathematics*. Reading (Mass.), London: Addison-Wesley, 1961.

[59] Halmos, P. R. *Naive Set Theory*. Princeton (N.J.): Van Nostrand, 1960.

[60] Hamming, R. W. Error Detecting and Error Correcting Codes. *Bell System Technical Journal*, 16, no. 2 (April 1950), pp. 147–60.

[61] Heath, F. W. and Gribble, M. W. Chain Codes and their Electronic Applications. *Proc. I.E.E. Monograph*, 392, 1961.

[62] Hersee, E. H. *A Simple Approach to Electronic Computers*. Blackie, 1960.

[63] Hohn, F. E. *Applied Boolean Algebra*. New York: Macmillan, 1960.

[64] Hollingdale, S. H. *High-speed Computing*. English Universities Press, 1960.

[65] Huffman, D. A. The Synthesis of Linear Sequential Coding Networks. In: *Information Theory*, ed. Colin Cherry. Butterworths Scientific Publications, 1956.

[66] Jeger, M. *Konstruktive Abbildungsgeometrie*. 2nd ed. Luzern: Räber, 1959.

[67] Johnson, D. A. and Glenn, W. H. *Exploring Mathematics on Your Own: Sets, Sentences and Operations*. St Louis [etc.]: Webster, 1960; London: Murray, 1963.

[68] Johnson, D. A. and Glenn, W. H. *Exploring Mathematics on Your Own: Topology*. St Louis [etc.]: Webster, 1960; London: Murray, 1963.

[69] Kelley, J. L. *Introduction to Modern Algebra.* Princeton (N.J.), London: Van Nostrand, 1960.

[70] Kemeny, J. G. *et al. Introduction to Finite Mathematics.* Englewood Cliffs (N.J.): Prentice-Hall, 1957.

[71] Kemeny, J. G. *et al. Finite Mathematical Structures.* Englewood Cliffs (N.J.): Prentice-Hall, 1959.

[72] Klein, F. *Elementary Mathematics from an Advanced Standpoint.* [Vol. 2]: *Geometry.* New York: Dover, 1939.

[73] Kutuzov, B. V. *Studies in Mathematics.* Vol. 4: *Geometry.* New Haven (Conn.): School Mathematics Study Group, 1960.

[74] Langer, S. *Introduction to Symbolic Logic.* 2nd ed. New York: Dover, 1953.

[75] Ledermann, W. *Introduction to the Theory of Finite Groups.* Oliver & Boyd, 1949.

[76] Levi, H. *Foundations of Geometry and Trigonometry.* Englewood Cliffs (N.J.): Prentice-Hall, 1960.

[77] Lieber, L. *Mits, Wits and Logic.* New York: Norton, 1947.

[78] Lieber, L. Modern Mathematics for Scientists and Engineers: Boolean Algebra. *Trans. N.Y. Acad. Sci.*, Ser. II, vol. 17, no. 4 (February 1955), pp. 331–8.

[79] Littlewood, D. E. *Skeleton Key of Mathematics.* Hutchinson Univ. Library, 1949.

[80] Livesley, R. K. *An Introduction to Automatic Digital Computers.* Cambridge Univ. Press, 1957.

[81] Mann, H. B. *Analysis and Design of Experiments.* New York: Dover, 1949.

[82] Mansfield, D. E. and Thompson, E. *Mathematics: A New Approach.* 6 vols. Chatto & Windus, 1962 (in progress).

[83] May, K. O. *Elements of Modern Mathematics.* Reading (Mass.): Addison-Wesley, 1959.

[84] Moisil, G. C. *Teoria algebrica a mecanismelor automate.* Bucharest: Editura Tehnica, 1959.

[85] Murphy, J. S. *Basics of Digital Computers.* (3 vols. in 1 or separate parts.) New York: J. F. Rider, 1958.

[86] National Physical Laboratory. *Modern Computing Methods.* 2nd ed. H.M.S.O., 1962.

[87] National Council of Teachers of Mathematics. *Insights into Modern Mathematics. 23rd Yearbook.* Washington: N.C.T.M., 1957.

[88] Newman, J. R. (ed.). *The World of Mathematics.* 4 vols. Allen & Unwin, 1961.

[89] Organization for European Economic Co-operation. *New Thinking in School Mathematics.* [Paris], OEEC, 1961.

[90] Organization for European Economic Co-operation. *Synopses for Modern Secondary School Mathematics.* [Paris], OEEC, 1961.

[91] Papy, G. *Premiers éléments de mathématique moderne.* Special order from l'Ecoles Normales Superieures, 72 rue Berkendael, Brussels 5.

[92] Papy, G. *Groupes.* Brussels: Presses Universitaires, 1961.

[93] Patterson, E. M. *Topology*. Oliver & Boyd, 1956.

[94] Peck, L. C. *Secret Codes, Remainder Arithmetic and Matrices*. Washington: National Council of Teachers of Mathematics, 1961.

[95] Petersen, W. W. Error Correcting Codes. *Scientific American*, 206, no. 2 (February 1962), pp. 96–108.

[96] Piaget, J. *Traité de logique*. Paris: Colin, 1949.

[97] Piaget, J. and Inhelder, B. *The Growth of Logical Thinking from Childhood to Adolescence*. Routledge & Kegan Paul, 1958.

[98] Piaget, J. *et al. L'enseignement des mathématiques*. Neuchâtel, Paris: Delachaux & Niestlé, 1955.

[99] Poe, E. A. *Tales of Mystery and Imagination: The Gold Bug*. Dent, 1946.

[100] Polya, G. *Mathematics and Plausible Reasoning*. 2 vols. Oxford Univ. Press, 1954.

[101] Quine, W. V. Paradox. *Scientific American*, 206, no. 4 (April 1962), pp. 84–96.

[102] Rademacher, H. and Toeplitz, O. *The Enjoyment of Mathematics*. Princeton (N.J.): Princeton Univ. Press, 1957.

[103] Redish, K. A. *An Introduction to Computational Methods*. English Universities Press, 1961.

[104] Rees, D. Note on a Paper by I. J. Good. *J. Lond. Math. Soc.* 21, no. 3 (May 1946), pp. 169–72.

[105] Roberts, J. B. *The Real Number System in Algebraic Setting*. San Francisco, London: Freeman, 1962.

[106] Rosenbloom, P. *Elements of Mathematical Logic*. New York: Dover, 1950.

[107] Rouse Ball, W. W. *Mathematical Recreations and Essays*. 11th ed. Macmillan, 1939.

[108] Sawyer, W. W.: *Prelude to Mathematics*. Harmondsworth: Penguin Books, 1955.

[109] Sawyer, W. W.: *A Concrete Approach to Abstract Algebra*. Freeman, 1959.

[110] Servais, W. *Seminar on the Modern Teaching of Geometry*. Aarhus University.

[111] School Mathematics Study Group. *Mathematics for High School*, and *Mathematics for Junior High School*. 30 parts. New Haven (Conn.): Yale University Press, 1960.

School Mathematics Study Group. Studies in Mathematics. 5 vols. New Haven (Conn.): S.M.S.G., 1959–60.

[112] Silvester, V. *Modern Ballroom Dancing*. 53rd ed. H. Jenkins, 1960.

[113] Stein, S. K.: The Mathematician as Explorer. *Scientific American*, 204, no. 5 (May 1961), pp. 149–58.

[114] Steinhaus, H. *Mathematical Snapshots*. Revised ed. New York: Oxford Univ. Press, 1960.

[115] Stoll, R. R. *Sets, Logic and Axiomatic Theories*. San Francisco, London: Freeman, 1961.

[116] Suppes, P. *Axiomatic Set Theory.* Princeton [etc.]: Van Nostrand, 1960.

[117] Suppes, P. *Introduction to Logic.* Princeton [etc.]: Van Nostrand, 1960.

[118] Sutton, O. G. *Mathematics in Action.* Bell, 1954.

[119] Thomas, H. H. *Further General Progress Papers.* Nelson, 1951.

[120] Thomsen, G. The Treatment of Elementary Geometry by a Group-calculus. *Mathematical Gazette*, 17, no. 224 (October 1933), pp. 230–42.

[121] Thurston, H. A. *The Number-system.* Blackie, 1956.

[122] Thurston, H. A. Celtic Interlacing Patterns and Topology. *Science News*, 33, pp. 50–62. Harmondsworth: Penguin Books, 1954.

[123] Thwaites, B. (ed.). *On Teaching Mathematics.* Oxford: Pergamon Press, 1961.

[124] Van der Waerden, B. L. *Modern Algebra*; trans. from 2nd revised ed. 2 vols, New York: Ungar, 1949–50.

[125] Vajda, S. *An Introduction to Linear Programming and the Theory of Games.* Methuen, 1960.

[126] Vajda, S. *Readings in Linear Programming.* Pitman, 1960.

[127] Wells, A. F. *The Third Dimension in Chemistry.* Oxford: Clarendon Press, 1956.

[128] Weyl, H. *Symmetry.* Princeton (N.J.): Princeton Univ. Press, 1952.

[129] Whitesitt, J. E.: *Boolean Algebra and its Applications.* Reading (Mass.), London: Addison-Wesley, 1961.

[130] Wilder, R. L. *Introduction to the Foundations of Mathematics.* New York: Wiley, 1952.

[131] Wilson, W. Proctor. Some Matrix Theorems. *Electronic and Radio Engineer*, 34, no. 6 (June 1957), pp. 229–31.

[132] Wooldridge, R. *An Introduction to Computing.* Oxford Univ. Press, 1962.

[133] Wylie, C. R. *101 Puzzles in Thought and Logic.* New York: Dover, 1957.

[134] Yaglom, I. M. *Geometric Transformations.* New York [etc.]: Random House, 1962.

[135] Yaglom, I. M. and Boltyanskii, V. G. *Convex Figures.* New York: Holt, Rinehart & Winston, 1961.

ADDENDA

[136] Belgium. Ministère de l'Instruction Publique. *Documentation: Cahier 5.* Brussels: The Ministry, 1956.

[137] Allen, R. G. D. *Basic Mathematics.* Macmillan, 1962.

[138] Moakes, A. J. *Numerical Mathematics: Exercises in Computing with a Desk Calculator.* London: Macmillan; New York: St Martins Press, 1963.

[139] Thomas, L. and Thomas, A. *Mathematics by Calculating Machine.* London: Cassell, 1963.

[140] St Dunstan's College Booklets. Directed by G. Matthews. London: Arnold (in the Press).

[141] Matthews, G. Matrices for the Million. *Mathematical Gazette*, 47, no. 359 (February 1963), pp. 2–8.

[142] Goodstein, R. L. *Boolean Algebra.* Oxford [etc.]: Pergamon Press, 1963.

[143] Stein, S. K. *Mathematics: the Man-made Universe, An Introduction to the Spirit of Mathematics.* San Francisco, London: Freeman, 1963.

INDEX

INDEX